T0260704

Dynamic
copula
methods
in finance

"Copulas address a central problem in financial modeling, namely how to describe the statistics of events which are related to two or more other events of interest. This important book provides a comprehensive and timely review of the theory and applications of copulas."

> Robert Elliott, Haskayne School of Business,University of Calgary

"Researchers and practitioners in the field of finance will welcome the appearance of Dynamic Copula Methods in Finance. In this innovative and well written book, the authors make a strong case for the application of convolution-based copulas in finance. The book features numerous illustrations and a wealth of examples, most of which concern applications to financial problems. Dynamic Copula Methods in Finance promises to be a valuable addition to the rapidly expanding literature on copula models in finance."

> Roger B. Nelsen, Professor Emeritus of Mathematics, Lewis & Clark College, Portland, Oregon

"Static copula models have been extensively used in finance for more than a decade. In this book the authors show how to apply copula methods to dynamic problems, setting the ground for a number of important financial applications, from derivatives pricing to risk management."

> Fabio Mercurio, Head of Quant Business Managers, Bloomberg LP, New York

About the authors

UMBERTO CHERUBINI is Associate Professor of Financial Mathematics at the University of Bologna, where he heads the Graduate Degree in Quantitative Finance. He is a fellow of the Financial Econometrics Research Center (FERC), a member of the Scientific Committees of Abiformazione – the professional education arm of the Italian banking association, and AIFIRM – the Italian Association of Financial RiskManagers. He has been consulting and teaching in the field of finance and risk management for more than ten years. Before joining academia he worked as an economist at the Economic Research Department of BCI Milan. He has published papers in finance and economics in international journals, and is co-author of six books on topics of risk management and financial mathematics, including *Fourier Transform Methods in Finance,* John Wiley & Sons, Ltd, 2009; and *Copula Methods in Finance,* John Wiley & Sons, Ltd, 2004.

FABIO GOBBI is a post-doctoral researcher at the University of Bologna. He has a PhD in Statistics from the University of Florence and his areas of research focus on probability and financial econometrics. This is his first book.

SABRINA MULINACCI is Associate Professor of Mathematical Methods for Economics and Finance at the University of Bologna, Italy. Prior to this Sabrina was Associate Professor of Mathematical Methods for Economics and Actuarial Sciences at the Catholic University of Milan. She has a PhD in Mathematics from the University of Pisa and has published a number of research papers in international journals on probability and mathematical finance. She is co-author of *Fourier Transform Methods in Finance*, John Wiley & Sons, Ltd, 2009.

SILVIA ROMAGNOLI is Assistant Professor of Mathematical Models for Economics and Actuarial and Financial Sciences at the University of Bologna. Her scientific research is mainly addressed to the applications of stochastic models to finance and insurance. She has published several research papers in international journals on mathematical finance.

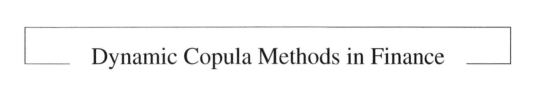

Dynamic Copula Methods in Finance

For other titles in the Wiley Finance series
please see www.wiley.com/finance

Dynamic Copula Methods in Finance

Umberto Cherubini
Fabio Gobbi
Sabrina Mulinacci
Silvia Romagnoli

A John Wiley & Sons, Ltd., Publication

This edition first published 2012
© 2012 John Wiley & Sons, Ltd

Registered office
John Wiley & Sons Ltd, The Atrium, Southern Gate, Chichester, West Sussex, PO19 8SQ, United Kingdom

For details of our global editorial offices, for customer services and for information about how to apply for
permission to reuse the copyright material in this book please see our website at www.wiley.com.

The right of the authors to be identified as the authors of this work has been asserted in accordance with the
Copyright, Designs and Patents Act 1988.

All rights reserved. No part of this publication may be reproduced, stored in a retrieval system, or transmitted, in any
form or by any means, electronic, mechanical, photocopying, recording or otherwise, except as permitted by the UK
Copyright, Designs and Patents Act 1988, without the prior permission of the publisher.

Wiley publishes in a variety of print and electronic formats and by print-on-demand. Some material included with
standard print versions of this book may not be included in e-books or in print-on-demand. If this book refers to
media such as a CD or DVD that is not included in the version you purchased, you may download this material at
http://booksupport.wiley.com. For more information about Wiley products, visit www.wiley.com.

Designations used by companies to distinguish their products are often claimed as trademarks. All brand names and
product names used in this book are trade names, service marks, trademarks or registered trademarks of their
respective owners. The publisher is not associated with any product or vendor mentioned in this book. This
publication is designed to provide accurate and authoritative information in regard to the subject matter covered. It is
sold on the understanding that the publisher is not engaged in rendering professional services. If professional advice
or other expert assistance is required, the services of a competent professional should be sought.

Library of Congress Cataloging-in-Publication Data

Dynamic copula methods in finance / Umberto Cherubini . . . [et al.].
    p. cm. — (The wiley finance series)
  Includes bibliographical references and index.
  ISBN 978-0-470-68307-1 (hardback)
1. Finance—Mathematical models.   I. Cherubini, Umberto
  HG106.D96 2011
  332.01′519233—dc23
                                        2011034154

A catalogue record for this book is available from the British Library.

ISBN 978-0-470-68307-1 (hardback) ISBN 978-1-119-95451-4 (ebk)
ISBN 978-1-119-95452-1 (ebk)   ISBN 978-1-119-95453-8 (ebk)

Set in 10/12pt Times by Aptara Inc., New Delhi, India

Contents

Preface

This book concludes five years of original research at the University of Bologna on the use of copulas in finance. We would like these results to be called the *Bologna school*. The problem tackled arises directly from financial applications and the fact that almost always in this field we are confronted with convolution problems along with non-normal distributions and non-linear dependence. More explicitly, almost always in finance we face the problem of evaluating the distribution of

$$X + Y$$

where X and Y may have arbitrary distributions and may be dependent on each other in quite a strange fashion. Very often, we may also be interested in the dependence of this sum on either X or Y. The *Bologna school* has studied the class of *convolution-based* copulas that is well suited to address this kind of problem. It is easy to see that this operates a restriction on the choice of copulas. In a sense, *convolution-based* copulas address a special *compatibility* problem, enforcing coherence in the dependence structure between variables and their sum. This compatibility issue is paramount and unavoidable for almost all the applications in finance. The first concept that comes to mind is the linear law of price enforced by the fundamental theorem of asset pricing: in order to avoid arbitrage, prices of complex products must be linear combinations of the primitive products constituting the replicating portfolio. In asset allocation, portfolios are also strictly linear concepts, even though they may include (and today they typically do) option-like and other non-linear products whose distribution is far from Gaussian and whose dependence on the other components of the portfolio is not Gaussian either. Moreover, trading and investment activities involve more and more exposures to credit risk that are non-Gaussian by definition: this was actually the very reason for copula function applications to finance in the first place. But, even in the case of credit, losses may be linked by the most complex dependence structure, but nevertheless they cumulate one after the other by a linear combination: computing cumulated losses is again a convolution problem. Finally, linear aggregation is crucial to understand the dynamics of markets. From this viewpoint, finance theory has developed under the main assumption of processes with independent increments: *convolution-based copulas* may allow us to considerably extend the set of possible market dynamics, allowing for general dependence structures between the price level at a given point in time and its increment (the return) in the following period: describing the distribution of the price at the end of a period is again a convolution problem.

The main message of this book is that copulas remain a flexible tool for applications in finance, but this flexibility is finite, and the *Bologna school* sets the frontier of this flexibility at the family of *convolution-based* copulas.

Chapters 1 and 2 review the general problem of dependence and correlation in finance. More particularly, Chapter 2 specializes the analysis to a review of the basic concepts of copulas, as they have been applied to financial problems until today. Chapters 3 and 4 introduce the theory of *convolution-based* copulas, and the concept of *C-convolution* within the mainstream of the Darsow, Nguyen, and Olsen (DNO) application of copulas to Markov processes. More specifically, Chapter 3 addresses theory and Chapter 4 deals with the application to econometrics. Chapters 5, 6, and 7 discuss applications of the approach in turn to the problems of: (i) evaluating multivariate equity derivatives; (ii) analyzing the credit risk exposure of a portfolio, (iii) aggregating Value-at-Risk measures across risk factors and business units. In all these chapters, we exploit the model to address dependence both in a spatial and temporal perspective. This twofold perspective is entirely new to these applications, and may easily be handled within the set of *convolution-based copulas*. Chapter 8 concludes by surveying other methodologies available in the mathematical finance and probability literature to set a dependence structure among processes: these approaches are mainly in continuous time, and raise the question, that we leave for future research, of whether they represent some or all the possible solutions that one would obtain by taking the continuous time limit of our model, which is defined in discrete time.

We conclude with thanks to our colleagues in the international community who have helped us during these years of work. Their support has been particularly precious, because our work is entirely free from government support. *Nemo propheta in patria.* As for comments on this manuscript, we would particularly like to thank, without implication, Xiaohong Chen, Fabrizio Durante, Marius Hofert, Matthias Scherer, Bruno Remillard, Paramsoothy Silvapulle, and an anonymous referee provided by John Wiley. And we thank our readers in advance for any comments they would like to share with us.

1
Correlation Risk in Finance

Over the last decade, financial markets have witnessed a progressive concentration of focus on correlation dynamics models. New terms such as *correlation trading* and *correlation products* have become the frontier topic of financial innovation. Correlation trading denotes the trading activity aimed at exploiting changes in correlation, or more generally in the dependence structure of assets and risk factors. Correlation products denote financial structures designed with the purpose of exploiting these changes. Likewise, the new term *correlation risk* in risk management is meant to identify the exposure to losses triggered by changes in correlation. Going long or short correlation has become a standard concept for everyone working in dealing rooms and risk management committees. This actually completes a trend that led the market to evolve from taking positions on the direction of prices towards taking exposures to volatility and higher moments of their distribution, and finally speculating and hedging on cross-moments. These trends were also accompanied by the development of new practices to transfer risk from one unit to others. In the aftermath of the recent crisis, these products have been blamed as one of the main causes. It is well beyond the scope of this book to digress on the economics of the crisis. We would only like to point out that the modular approach which has been typical of financial innovation in the *structured finance* era may turn out extremely useful to ensure the efficient allocation of risks among the agents. While on the one hand the use of these techniques without adequate knowledge may represent a source risk, avoiding them for sure represents a distortion and a source of cost. Of course, accomplishing this requires the use of modular mathematical models to split and transfer risk. This book is devoted to such models, which in the framework of dependence are called *dependence functions* or *copula functions*.

1.1 CORRELATION RISK IN PRICING AND RISK MANAGEMENT

In order to measure the distance between the current practice of markets and standard textbook theory of finance, let us consider the standard static portfolio allocation problem. The aim is to maximize the expected utility of wealth W at some final date T using a set of risky assets, $S_i, i = 1, \ldots, m$. Formally, we have

$$\mathbb{E}_{\mathbb{P}}\left[\mathbb{U}\left(R_f + \sum_{i=1}^{m} w_i(R_i - R_f)\right)\right],$$

where $R_i = \ln(S_i(T)/S_i(0))$ are the log-returns on the risky assets and R_f is the risk-free rate. The asset allocation problem is completely described by two functions: (i) the utility function $\mathbb{U}(.)$, assumed strictly increasing and concave; (ii) the joint distribution function of the returns $\mathbb{P}$. While we could argue in depth about both of them, throughout this book the focus will be on the specification of the joint distribution function. In the standard textbook problem, this is actually kept in the background and returns are assumed to be jointly normally distributed, which leads to rewriting the expected utility in terms of a mean–variance problem.

Nowadays, real-world asset management has moved miles away from this textbook problem, mainly for two reasons: first, investments are no longer restricted to linear products, such as stocks and bonds, but involve options and complex derivatives; second, the assumption that the distribution of returns is Gaussian is clearly rejected by the data. As a result, the expected utility problem should take into account three different dimensions of risk: (i) directional movements of the market; (ii) changes in volatility of the assets; (iii) changes in their correlation. More importantly, there is also clear evidence that changes in both volatility and correlation are themselves correlated with swings in the market. Typically, both volatility and correlation increase when the market is heading downward (which is called the *leverage effect*). It is the need to account for these new dimensions of risk that has led to the diffusion of derivative products to hedge against and take exposures to both changes in volatility and changes in correlation. In the same setting, it is easy to recover the other face of the same problem encountered by the pricer. From his point of view, the problem is tackled from the first-order conditions of the investment problem:

$$\mathbb{E}_{\mathbb{P}}\left[\mathbb{U}'\left(R_f + \sum_{i=1}^{m} w_i(R_i - R_f)\right)(R_i - R_f)\right] = 0 = \mathbb{E}_{\mathbb{Q}}\left[R_i - R_f\right],$$

where the new probability measure $\mathbb{Q}$ is defined after the Radon–Nikodym derivative

$$\frac{\partial \mathbb{Q}}{\partial \mathbb{P}} = \frac{\mathbb{U}'\left(R_f + \sum_{i=1}^{m} w_i(R_i - R_f)\right)}{\mathbb{E}_{\mathbb{P}}\left[\mathbb{U}'\left(R_f + \sum_{i=1}^{m} w_i(R_i - R_f)\right)\right]}.$$

Pricers face the problem of evaluating financial products using measure $\mathbb{Q}$, which is called the *risk-neutral measure* (because all the risky assets are expected to yield the same return as the risk-free asset), or the *equivalent martingale measure* (EMM, because $\mathbb{Q}$ is a measure equivalent to $\mathbb{P}$ with the property that prices expressed using the risk-free asset as *numeraire* are martingale). An open issue is whether and under what circumstances volatility and correlation of the original measure $\mathbb{P}$ are preserved under this change of measure. If this is not the case, we say that volatility and correlation risks are priced in the market (that is, a risk premium is required for facing these risks). Under this new measure, the pricers face problems which are similar to those of the asset manager, that is evaluating the sensitivity of financial products to changes in the direction of the market (long/short the asset), volatility (long/short volatility) and correlation (long/short correlation). They face a further problem, though, that is going to be the main motivation of this book: they must ensure that prices of multivariate products are consistent with prices of univariate products. This consistency is part of the so-called *arbitrage-free* approach to pricing, which leads to the martingale requirement presented above. In the jargon of statisticians, this consistency leads to the term *compatibility*: the risk-neutral joint distribution $\mathbb{Q}$ has to be compatible with the marginal distributions $\mathbb{Q}_i$.

Like the asset manager and the pricer, the risk manager also faces an intrinsically multivariate problem. This is the issue of measuring the exposure of the position to different risk factors. In standard practice, he transforms the financial positions in the different assets and markets into a set of exposures (*buckets*, in the jargon) to a set of risk factors (*mapping process*). The problem is then to estimate the joint distribution of losses on these exposures and define a risk measure on this distribution. Typical measures are *Value-at-Risk* (VaR) and *Expected Shortfall* (ES). These measures are multivariate in the sense that they must account for correlation among the losses, but there is a subtle point to be noticed here, which makes this practice obsolete with respect to structured finance products, and correlation products in particular.

A first point is that these products are non-linear, so that their value may change even though market prices do not move but their volatilities do. As for volatility, the problem can be handled by including a *bucket* of volatility exposures for every risk factor. But there is a crucial point that gets lost if correlation products are taken into account. It is the fact that the value of these products can change even if neither the market prices nor their volatilities move, but simply because of a change in correlation. In fact, this exposure to correlation among the assets included in the specific product is lost in the mapping procedure. Correlation risk then induces risk managers to measure this dimension of risk on a product-by-product basis, using either historical simulation or stress-testing techniques.

1.2 IMPLIED VS REALIZED CORRELATION

A peculiar feature of applications of probability and statistics to finance is the distinction between historical and implied information. This duality, that is found in many (if not all) applications in univariate analysis, shows up in the multivariate setting as well. On the one side, standard time series data from the market enable us to gauge the relevance of market co-movements for investment strategies and risk management issues. On the other side, if there exist derivative prices which are dependent on market correlation, it is possible to recover the degree of co-movement credited by investors and financial intermediaries to the markets, and this is done by simply inverting the prices of these derivatives. Of course, recovering implied information is subject to the same flaws as those that are typical of the univariate setting. First, the possibility of neatly backing out this information may be limited by the market incompleteness problem, which has the effect of introducing a source of noise into market prices. Second, the distribution backed out is the risk-neutral one and a market price of risk could be charged to allow for the possibility of correlation changes. These problems are indeed compounded and in a sense magnified in the multivariate setting, in which the uncertainty concerning the dependence structure among the markets adds to that on the shape of marginal distributions.

Unfortunately, there are not many cases in which correlation can be implied from the market. An important exception is found in the FOREX market, because of the so-called *triangular arbitrage* relationship. Consider the Dollar/Euro ($e_{US,E}$), the Euro/Yen ($e_{E,Y}$) and the Dollar/Yen ($e_{US,Y}$) exchange rates. Triangular arbitrage requires that

$$e_{US,E} = e_{E,Y} e_{US,Y}.$$

Taking logs and denoting by $\sigma_{US,E}$, $\sigma_{E,Y}$, and $\sigma_{US,Y}$ the corresponding implied volatilities, we have that

$$\sigma_{US,E}^2 = \sigma_{E,Y}^2 + \sigma_{US,Y}^2 + 2\rho\sigma_{E,Y}\sigma_{US,Y},$$

from which

$$\rho = \frac{\sigma_{US,E}^2 - \sigma_{E,Y}^2 - \sigma_{US,Y}^2}{2\sigma_{E,Y}\sigma_{US,Y}}$$

is the implied correlation between the Euro/Yen and the Dollar/Yen priced by the market.

1.3 BOTTOM-UP VS TOP-DOWN MODELS

For all the reasons above, estimating correlation, either historical or implied, has become the focus of research in the last decade. More precisely, the focus has been on the specification of the joint distribution of prices and risk factors. This has raised a first strategic choice between two opposite classes of models, that have been denoted *top-down* and *bottom-up* approaches. In all applications, pricing of equity and credit derivatives, risk management aggregation and allocation, the first choice is then to fit all markets and risk factors with a joint distribution and to specify in the process both the marginal distributions of the risk factors and their dependence structure. The alternative is to take care of marginal distributions first, and of the dependence structure in a second step. It is clear that copula functions represent the main tool of the latter approach. It is not difficult to gauge what the pros and cons of the two alternatives might be. Selecting a joint distribution fitting all risks may not be easy, beyond the standard choices of the normal distribution for continuous variables and the Poisson distribution for discrete random variables. If one settles instead for the choice of non-parametric statistics, even for a moderate number of risk factors, the implementation runs into the so-called *curse of dimensionality*. As for the advantages, a top-down model would make it fairly easy to impose restrictions that make prices consistent with the equilibrium or no-arbitrage restrictions. Nevertheless, this may come at the cost of marginal distributions that do not fit those observed in the market. Only seldom (to say never) does this poor fit correspond to arbitrage opportunities, while more often it is merely a symptom of *model risk*. On the opposite side, the bottom-up model may ensure that marginal distributions are properly fitted, but it may be the case that this fit does not abide by the consistency relationships that must exist among prices: the most well known example is the no-arbitrage restriction requiring that prices of assets in speculative markets follow martingale processes. The main goal of this book is actually to show how to impound restrictions like these in a bottom-up framework.

1.4 COPULA FUNCTIONS

Copula functions are the main tool for a bottom-up approach. They are actually built on purpose with the goal of pegging a multivariate structure to prescribed marginal distributions. This problem was first addressed and solved by Abe Sklar in 1959. His theorem showed that any joint distribution can be written as a function of marginal distributions:

$$F(x_1, x_2, \ldots, x_n) = C(F_1(x_1), F_2(x_2), \ldots, F_n(x_n))$$

and that the class of functions $C(.)$, denoted copula functions, may be used to extend the class of multivariate distributions well beyond those known and usually applied. To quote the dual approach above, the former result allows us to say that any top-down approach may be written in the formalism of copula functions, while the latter states that copulas can be applied in a bottom-up approach to generate infinitely many distributions. A question is whether this multiplicity may be excessive for financial applications, and this whole book is devoted to that question.

Often a more radical question is raised, whether there is any advantage at all to working with copulas. More explicitly, one could ask what can be done with copulas that cannot be done with other techniques. The answer is again the essence of the bottom-up philosophy. The crucial point is that in the market, we are used to observing marginal distributions. All the information that we can collect is about marginals: the time series of this and that price, and the implied distribution of the underlying asset of an option market for a given exercise date. We can couple time series of prices or of distributions together and study their dependence, but only seldom can we observe multivariate distributions. For this reason, it is mandatory that any model be made consistent with the univariate distributions observed in the market: this is nothing but an instance of that procedure pervasively used in the markets and called *calibration*.

1.5 SPATIAL AND TEMPORAL DEPENDENCE

To summarize the arguments of the previous section, it is of the utmost importance that multivariate models be consistent with univariate observed prices, but this consistency must be subject to some rules and cannot be set without limits. These limits were not considered in standard copula functions applications to finance problems. In these applications the term *multivariate* was used with the meaning that several different risk factors at a given point in time were responsible for the value of a position at that time. This concept is called *spatial dependence* in statistics and is also known as *cross-section* dependence in econometrics. Copula functions could be used in full flexibility to represent the consistency between the price of a multivariate product at a given date and the prices of the constituent products observed in the market. However, the term multivariate could be used with a different meaning, that would make things less easy. It could in fact refer to the dependence structure of the value of the same variable observed at different points in time: this is actually defined as a *stochastic process*. In the language of statistics, the dependence among these variables would be called *temporal dependence*. Curiously, in econometric applications copula functions have mainly been intended in this sense. If copulas are used in the same sense in derivative pricing problems, the flexibility of copulas immediately becomes a problem: for example, one would like to impose restrictions on the dynamics to have Markov processes and martingales, and only a proper specification of copulas could be selected to satisfy these requirements.

Even more restrictions would apply in an even more general setting, in which a multivariate process would be considered as a collection of random variables representing the value of each asset or risk factor at different points in time. In the standard practice of econometrics, in which it is often assumed that relationships are linear, this would give rise to the models called *vector autoregression* (VAR). Copula functions allow us to extend these models to a general setting in which relationships are allowed to be non-linear and non-Gaussian, which is the rule rather than the exception of portfolios of derivative products.

1.6 LONG-RANGE DEPENDENCE

These models that extend the traditional VAR time series with a specification in terms of copula functions are called *semi-parametric* and the most well known example is given by the so-called SCOMDY model (Semiparametric Copula-based Multivariate Dynamic). By taking the comparison with linear models one step further, these models raise the question of the behavior of the processes over long time horizons. We know from the theory of time series that a univariate process y_t modeled with the dynamics

$$y_t = \omega + \epsilon_t + \alpha_1 \epsilon_{t-1} + \ldots + \alpha_q \epsilon_{t-q} + \beta_1 y_{t-1} + \ldots + \beta_p y_{t-p},$$

where ω, $\alpha_1, \ldots, \alpha_p$ and $\beta_1, \ldots, \beta_q$ are constant parameters, $\epsilon_{t-i} \sim N(0, \sigma)$ and $Cov(\epsilon_{t-i}, \epsilon_{t-j}) = 0, i \neq j$ is called an ARMA(p,q) model (Autoregressive Moving Average

Process), that could be extended to a multiple set of processes called VARMA. In particular, the MA part of the process is represented by the dependence of y_t on the past q innovations ϵ_{t-i} and the AR part is given by its dependence on the past p values of the process itself. If we focus on the autoregressive part, we know that in cases in which the characteristic equation of the process

$$z - \beta_1 z - \ldots - \beta_p z^p = 0$$

has solutions strictly inside the unit circle, the process is said to be *stationary* in mean. To make the meaning clear, let us just focus on the simplest AR(1) process:

$$y_t = \omega + \epsilon_t + \beta_1 y_{t-1}.$$

It is easy to show that if $\beta_1 < 1$ by recursive substitution of y_{t-i-1} into y_{t-i}, we have

$$\mathbb{E}(y_t) = \omega \sum_{i=0}^{\infty} \beta_1^i + \sum_{i=0}^{\infty} \beta_1^i \mathbb{E}(\epsilon_{t-i}) = \frac{\omega}{1 - \beta_1}$$

and

$$VAR(y_t) = \sum_{i=0}^{\infty} \mathbb{E}(\beta_1^i \epsilon_{t-i})^2 = \frac{\sigma^2}{1 - \beta_1^2},$$

where we have used the moments of the distribution of ϵ_{t-i}. Notice that if instead it is $\beta_1 = 1$, the dynamics of y_t is defined by

$$y_t = \omega + \epsilon_t + y_{t-1}$$

and neither the mean nor the variance of the unconditional distribution are defined. In this case the process is called *integrated* (of order 1) or *difference stationary*, or we say that the process contains a *unit root*. The idea is that the first difference of the process is stationary (in mean). The distinguishing feature of these processes is that any shock affecting a variable remains in its history forever, a property called *persistence*. As an extension, one can conceive that several processes may be linear combinations of the same persistent shock y_t, that is also called the *common stochastic trend* of the processes. In this case we say that the set of processes constitutes a *co-integrated* system. More formally, a set of processes is said to constitute a co-integrated system if there exists at least one linear combination of the processes that is stationary in mean.

In another stream of literature, another intermediate case has been analyzed, in which the process is said to be *fractionally* integrated, so that the process is made stationary by taking fractional differences: the long-run behavior of these processes is denoted *long memory*. In Chapter 4 we shall give a formal definition of long memory (due to Granger, 2003) and we will discuss the linkage with a copula-based stochastic process. As for the contribution of copulas to these issues, notice that while most of the literature on unit roots and persistence vs stationary models has developed under the maintained assumption of Gaussian innovations, the use of copula functions extends the analysis to non-Gaussian models. Whether these models can represent a new specification for the long-run behavior of time series remains an open issue.

1.7 MULTIVARIATE GARCH MODELS

Since copula functions are a general technique to address multivariate non-Gaussian systems, one of the alternative approaches that represents the fiercest challenge to them is the multivariate version of conditional heteroscedasticity models. The idea is to assume that the Gaussian dynamics, or a dynamics that can be handled with tractable distributions, can be maintained if it is modeled conditionally on the dynamics of volatility. So, in the univariate case a shock has the twofold effect of changing the return of a period and the volatility of the following period. Just like in the univariate case, a possible choice to model the non-normal joint distribution of returns is to assume that they could be conditionally normal. A major problem with this approach is that the number of parameters to be estimated can become huge very soon. Furthermore, restrictions are to be imposed to ensure that the covariance matrix is symmetric and positive definite. In particular, imposing the latter restriction is not very easy. The most general model of covariance matrix dynamics is specified by arranging all the coefficients in the matrix in a vector. The most well known specification is, however, that called BEKK (from the authors: Baba, Engle, Kraft, and Kroner, 1990). Calling $\mathbf{V}_t$ the covariance matrix at time t, this specification reads

$$\mathbf{V}_t = \Omega'\Omega + \mathbf{A}'\epsilon'_{t-1}\epsilon_{t-1}\mathbf{A} + \mathbf{B}'\mathbf{VB},$$

where Ω, $\mathbf{A}$ and $\mathbf{B}$ are n-dimensional matrices of coefficients. Very often, special restrictions are imposed on the matrices in order to reduce the dimension of the estimation problem. For example, typical restrictions are to assume the matrices $\mathbf{A}$ and $\mathbf{B}$ are diagonal, so limiting the flexibility of the representation.

In order to reduce the dimensionality of the problem, the typical recipe used in statistics is to resort to data compression methods: principal component analysis and factor analysis. Both these approaches have been applied to the multivariate GARCH problem. Engle, Ng and Rothschild (1990) resort to a factor GARCH representation in which the dynamics of the common factors are modeled as a multivariate factor model. This way, the dimension of the estimation problem is drastically reduced. Alexander (2001) proposes the so-called orthogonal GARCH model. The idea is to use principal component analysis to diagonalize the covariance matrix and to estimate a GARCH model on the diagonalized model. Eigenvectors are then used to reconstruct the variance matrix. The maintained assumption in this model is of course that the same linear transformation diagonalizes not only the unconditional variance matrix, but also the conditional one.

Besides data compression methods, a computationally effective alternative would be to separate the specification of the marginal distributions from the dependence structure. This approach to the specification of the system, that very closely resembles copula functions, was proposed by Engle (2002). The model is known as *Dynamic Conditional Correlation* (DCC). The idea is to separate the specification of a multivariate GARCH model into a two-step procedure in which univariate marginal GARCH processes are specified in the first step, and the dependence structure in the second step. More formally, the approach is based on standardized returns, defined as

$$\eta_{j,t} = \frac{r_{j,t}}{\sigma_{j,t}},$$

which are assumed to have standard normal distribution. Now, notice that the pairwise conditional correlation of return j and k is

$$\rho_{jk,t} = \frac{\mathbb{E}_t[r_{j,t} r_{k,t}]}{\sqrt{\mathbb{E}_t[r_{j,t}^2] \mathbb{E}_t[r_{k,t}^2]}} = \mathbb{E}_t[\eta_{j,t} \eta_{k,t}].$$

Engle (2002) proposes to model such conditional correlations in an autoregressive framework:

$$\chi_{jk,t} = \bar{\rho}_{jk} + \alpha \left(\eta_{j,t} \eta_{k,t} - \bar{\rho}_{jk} \right) + \beta \left(\chi_{jk,t-1} - \bar{\rho}_{jk} \right),$$

where $\bar{\rho}_{jk}$ is the steady-state value of the correlation between returns j and k.

The estimator proposed for conditional correlation is

$$\hat{\rho}_{jk,t} = \frac{\mathbb{E}_t[\chi_{j,t} \chi_{k,t}]}{\sqrt{\mathbb{E}_t[\chi_{j,t}^2] \mathbb{E}_t[\chi_{k,t}^2]}}.$$

Once written in matrix form, the DCC model can be estimated in a two-stage procedure similar to the IFM (inference functions for margins) method that is typical of copula functions. The model is also well suited to generalization beyond the Gaussian dependence structure, applying copula function specifications as suggested in Fengler *et al.* (2010).

1.8   COPULAS AND CONVOLUTION

In this book, we propose an original approach to the use of copulas in financial applications. Our main task is to identify the limits that must be imposed on the freedom of selecting whatever copula for whatever application. The motivation behind this project is that the feeling of complete flexibility and freedom that from the start was considered to be the main advantage of this tool in the end turned out to be its major limitation. As we are going to see, there is something of a paradox in this: the major flaw in the use of copulas in finance, which in the first place were applied to allow for non-linear relationships, is that it is quite difficult to apply them to linear combinations of variables. We could say that a sort of *curse of linearity* is haunting financial econometrics. To put it in more formal terms, in almost all applications in finance we face the problem of determining the distribution of a sum of random variables. If random variables are assumed to be independent, this distribution is called the *convolution* of the distributions of those variables. In a setting in which the variables are not independent, this convolution concept has to be extended. The extension of this concept to the general case of dependent variables is the innovation at the root of this book: we define and call *C-convolution* the distribution of the sum of variables linked by a copula function $C(.)$. We also show that once this convolution concept is established, we immediately obtain the copula functions that link the distribution of variables which are part of a sum and the convolution itself. It is our claim that this convolution restriction is the main limitation that must be imposed on the selection of copula functions if one wants to obtain well-founded financial applications. Throughout this book we provide examples of this consistency requirement in all typical fields of copula applications, namely multivariate equity and credit derivatives and risk aggregation and allocation.

As a guide to the reader, the book is made up of three parts. Chapter 2 recalls the main concepts of copula function theory and collects the latest results, limiting the review to the standard way in which copula functions are applied in finance. In Chapters 3 and 4 we construct the theory and the econometrics of convolution-based copulas. In Chapters 5, 6 and 7 we show that limiting the selection of copulas to members of this class enables us to solve many of the consistency problems arising in financial applications. In Chapter 8 we cast a bridge to future research by addressing the problem of convergence of our copula model, which is developed in a discrete time setting, to continuous time: the issue is open, and for the time being we cannot do any better than review existing copula models that were built in continuous time or that are able to deal with the convolution restriction.

To provide further motivation to the reader, we anticipate here examples in which the convolution restriction arises in applied work, in equity, credit and risk measurement. Consider multivariate equity or FOREX derivatives expiring at different dates. One of the current arguments against the use of copula functions is that they make it difficult to set price consistency between different dates. Indeed, both the efficient market hypothesis and the martingale-based pricing techniques require specific restrictions to the dynamics of market prices. These restrictions, which for short mean that price increments must be unpredictable, imply a restriction in the copula functions that may be used to represent the temporal dependence of the price process: more explicitly, future levels cannot be predicted using the dependence between increment and level. This restriction is exactly of the convolution type: the dependence of the price at different points in time must be consistent with the dependence between a price and its increment in the period, and the latter must be designed to fulfill the unpredictability requirement above. Let us now come to credit derivatives, and consider a typical term structure problem: assume you buy insurance on a basket of credit exposures for five years and sell the same insurance on the same basket for seven years. Could the prices of the two insurance contracts be chosen arbitrarily and independently of each other? The answer is clearly no, because the difference in the two prices is the value today of selling insurance for two years starting five years from now: put in other terms, the distribution of losses in seven years is the convolution of the losses in the first five years and those in the following two years; ignoring this convolution restriction could leave room for arbitrage possibilities. We may conclude with the most obvious risk management application: your firm is made up of a desk that invests in equity products and another one that invests in credit products. For each of the desks you may come up with a risk measure based on the profit and loss distribution. How much capital would you require for the firm as a whole? One more time, the answer to this standard question is convolution.

2

Copula Functions: The State of The Art

This chapter is devoted to a bird's-eye review of the basic concepts of the theory of *dependence functions* or *copula functions*, as they are more usually called. Experts in the field may skip this reading, or, if they do not, they are warned it could be boring. People who have never been exposed to concepts such as non-parametric association measures or tail dependence indexes are instead welcome to go through the chapter, and they will find almost all the basic tools they need to deal with the rest of the book. The menu includes a starter of basic concepts on copula functions, with some examples and their basic properties, a main course of the main families of copulas with a discussion of dependence measures on the side, and dessert of advanced modeling problems of copula functions. This will suffice as an introduction to the meals that will follow in the coming chapters. However, for those who want to undertake a career as a *chef* and get their own restaurant started, we suggest the classical books by Nelsen (2006) and Joe (1997) as mandatory readings.

2.1 COPULA FUNCTIONS: THE BASIC RECIPE

What ordinary people know about copula functions is that they are used to separate the specification of marginal distributions from the dependence structure. It is not difficult to take a deeper insight into this result and to become somewhat of an expert in the field. The key result upon which all of this theory is built is the well-known theorem of *probability integral transformation* of continuous random variables, or *Fisher transform*. All it says is that if one takes whatever random variable X with continuous distribution F_X (that is the function that denotes $\mathbb{P}(X \leq x) = F_X(x)$) and computes $U \equiv F_X(X)$, the variable U obtained as a result has uniform distribution in the unit interval (that is $\mathbb{P}(U \leq u) = u$, for all $u \in [0, 1]$). This is true for every random variable endowed with a continuous distribution (the only kind of distribution that will be addressed throughout this book), such as Y with distribution F_Y. So, $V \equiv F_Y(Y)$ will also be uniformly distributed. A question arises quite naturally: what about the joint distribution $F(x,y)$? Can we apply the probability integral transformation theorem to it? The answer is a resounding no, and one has always to keep in mind that the theorem holds for univariate distributions only. Nevertheless, the theorem may help to rewrite the joint distribution in an interesting way. Assume we may define the inverse function of both F_X and F_Y: notice that this can always be done because distribution functions are non-decreasing, even though it may happen that such an inverse is not unique in points at which some of the distribution functions are represented by horizontal lines. In order to avoid this problem, we can use the following concept of *generalized inverse* $F^{-1} : (0, 1) \rightarrow \mathbb{R}$ as

$$F^{-1}(w) = \inf \{l \in \mathbb{R} : F(l) \geq w\}, \tag{2.1}$$

that coincides with the standard inverse function when this exists. So, why not use this inverse function in the joint distribution? We get

$$F(X, Y) = F(F_X^{-1}(U), F_Y^{-1}(V)) \equiv C(U, V),$$

and we have rewritten the joint distribution of X and Y in terms of the variables U and V that are uniformly distributed, i.e.

$$F(x, y) = F(F_X^{-1}(u), F_Y^{-1}(v)) \equiv C(u, v), \qquad (2.2)$$

where $u = F_X(x)$ and $v = F_Y(y)$. The function that links the uniform variables to the joint distribution is called the *dependence function* or *copula function*. Notice that we could have done the same with any n-dimensional distribution. So, every joint distribution can be written as a function of n uniform distributions. An interesting question is how we can reverse the construction. That is: which requirements are to be met by the function $C(u,v)$ for it to deliver a well-defined joint distribution? Indeed, these requirements are quite light. First, for the function to return a probability it has to have a range in the unit interval (so, for example, a function like $C(u, v) = u + v$ would not do the job). Second, if the probability of $\{X \leq x\}$ is 0 ($F_X(x) = 0$), then we would like the joint probability $F(x,y)$ to be equal to zero as well. The same should of course hold for any y such that $F_Y(y) = 0$. Thus, if one of the arguments of $C(u,v)$ is equal to zero, then the function should return zero. Third, if $\{X \leq x\}$ has probability 1, we want the joint probability of X and Y evaluated at (x, y) to be equal to $F_Y(y)$, and symmetrically if $F_Y(y) = 1$ we want to get $F(x, y) = F_X(x)$. In plain words, if we ask what is the joint probability of two events and we know that one event will take place for sure, the joint probability will be equal to the probability of the other event. The last requirement is the most involved to explain and to prove in copula function theory. Consider the analogy with univariate probability theory. The probability that a variable will be in a given interval (a segment) cannot be negative, and we have $F_X(x_1) - F_X(x_2) \geq 0$ for $x_1 > x_2$. In a multivariate setting, we want to set the same requirement, but this time it will not apply to a segment, but to the *volume* of an n-dimensional figure (a rectangle in the two-dimensional case). So, in two dimensions for $x_1 > x_2$ and $y_1 > y_2$ we will have

$$F(x_1, y_1) - F(x_1, y_2) - F(x_2, y_1) + F(x_2, y_2) \geq 0. \qquad (2.3)$$

Substituting equation (2.2) into the requirement above completes the properties that describe a proper copula function. We formalize them below for future reference. We stick to the bivariate case for the time, but extension to higher dimensions is only a matter of notation: the very meaning of requirements remains the same.

Definition 2.1.1. A copula is a function $C : [0, 1]^2 \to [0, 1]$ satisfying the following requirements:

1. *Grounded*: $C(0, v) = C(u, 0) = 0$.
2. *Uniform marginals*: $C(u, 1) = u, C(1, v) = v$.
3. *2-Increasing*: $C(u_1, v_1) - C(u_1, v_2) - C(u_2, v_1) + C(u_2, v_2) \geq 0$ for $u_1 > u_2$ and $v_1 > v_2$.

The term *2-increasing* (*n-increasing* in the general case) is also referred to as the non-negativity condition on the volume. So, we have a sketch of the proof of what is known as the *Sklar theorem*, which is the basic pillar from which the theory of copula functions has grown up. In a nutshell, it means that every joint distribution can be written as a copula function taking the univariate distributions (technically, the *marginal* distributions) as arguments, and the other way around: including marginal distributions in a copula function yields a joint distribution. It is clear that the main advantage of this is that the specification of the joint distribution can be separated from the specification of marginal distributions. Then, we may conceive two models with the same marginal distributions and different joint distribution.

One may ask: where would the difference come from? It is obvious to conclude that the only difference would be in the way the variables co-move together, that is in their dependence structure. Copulas collect all the information concerning such co-movements.

The first copula that comes to mind corresponds to the dependence case we are taught in introductory courses in statistics, that is the case of no dependence at all. In this case the joint distribution is the product of the marginal ones: in fact, it is easy to prove that $C_{\perp}(u, v) \equiv uv$ is a copula function, and it is actually known as the *product copula*. It is not difficult to figure out what the copula should look like for the case in which the variables are perfectly dependent on each other. Actually, if two events are perfectly dependent, one can say that one happens whenever the other takes place, or the other way around: the latter would take place only if the former does. So, the probability of both the events taking place must be the minimum of the marginal probabilities, and in fact, it is easy to prove that $C_{max}(u, v) \equiv \min(u, v)$ is a copula. It is called the *maximal copula*. It is just as simple as this to prove that if two variables are perfectly negatively dependent, the copula function is $C_{min}(u, v) \equiv \max(u + v - 1, 0)$. In particular, this can be obtained by applying (2.6) below:

$$W(u, v) = u - M(u, 1 - v) = u - \min(u, v) = \max(u + v - 1, 0).$$

This is also called the *minimal copula*. The terms *minimal* and *maximal* immediately suggest that the value of a copula function increases with the dependence, and actually you may check that the product copula is in the middle. Based on these three copulas we can set up the first copula family, the easiest one.

Example 2.1.1. *Weighted averages of the maximum copula, the minimum copula and the product copula, make up what is called the* Fréchet *family of copulas. Formally, the Fréchet family of copulas is defined as*

$$C(u, v) \equiv \alpha \min(u, v) + (1 - \alpha - \beta)uv + \beta \max(u + v - 1, 0), \tag{2.4}$$

with $\alpha, \beta \geq 0$ and $\alpha + \beta \leq 1$. A special case that is often used in financial applications is that with $\beta = 0$:

$$C(u, v) \equiv \alpha \min(u, v) + (1 - \alpha)uv, \tag{2.5}$$

which is known as a mixture copula.

Remark 2.1.1. If $C_{X,Y}$ is the copula associated with the pair of random variables (X, Y), it is interesting (also because it will often be used in the sequel) to find what is the dependence structure that links the pair $(X, -Y)$. This can be trivially obtained by the following:

$$
\begin{aligned}
C_{X,-Y}(F_X(x), F_{-Y}(y)) &= \mathbb{P}(X \leq x, -Y \leq y) \\
&= \mathbb{P}(X \leq x) - \mathbb{P}(X \leq x, Y \leq -y) \\
&= \mathbb{P}(X \leq x) - C_{X,Y}(F_X(x), F_Y(-y)) \\
&= \mathbb{P}(X \leq x) - C_{X,Y}(F_X(x), 1 - F_{-Y}(y)).
\end{aligned}
$$

It follows that

$$C_{X,-Y}(u, v) = u - C_{X,Y}(u, 1 - v) \tag{2.6}$$

and, similarly,

$$C_{-X,Y}(u, v) = v - C_{X,Y}(1 - u, v). \tag{2.7}$$

The concept of a copula function extends to the n-dimensional case.

In order to specify this increasing property, we have first to introduce the definition of volume. Let $[\mathbf{u}, \mathbf{v}] = \prod_{i=1}^{n}[u_i, v_i]$ be the rectangle with $u_i \leq v_i$. The quantity

$$V_{[\mathbf{u},\mathbf{v}]} = \sum_{V \in R} sign(V)C(V)$$

is called the *volume* associated with the rectangle $[\mathbf{u}, \mathbf{v}]$. The summation above is over all the vertices $V = (z_1, \ldots, z_n)$ of the rectangle, where $z_i = u_i$ or v_i and

$$sign(V) = \begin{cases} -1, & \text{if the number of } u_i \text{ is odd,} \\ 1, & \text{otherwise.} \end{cases}$$

Definition 2.1.2. An n-copula is a function $C(u_1, \ldots, u_n)$ from $[0, 1]^n \rightarrow [0, 1]$ satisfying the following requirements:

1. *Grounded*: $C(u_1, \ldots, u_{i-1}, 0, u_{i+1}, \ldots, u_n) = 0$ for all i and all $u_1, \ldots, u_n$.
2. *Copula marginal*: $C(u_1, \ldots, u_{i-1}, 1, u_{i+1}, \ldots, u_n)$ is an $(n-1)$-copula for all i.
3. n -*Increasing*: $V_{[\mathbf{u},\mathbf{v}]} \geq 0$ for all rectangles $[\mathbf{u}, \mathbf{v}]$.

Remark 2.1.2. Empirical copula.   Here, we present the notion of the empirical copula introduced by Deheuvels (1979, 1981). Let $X = (X_t^1, \ldots, X_t^n)_{t \geq 0}$ be an i.i.d. sequence of n-dimensional vectors with continuous joint distribution F and continuous margins F_i. Let $\{x_{(t)}^1, \ldots, x_{(t)}^n\}$ be the order statistic and let $\{r_{(t)}^1, \ldots, r_{(t)}^n\}$ be the rank statistic of the sample. The Deheuvels' empirical copula defined on the lattice

$$l = \left\{ \left(\frac{t_1}{T}, \ldots, \frac{t_n}{T} \right) : t_i = 0, 1, \ldots, T, 1 \leq i \leq n \right\}$$

is the following function:

$$\hat{C}\left(\frac{t_1}{T}, \ldots, \frac{t_n}{T} \right) = \frac{1}{T} \sum_{t=1}^{T} \prod_{i=1}^{n} \mathbf{1}_{\{r_t^i \leq t_i\}}.$$

2.2 MARKET CO-MOVEMENTS

Having defined copulas and built the first copula family, we may get started with a first application to get a better understanding of the use of copula functions in finance. Consider the simplest pricing problem, that we will follow throughout this chapter as a term of reference.

Example 2.2.1. *Bivariate digital put option*. *Consider a European option paying a fixed sum if market A (say the S&P 500 index) and market B (say the Eurostoxx 50 index) are both below the corresponding strike prices K_A and K_B at time T.*

We want to *mark-to-market* this digital option. What are the ingredients of the price? Certainly there are option markets in which contracts for Eurostoxx and S&P options are actively traded. From these markets we may definitely recover consistent prices for univariate digital options. Whatever option pricing model we use, parametric or non-parametric, Breeden and Litzenberger (1978) provides a solution to the problem. So, we can first *mark-to-market* the prices of the univariate options. Call these univariate put options prices DP_i, with $i = A, B$. We want the price of the bivariate digital put option to be consistent with these prices.

So, we want

$$MDP(K_A, K_B) = v(t, T)f(DP_A, DP_B),$$ (2.8)

where MDP is the price of the bivariate option and $v(t,T)$ is the risk-free discount factor. It is not difficult to understand that addressing the problem of market consistency between the bivariate product and the univariate ones inevitably leads us to define a relationship between this function $f(x,y)$ and a copula function. The standard martingale approach to pricing in fact requires that $DP_i = v(t, T)Q_i(K_i), i = A, B$ where $Q_i(K_i) \equiv \mathbb{P}(S_i(T) \leq K_i)$ denotes the probability that each price S_i will be under the corresponding strike on the exercise date. Arbitrage-free valuation requires the existence of such a probability measure, which is called a *risk-neutral* or *equivalent martingale measure*. By the same token, this valuation approach would require the price of the bivariate digital option to be

$$\begin{aligned} MDP(t) &= v(t, T)\mathbb{P}(S_A(T) \leq K_A, S_B(T) \leq K_B) \\ &= v(t, T)C(Q_A(K_A), Q_B(K_B)) \\ &= v(t, T)C\left(\frac{DP_A}{v(t, T)}, \frac{DP_B}{v(t, T)} \right), \end{aligned}$$ (2.9)

and the connection with the function $f(x,y)$ above is so explicit that it does not deserve any further comment. So, using copulas ensures that the price we come up with for our multivariate derivative is consistent with the *mark-to-market* value of the univariate assets. We then have a tool to map every change of value of each and every underlying asset into a consistent change of value of the multivariate one.

Let us now address the crucial issue on which correlation trading is based. What contribution does the copula function make to the price? It may be the case that the price of each and every underlying asset does not change by a penny, while the value of the multivariate derivative can change substantially: how come? The answer is that the copula function represents the dependence structure among the underlying assets: for this reason, using a copula function or another one, or changing the parameter of a copula function affects the price of multivariate options. It must always be kept in mind that if a borrower or a structurer issues an option written on a basket of underlying assets, instead of a basket of options on that asset, it does so in order to take advantage, or allow someone to take advantage, of the dependence structure among the underlying assets and its changes. That is why these products are called *correlation products* and the business of dealing in them is called *correlation trading*.

Copulas are then useful because they single out the co-movement part among the assets from the movements of each one of them. Actually, it is intuitive to grasp that if two copulas can give different values to the same pair (or set) of marginal distributions, they must include all information about dependence. Moreover, it should be possible to represent the contribution of dependence to the price in terms of copulas, or their parameters. Using again our reference problem, one first could check whether

$$MDP(t) \geq v(t, T)Q_A(K_A)Q_B(K_B),$$ (2.10)

meaning that one could ask whether the value of the product is greater than it should be under the assumption of independence of the underlying assets. This corresponds to a concept that in statistics is known as *Positive Quadrant Dependence (PQD)*. It is still a qualitative concept, but it speaks out on the degree of dependence that is priced in the product. A second question concerns the maximum and minimum values the product could attain, and the answer

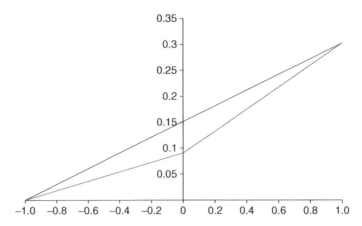

Figure 2.1 Bivariate digital put prices with respect to Spearman's ρ for marginals both equal to 0.3. Horizontal axis: Kendall's *tau*, vertical axis: copula values.

is obviously the maximal and minimal copula, respectively:

$$v(t, T)\max(Q_A(K_A) + Q_B(K_B) - 1, 0) \leq MDP(t) \leq v(t, T)\min(Q_A(K_A), Q_B(K_B)).$$
$$(2.11)$$

This is simply the financial restatement of a well-known concept in statistics, called the *Fréchet bounds*. Finally, if you want to come up with a more precise evaluation of the product sensitivity to changes in the dependence structure, within these extreme dependence cases, you can try the first example of a copula family that we just laid out. So, the price of this product under the *Fréchet family* of copulas will be

$$MDP(t) = v(t, T)(\beta C_{\min}(Q_A(K_A), Q_B(K_B))$$
$$+ (1 - \alpha - \beta)C_\perp(Q_A(K_A), Q_B(K_B)) + \alpha C_{\max}(Q_A(K_A), Q_B(K_B))). \quad (2.12)$$

Figure 2.1 depicts the different values that the product can take for given prices of the univariate digital options (that is, for given marginal distribution). The assumption is that the univariate digital put options are both worth 0.3. Given these marginals, the market price can take on the values comprised in the triangle. Notice that the impact of dependence is most relevant for a contract around the *at-the-money* (ATM) area. Just as for univariate options, where the impact of volatility changes fades away for deep *in-the-money* (ITM) and far *out-of-the-money* (OTM) contracts, the same happens here: the value of the product will converge to $v(t,T)$ or to 0 across the board as a contract is deep ITM or far OTM, respectively.

2.3 DELTA HEDGING MULTIVARIATE DIGITAL PRODUCTS

Let us consider a multivariate digital put with strikes $K_i, i = 1, \ldots, n$ whose price is

$$MDP_{X_1, \ldots, X_n}(K_1, \ldots, K_n) = C(u_1, \ldots, u_n),$$

where C is the copula function representing the dependence structure between the underlying assets and $u_i = F_{X_i}(K_i), i = 1, \ldots, n$. We observe that we cannot replicate this multivariate product by linearizing its payoff as a sum of univariate digital puts, since the multivariate product involves the dependence structure between the underlying assets.

By a Taylor series expansion approximated at first order, we recover the fundamental of the delta-hedging approach

$$\Delta MDP_{X_1,\ldots,X_n}(K_1,\ldots,K_n) = \sum_{i=1}^{n} D_i C(u_1,\ldots,u_n)\Delta DP_i,$$

where $D_i C$ stands for the partial derivative of copula C with respect to the ith argument. Then we can say that this product can be replicated by a portfolio of the underlying assets in a proportion corresponding to their delta, defined as

$$\Delta^{MDP} = \begin{bmatrix} D_1 C(u_1,\ldots,u_n) \\ \vdots \\ D_n C(u_1,\ldots,u_n) \end{bmatrix}.$$

Let us consider a univariate corridor whose payoff may be linearized, implying that it may be replicated by a portfolio of univariate digital puts

$$DCor(X) = \mathbb{P}(X \in [a,b])$$
$$= DP_X(b) - DP_X(a).$$

In the multivariate setting we cannot linearize the payoff in the same way. For example, the price of the bivariate corridor is

$$MDCor(X_1, X_2) = \mathbb{P}(a_1 \leq X_1 \leq b_1, a_2 \leq X_2 \leq b_2)$$
$$= \mathbb{P}(X_1 \leq b_1, X_2 \leq b_2) - \mathbb{P}(X_1 \leq b_1, X_2 \leq a_2)$$
$$-\mathbb{P}(X_1 \leq a_1, X_2 \leq b_2) + \mathbb{P}(X_1 \leq a_1, X_2 \leq a_2)$$
$$= V_C\left(\begin{bmatrix} F_{X_1}(a_1) \\ F_{X_2}(a_2) \end{bmatrix} \times \begin{bmatrix} F_{X_1}(b_1) \\ F_{X_2}(b_2) \end{bmatrix}\right),$$

where $V_C(\mathbf{S})$ is the volume of the copula function on the box $\mathbf{S}$. Then the bivariate product is replicable by a portfolio called the *hedging portfolio*, whose value is

$$HP = \Delta_{1,a_1} DP_1(a_1)_1 + \Delta_{2,a_2} DP_2(a_2) + \Delta_{1,b_1} DP_1(b_1) + \Delta_{2,b_2} DP_2(b_2)$$

where Δ_i, a_i and Δ_i, b_i is the partial derivatives of volume with respect to its arguments.

Example 2.3.1. *Bivariate digital put on exchange rates. We consider a bivariate digital put on the exchange rates euro/yen and euro/pound with strike equal to 0.8 times the starting historical value, denoted as $X_0^i, i = 1, 2$, respectively. From the historical data, we estimate the empirical c.d.f. of the exchange rates and then the marginal probabilities.*

We compute the empirical bivariate copulas of the exchange rates and estimate a Kendall $\tau = 0.031$ corresponding to a Clayton copula function parameter $\theta = 0.064$.

The price of the bivariate digital put corresponds to the value of a Clayton copula:

$$MDP_{EU/YEN,EU/POUND}(0.8X_0^1, 0.8X_0^2)$$
$$= C(F_{EU/YEN}(0.8X_0^1), F_{EU/POUND}(0.8X_0^2))$$
$$= 0.0834012.$$

To hedge this product we must construct a portfolio composed of the underlying exchange rates in proportion corresponding to the deltas:

$$\Delta_i = D_i C(F_{EU/YEN}(0.8X_0^1), F_{EU/POUND}(0.8X_0^2)), \quad i = 1, 2,$$

where $D_i C$ stands for the partial derivative of the copula function representing the price with respect to the ith argument.

As observed before, this product cannot be replicated with long positions in univariate digital puts. In fact, the price of this portfolio is:

$$
\begin{aligned}
P &= DP_{EU/YEN}(0.8X_0^1) + DP_{EU/POUND}(0.8X_0^2) \\
&= 0.55.
\end{aligned}
$$

Example 2.3.2. *Bivariate digital corridor on exchange rates. We consider a bivariate digital corridor on the exchange rates euro/yen and euro/pound and we choose a corridor, on the basis of the comparison of the empirical distributions of the exchange rates normalized with respect to the historical starting value, equal to [0.9, 1.1] of the starting values of the historical data, denoted as X_0^i, $i = 1, 2$, respectively. From the historical data, we estimate the empirical c.d.f. of the exchange rates and then the marginal probabilities corresponding to the corridor.*

The price of the bivariate digital corridor corresponds to the volume of a copula (here we work with a Clayton copula) in the box defined by the corridor:

$$
\begin{aligned}
MDCor(EU/YEN, EU/POUND) &= \mathbb{P}\left(\frac{EU/YEN}{X_0^1} \leq 1.1, \frac{EU/POUND}{X_0^2} \leq 1.1\right) \\
&\quad -\mathbb{P}\left(\frac{EU/YEN}{X_0^1} \leq 1.1, \frac{EU/POUND}{X_0^2} \leq 0.9\right) \\
&\quad -\mathbb{P}\left(\frac{EU/YEN}{X_0^1} \leq 0.9, \frac{EU/POUND}{X_0^2} \leq 1.1\right) \\
&\quad +\mathbb{P}\left(\frac{EU/YEN}{X_0^1} \leq 0.9, \frac{EU/POUND}{X_0^2} \leq 0.9\right) \\
&= V_C\left(\begin{bmatrix} F_{\frac{EU/YEN}{X_0^1}}(0.9) \\ F_{\frac{EU/POUND}{X_0^2}}(0.9) \end{bmatrix} \times \begin{bmatrix} F_{\frac{EU/YEN}{X_0^1}}(1.1) \\ F_{\frac{EU/POUND}{X_0^2}}(1.1) \end{bmatrix}\right) \\
&= V_C\left(\begin{bmatrix} 0.693 \\ 0.9225 \end{bmatrix} \times \begin{bmatrix} 1 \\ 1 \end{bmatrix}\right) \\
&= 0.0394,
\end{aligned}
$$

where the marginals into the box correspond to the empirical c.d.f. at the percentile defining the corridor and the parameter of the Clayton copula function is estimated from historical data ($\theta = 0.064$). We can see that the probability of having all the exchange rates into the corridor is more than 3%.

We can observe that in the univariate case, the digital corridor may be replicated by a long position on a digital put with strike equal to the superior bound of the corridor and a short position on a digital put with strike equal to the inferior bound of the corridor. Anyway, in the multivariate case this replication approach does not run. In fact, the bivariate product under examination cannot be replicated with a long position in a bivariate digital put with strike equal to the superior extremes of the volume box and a short position on a bivariate digital

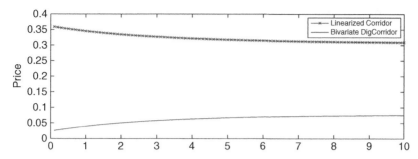

Figure 2.2 Bivariate digital corridor vs linearized corridor prices with respect to the dependence parameter of the Clayton copula function (horizontal axis).

put with strike equal to the inferior extremes of the volume box. The price of this portfolio that we call the linearized corridor is:

$$LCor = C(F_{EU/YEN}(1.1X_0^1), F_{EU/POUND}(1.1X_0^2)) - C(F_{EU/YEN}(0.9X_0^1), F_{EU/POUND}(0.9X_0^2))$$
$$= MDPut_{EU/YEN,EU/POUND}(1.1X_0^1, 1.1X_0^2) - MDPut_{EU/YEN,EU/POUND}(0.9X_0^1, 0.9X_0^2)$$
$$= 0.3451.$$

If we consider a shift of the dependence parameter of 10% each time, the prices of the corridor and of the linearized corridor move as depicted in Figure 2.2.

For replicating purposes we need to compute the delta of the corridor product:

$$\Delta^{Cor} = DV_C \left(\begin{bmatrix} 0.693 \\ 0.9225 \end{bmatrix} \times \begin{bmatrix} 1 \\ 1 \end{bmatrix} \right)$$
$$= \begin{bmatrix} 0.0019 \\ 0.0014 \end{bmatrix},$$

where DV_C stands for the vector of the partial derivatives of the volume with respect to the arguments of the copula function.

In the same spirit, we can also replicate the linearized corridor with a portfolio represented by the following delta:

$$\Delta^{LCor} = DC(F_{EU/YEN}(1.1X_0^1), F_{EU/POUND}(1.1X_0^2))$$
$$- DC(F_{EU/YEN}(0.9X_0^1), F_{EU/POUND}(0.9X_0^2))$$
$$= \begin{bmatrix} 0.0804 \\ 0.3217 \end{bmatrix}.$$

where DC is the vector of the partial derivatives of the copula C with respect to the arguments of the function.

2.4 LINEAR CORRELATION

Since copulas are linked to the dependence structure, they must be related to dependence measures. Let us start with the most usual one, that we are taught at school, which is correlation. Actually, we would rather say *linear correlation* throughout this book. The term *linear* should not be forgotten, since this measure is explicitly designed and well suited to represent *linear* relationships. We will soon see that this is a relevant limitation of this measure. Technically, the correlation measure we typically use is called the *Pearson correlation* measure. It is the covariance figure divided by the product of standard deviations. Just focus on the covariance figure, by assuming the data to be normalized to unit variance. Covariance can be written as

$$\int \int (F(x, y) - F_X(x)F_Y(y))\,dxdy, \tag{2.13}$$

where the integral is taken over the whole support of the variables X and Y. This is the crucial point. The integral is taken over the domain of the variables and depends not only on the joint distribution, but also on the marginal ones. For this reason, the measure will depend on all of them. This is a problem because it turns out that it may reach a level lower than 1 even if $F(X, Y) = \min(F_X(X), F_Y(Y))$, so that the variables are perfectly dependent (and you would expect it to be 1). By the same token, it may happen that you record a value higher than -1 even if the variables are perfectly negatively associated. If you want to check an extreme case yourself, take X with standard normal distribution, and take $Y \equiv X^2$. Are X and Y independent? Certainly not at all, they are indeed perfectly dependent. See what happens if you measure the covariance. You get

$$\mathbb{E}[XY] - \mathbb{E}[X]\mathbb{E}[Y] = \mathbb{E}[X^3] - 0 = 0, \tag{2.14}$$

by the assumption on X. What you have learned is that linear correlation is not good at measuring dependence structures that are not linear. Looking at the example, you could also put it another way. Linear correlation does not work for different probability distributions (one standard normal and the other chi-square, in the case at hand).

So, without getting to extreme cases, linear correlation is a measure for which you do not know the upper and lower bounds. For this reason you have to be careful when you use it in pricing, and when your counterpart wants to convince you in favor or against it. For a general review of the pitfalls of correlation, you are referred to Embrechts *et al.* (2002). To get an idea of how much the dependence structure measure corresponding to a price can be biased, you may take a look at the experiments in Cherubini and Luciano (2002). To conclude, in a non-Gaussian world you must consider either scaling linear correlation with respect to its maximum bounds (that change with the marginal distributions) or substituting it with other measures that ensure, when you have perfect dependence, you actually get 1 and -1 values.

2.5 RANK CORRELATION

By definition, non-parametric measures can be used independently of the marginal distributions, and they possess the desirable property of yielding 1 if the variables are perfectly dependent, and -1 if they are perfectly negatively associated. To construct the first and most widely used such measure, just consider the following intuitive idea. Why not apply the *Pearson correlation measure* to the marginal probability measures instead of the raw variables? By the integral transformation theorem we know that such distributions are in fact uniform, and the joint one is a copula. So, we have

$$\rho_S \equiv \frac{\mathbb{E}[UV] - \mathbb{E}[U]\mathbb{E}[V]}{\sqrt{Var(U)Var(V)}}, \tag{2.15}$$

where ρ_S is called *Spearman's ρ* (or *rank correlation*). Since U and V are uniformly distributed, we have $\mathbb{E}[U] = \mathbb{E}[V] = 1/2$ and $Var(U) = Var(V) = 1/12$. We may then rewrite the above equation as

$$\rho_S = 12 \int_0^1 \int_0^1 C(u, v) du dv - 3. \tag{2.16}$$

Now, the integral is in the unit square no matter what the marginal distributions are, and they do not affect the measure. Notice that the value only depends on the copula function. You may verify that you get 1 if $C(u, v) = \min(u, v)$ and -1 if $C(u, v) = \max(u + v - 1, 0)$. Also, you will get 0 by setting $C(u, v) = uv$.

The algorithm to compute *rank correlation* is straightforward: just change the variables to probabilities (*ranks*) and compute the Pearson correlation among them. Notice that in financial applications these probabilities could be represented either by ranks or by implied probabilities, as the following example shows.

Example 2.5.1. *Assume we want to price the bivariate digital put option above. We have the time series of option prices and the smiles. In order to come up with a price consistent with such a history, we may resort to historical simulation. We complete the following steps:*

- *Use past values from the option market to compute a sequence of values of the digital options for markets A and B.*
- *Compute the Pearson correlation figure between the two sequences.*
- *Select a copula function and calibrate it to the dependence coefficient estimated.*
- *Use the copula function selected to compute the price of the bivariate digital option.*

In the above algorithm the most difficult step is probably that of calibrating the copula function to the rank correlation measure. This may have to be done by numerical integration. Luckily, in some cases the integral can be computed in closed form. This is the case with our Fréchet family of copulas (see Example 2.1.1), for which we have the very simple expression (that should have been expected from the linearity property of integration)

$$\rho_S = \alpha - \beta. \tag{2.17}$$

In this case you have even more degrees of freedom than those you need. You may, in fact, restrict the search to the *mixture copula* case, so that you set $\beta = 0$. This gives you the price

$$MDP(t) = v(t, T)((1 - \rho_S)C_\perp + \rho_S C_{\max}). \tag{2.18}$$

Looking back at the triangle in Figure 2.1. It is easy to check that this price represents a bound with respect to the *Fréchet family*. Actually, it is the price corresponding to one of the edges of the triangle. It is very easy to come up with an upper price within the Fréchet family: that would be the opposite edge, corresponding to the condition $\alpha + \beta = 1$. So, this bound is

$$MDP(t) = v(t, T) \left(\frac{1 - \rho_S}{2} C_{\min} + \frac{1 + \rho_S}{2} C_{\max} \right). \tag{2.19}$$

Another measure of association that is commonly used in non-parametric statistics is *Kendall's* τ. This refers to the probability of extracting concordant pairs from the distribution (that is, $u_1 > v_1$ and $u_2 > v_2$) with respect to discordant pairs ($u_1 > v_1$ and $u_2 < v_2$ or vice versa). The measure would yield zero if the probability of extracting a concordant pair is equal to that of a discordant pair. We refer the reader to Nelsen (2006) for a detailed study of this, and other non-parametric measures. Here we just report the representation that links the *Kendall's*

τ figure to copula functions:

$$\tau = 4 \int_0^1 \int_0^1 C(u, v) dC(u, v) - 1$$
$$= 4\mathbb{E}[C(U, V)] - 1. \tag{2.20}$$

The equation provides a suggestive relationship between the *Kendall's* τ measure and the expected value of the copula function.

2.6 MULTIVARIATE SPEARMAN'S RHO

The following are three d-dimensional population versions of Spearman's rho:

$$\rho_1 = h(d) \left(2^d \int_{[0,1]^d} C(\mathbf{u}) d\mathbf{u} - 1 \right),$$

$$\rho_2 = h(d) \left(2^d \int_{[0,1]^d} \Pi(\mathbf{u}) dC(\mathbf{u}) - 1 \right),$$

$$\rho_3 = h(2) \left(4 \sum_{k<l} \binom{d}{2}^{-1} \int_{[0,1]^2} C_{kl}(u, v) du dv - 1 \right),$$

where $\mathbf{u} = (u_1, \ldots, u_d)$, $h(x) = \frac{1+x}{2^x-(1+x)}$, and $C_{kl}(u, v)$ refers to the bivariate marginal copula of C which corresponds to the kth and lth margin. The quantity ρ_1 was originally considered in Ruymgaart and van Zuijlen (1978), ρ_2 appeared first in Joe (1990), whereas ρ_3 is known as the population version of the weighted average pairwise Spearman's ρ given in Kendall (1970). The motivation of ρ_1 and ρ_2 becomes clear if we consider the following interpretation of the two-dimensional Spearman's ρ. Let (X, Y) be a two-dimensional random vector with copula C, since

$$\rho = \frac{Cov(U, V)}{\sqrt{Var(U)}\sqrt{Var(V)}} = 12 \int_{[0,1]^2} C(u, v) du dv - 3,$$

where (U, V) have joint distribution function C, we can write

$$\rho = \frac{\int_{[0,1]^2} C(u, v) du dv - \int_{[0,1]^2} \Pi(u, v) du dv}{\int_{[0,1]^2} M(u, v) du dv - \int_{[0,1]^2} \Pi(u, v) du dv} = \frac{\int_{[0,1]^2} uv dC(u, v) - \int_{[0,1]^2} uv d\Pi(u, v)}{\int_{[0,1]^2} uv dM(u, v) - \int_{[0,1]^2} uv d\Pi(u, v)},$$

because $\int_{[0,1]^2} M(u, v) du dv = \frac{1}{3}$ and $\int_{[0,1]^2} \Pi(u, v) du dv = \frac{1}{4}$. Then, ρ can be interpreted as the normalized average distance between the copula C and the independence copula Π. A natural d-dimensional extension of ρ is then

$$\frac{\int_{[0,1]^d} C(\mathbf{u}) d\mathbf{u} - \int_{[0,1]^d} \Pi(\mathbf{u}) d\mathbf{u}}{\int_{[0,1]^d} M(\mathbf{u}) d\mathbf{u} - \int_{[0,1]^d} \Pi(\mathbf{u}) d\mathbf{u}},$$

and it is straightforward to show that

$$\frac{\int_{[0,1]^d} C(\mathbf{u}) d\mathbf{u} - \int_{[0,1]^d} \Pi(\mathbf{u}) d\mathbf{u}}{\int_{[0,1]^d} M(\mathbf{u}) d\mathbf{u} - \int_{[0,1]^d} \Pi(\mathbf{u}) d\mathbf{u}} = h(d) \left(2^d \int_{[0,1]^d} C(\mathbf{u}) d\mathbf{u} - 1 \right) = \rho_1.$$

An analogous extension can be shown for ρ_2. Observe that for $d = 2$ we have $\rho_1 = \rho_2 = \rho_3$. Now consider a copula C_0, which is a convex combination of the marginal copulas corresponding to the kth and lth margin of $C(\mathbf{u})$, that is

$$C_0(\mathbf{u}) = \sum_{k<l} \binom{d}{2}^{-1} C_{kl}(u_k, v_l) \prod_{i \neq k,l} u_i.$$

In this case,

$$\rho_3(C) = h(2) \left(2^d \int_{[0,1]^d} C_0(\mathbf{u})d\mathbf{u} - 1 \right) = \frac{h(2)}{h(d)} \rho_1(C_0).$$

The coefficient ρ_3 has certain disadvantages as a multidimensional measure of association since it depends on the bivariate copulas only. For example, consider the three-dimensional Farlie–Gumbel–Morgenstern copula $C(u_1, u_2, u_3) = u_1 u_2 u_3 [1 + \theta(1 - u_1)(1 - u_2)(1 - u_3)]$, $|\theta| \leq 1$. This copula possesses independent bivariate margins, therefore $\rho_3 = 0$ for all θ. However, $\rho_1 = \frac{\theta}{6^3}$.

2.7 SURVIVAL COPULAS AND RADIAL SYMMETRY

We now introduce a concept that will be paramount in pricing applications. Take the joint probability distribution $F(x, y) \equiv \mathbb{P}(X \leq x, Y \leq y)$. The question is: what is the joint probability $\overline{F}(x, y) \equiv \mathbb{P}(X > x, Y > y)$? Obviously, it cannot be random. Actually, $\overline{F}(x, y)$ and $F(x,y)$ have to be linked by the following relationship:

$$\overline{F}(x, y) = \mathbb{P}(X > x) + \mathbb{P}(Y > y) - 1 + \mathbb{P}(X \leq x, Y \leq y). \tag{2.21}$$

The probability $\overline{F}(x, y)$ is called the *survival probability* (the definition comes from actuarial science). Of course, Sklar's theorem also applies to the survival probability. That is, we can recover a copula function $\hat{C}(u, v)$ such that

$$\overline{F}(x, y) = \hat{C}\left(1 - F_X(x), 1 - F_Y(y)\right). \tag{2.22}$$

This is called the *survival copula* corresponding to $C(u,v)$. Using the Fisher transform and Sklar's theorem in (2.21), it is easy to define

$$\hat{C}(u, v) = u + v - 1 + C(1 - u, 1 - v). \tag{2.23}$$

The above relationship is essential to ensure *put–call parity* among multivariate assets, and it will occur many times throughout the book. The following example explains the link.

Example 2.7.1. *Bivariate digital call option.* *Consider a European option paying a fixed sum if market A (say the S&P 500 index) and market B (say the Eurostoxx 50 index) are both above the corresponding strike prices K_A and K_B at time T. Assume we already know the prices of the corresponding bivariate digital put option price $MDP(K_A, K_B)$ from equation (2.9), and we know the prices of the univariate digital call options $DC_A(K_A)$ and $DC_B(K_B)$. Consider the following strategy:*

* *long one unit $DC_A(K_A)$, one unit $DC_B(K_B)$, and one unit $MDP(K_A, K_B)$;*
* *short one unit of risk-free asset.*

It is easy to check that the strategy provides perfect static replication of the bivariate digital call MDC(K_A, K_B). To see this, consider the three possible scenarios:

- *Both $S_A \geq K_A$ and $S_B \geq K_B$, then, both the univariate digital options will end in-the-money, and upon paying the one unit funding in the risk-free asset will leave one unit of payoff.*
- *Both $S_A < K_A$ and $S_B < K_B$, then only the bivariate digital put option will earn the payoff, and that will enable facing the funding.*
- *Only one of the events $S_A \geq K_A$ or $S_B \geq K_B$ occurs, in which case only one of the univariate digital options will earn a payoff, and this will be passed through to the funding.*

So, we have proved that

$$
\begin{aligned}
MDC(K_A, K_B) &= DC_A(K_A) + DC_B(K_B) - v(t, T) + MDP(K_A, K_B) \\
&= v(t, T)(\overline{Q}_A(K_A) + \overline{Q}_B(K_B) - 1 + C(1 - \overline{Q}_A(K_A), 1 - \overline{Q}_B(K_B)) \\
&= v(t, T)\hat{C}\left(\frac{DC_A}{v(t, T)}, \frac{DC_B}{v(t, T)}\right),
\end{aligned}
\tag{2.24}
$$

where we define $\overline{Q}_i(K_i) \equiv \mathbb{P}(S_i \geq K_i)$, $i = A, B$, the probability being computed under the equivalent martingale measure.

The above example shows that the relationship between each copula and its survival counterpart actually enforces a no-arbitrage relationship between multivariate call and put options. Actually, if we were to observe traded prices for the bivariate digital options above we could recover from the market the *implied* dependence measure, just like we recover implied volatility, or implied probability from univariate option prices. Just like in the univariate case, we could recover a *smile effect*, with the dependence measure changing with the degree of "moneyness" of the product. We could also observe *skews*, that is a different degree of dependence for products paying in upward scenarios, with respect to those paying in downward ones. And, like in the univariate problem, the shape of the smile or skew in volatility depends on symmetry or asymmetry of the distribution; in the multivariate case the same symmetry concept will apply to copulas. More precisely, this symmetry concept is known as *radial symmetry* and is defined as the equality of a copula function and its survival version. That is, a copula is said to be radially symmetric if for all u and v, we have

$$
\hat{C}(u, v) = C(u, v).
\tag{2.25}
$$

Notice that this implies $MDP(x, y) = MDC(x, y)$ and a symmetric smile of the dependence measures backed out from such prices.

The same arguments above extend to products with a larger number of underlying assets, even though computing the survival version of a large-dimensional copula function may be computationally demanding. Handling the problem is actually the same as computing the volume of the copula between two points, as we are going to show below.

2.8 COPULA VOLUME AND SURVIVAL COPULAS

There is a tight connection between the task of computing the volume of a copula function and that of computing its survival version. The link is immediately clear in a two-variable setting.

Consider the volume of a copula function between two points $u_1 > u_2$ and $v_1 > v_2$:

$$C(u_1, v_1) - C(u_1, v_2) - C(u_2, v_1) + C(u_2, v_2).$$

Now take the case $u_1 = v_1 = 1$. By the uniform marginal property of copula functions we have

$$1 - v_2 - u_2 + C(u_2, v_2).$$

Now add and subtract 1 and define $\hat{v}_2 = 1 - v_2$ and $\hat{u}_2 = 1 - u_2$:

$$\hat{v}_2 + \hat{u}_2 - 1 + C(1 - \hat{u}_2, 1 - \overline{v}_2) = \hat{C}(\hat{u}_2, \hat{v}_2).$$

So, we have proved that computing the survival version of a copula C at the point (u, v) amounts to computing the value of the volume of that copula between the point (u, v) and the vertex $(1, 1)$.

Computing survival copulas is then the same combinatorial problem as computing the volume of the copula between two points. Unfortunately, the problem may get very involved for higher dimensions. Cherubini and Romagnoli (2009) propose an algorithm to solve the problem in n dimensions, with $n \leq 20$. We report it below.

Remark 2.8.1. Computation of the survival copula functions. Assume we want to compute the survival version of the copula $C(w_1, w_2, \ldots, w_n)$. We use a vector notation, denoting by **w** the vector including the coordinates of the copula: $\mathbf{w} = \{w_1, w_2, \ldots, w_n\}$. We also use a vector **c**, which we update at each recursion.

Algorithm (pseudo code)

1. Initialize the vector $\mathbf{c} = [c_1, c_2, \ldots, c_n] = \mathbf{e}$, the unit vector in n dimensions
2. Set $\hat{C} = 0$
3. For $i = 0$ to $2^n - 1$

   Change i into a binary number and arrange the elements of such number in the vector $\mathbf{p} = [p_1, p_2, \ldots, p_n]$

       $k = 0$

       For $j = 1$ to $f(\log_2(i + 1))$

           if $p_j = 1$ then

               $c_j = w_j$

               $k = k + 1$

           end if

       next j

   $\hat{C} = \hat{C} + (-1)^k C(\mathbf{c})$

   next i
4. Return $\hat{C}$

where $f(\log_2(i + 1))$ denotes the function reporting the upper integer value of $\log_2(i + 1)$.

As for the scalability of the algorithm above, the authors find that good performances in terms of computing time can be achieved even in Matlab for dimensions up to 20. Beyond this dimension, the problem becomes unfeasible because of the physical limits of memory. Anyway, this is quite good for most equity applications, which do not exceed this dimension (the typical size of basket equity derivatives is between 5 and 15 underlying assets). So, in the

Table 2.1 Clayton dependence parameter estimated from the historical data

Stock index	Nikkei 225	FTSE 100	Nasdaq 100	Eurostoxx
Nikkei 225	–	2.4422	1.3740	0.129
FTSE 100	2.4422	–	2.6487	0.1345
Nasdaq 100	1.3740	2.6487	–	0.0932
EuroStoxx	0.129	0.1345	0.0932	–

equity market, the put–call parity of these products can easily be handled even with standard computing power.

Example 2.8.1. *Multivariate digital put and call on stock index. Let us consider a digital put option on four stock indexes, i.e. the Nikkei 225, FTSE 100, Nasdaq 100, and Eurostoxx. To solve this pricing problem, we choose to represent the dependence by a Clayton copula function. We estimated the dependence parameters reported in Table 2.1 from the historical data for the period from January 2002 to September 2010.*

The prices of the digital put and call, for an average estimated Clayton dependence parameter equal to 1.1369 and for a strike equal to 0.67 of the starting value of the historical data, denoted as X_0^i, $i = 1, 2, 3, 4$, are

$$MDP_{Nikkei,FTSE,Nasdaq,EStoxx}(0.67X_0^1, 0.67X_0^2, 0.67X_0^3, 0.67X_0^4)$$
$$= C(u_{Nikkei}, u_{FTSE}, u_{Nasdaq}, u_{EStoxx})$$
$$= C(0.594, 0.3, 0.9, 0.482)$$
$$= 0.2021$$
$$MDC_{Nikkei,FTSE,Nasdaq,EStoxx}(0.67X_0^1, 0.67X_0^2, 0.67X_0^3, 0.67X_0^4)$$
$$= \overline{C}(u_{Nikkei}, u_{FTSE}, u_{Nasdaq}, u_{EStoxx})$$
$$= 0.0547,$$

where $\overline{C}$ is the survival Clayton copula function. In Figure 2.3 we represent the prices of the digital call and put for different values of the Clayton dependence parameter.

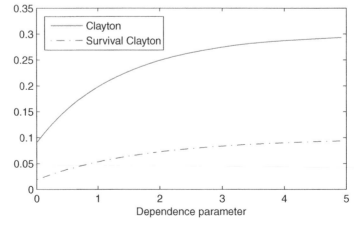

Figure 2.3 Digital put and call prices with respect to the Clayton dependence parameter.

2.9   TAIL DEPENDENCE

In univariate settings, non-normality of returns is often referred to the so-called *fat tails* phenomenon, meaning that the probability of extreme events is larger than expected according to the normal distribution. When the problem is cast in a multidimensional setting, a new feature comes into play, and it has to do with the association of extreme events. It is indeed relevant whether large downward and upward price movements are likely to occur at the same time in different markets, or whether they are independent from each other. This concept is measured by the so-called *tail dependence* indexes. Again, by using copulas this measure can be recovered without any reference to marginal distributions. The result follows from a straightforward application of *Bayes' theorem*. We want to assess what is the probability of observing an extreme price movement in market A (say a movement with very low probability p), given that an event with the same low probability has occurred in market B. This conditional probability may be written as

$$\lambda(p) = \frac{C(p, p)}{p}. \tag{2.26}$$

Tail dependence is measured by taking the limit of this conditional distribution with p tending to 0. We then define

$$\lambda_L \equiv \lim_{p \to 0^+} \lambda(p) = \lim_{p \to 0^+} \frac{C(p, p)}{p}. \tag{2.27}$$

The measure λ_L denotes the *lower tail index*, and represents the association of simultaneous major crashes in the markets. By the same token, the *upper tail index* is recovered from the survival copulas, computing

$$\lambda_U \equiv \lim_{w \to 1^-} \lambda(p) = \lim_{p \to 1^-} \frac{1 - 2p + C(p, p)}{1 - p}. \tag{2.28}$$

Notice that even in this case copulas have delivered the flexibility required. We can then model joint distributions with fat tails for the marginal distributions and no tail dependence: this means that each market would have a higher probability than those expected for normal distributions, but they will not be expected to crash or soar together. Alternatively, we could model distributions with Gaussian marginal probabilities and tail dependence: in this case markets will experience extreme events at the same time. Finally, we are free to model upper and lower tail indexes independently. So, even with respect to tails, we can choose symmetric or asymmetric distributions. In fact, it may easily be checked that if a copula function is radially symmetric, then $\lambda_L = \lambda_U$.

2.10   LONG/SHORT CORRELATION

It is by now clear that when buying or selling multivariate contracts involving a set of underlying assets one wants to exploit changes in their dependence structure. A question to be addressed is: in which direction? The party of the contract that takes advantage of an increase in dependence is typically said to be *long correlation*. The party that is instead negatively affected by an increase in correlation is said to be *short correlation*.

There is a very useful rule of thumb that can be used to assess whether one is long or short correlation. We could call it the AND/OR rule. Whenever, under a contract, you are getting paid if one event AND other events take place, you are *long correlation*. The reason is that under the no-arbitrage assumption the price is given by the joint probability of the events and if the degree of dependence among the events increases, the copula and the joint distribution increase as well. The bivariate digital options in the examples above are cases of *long correlation* positions (for the buyer, of course). The question that remains open is what about the short positions (the OR contracts).

We are now going to see that in fact OR operators can be described using formulas in which the copula function enters with a minus sign in front of it. Constructions like these, corresponding to OR statements, bear the name *co-copulas*:

$$\check{C}(u, v) = 1 - C(u, v),$$

and *dual of a copula:*

$$\check{C}(u, v) = u + v - C(u, v),$$

where $C(u,v)$ is a copula function.

We are now going to see some examples of products for which the payment is contingent on an OR statement, and we will check how their replicating portfolios actually generate either a *co-copula* or the *dual* of a copula.

Example 2.10.1. *Digital* **best-of** *options. We now modify slightly the examples of the bivariate European call or European put digital options above. Take the digital call option as in Example 2.7.1. This was paying one unit of cash if both the underlying assets, namely market A (the S&P 500) and market B (the Eurostoxx 50) were above the strike levels K_A and K_B. Notice that we could also have called this product a* digital call on the minimum *of the two assets, or else a* digital worst-of call option. *Consider now the case of a product that pays one unit of cash if at least one of the two markets is above the corresponding strike. By the same token, we could also call this product a* digital call on the maximum *or a* digital best-of call option. *Let us now see how to replicate this option. Remember that we denote by $DC_A(K_A)$ and $DC_B(K_B)$ the two univariate digital call options, and by $MDC(K_A, K_B)$ the bivariate digital call. Now consider the following strategy:*

- *long one unit of $DC_A(K_A)$ and $DC_B(K_B)$,*
- *short one unit of $MDC(K_A, K_B)$.*

It is easy to check that this strategy provides perfect replication of the digital best-of *option. Consider the three possible scenarios:*

- *Both markets end above the corresponding strike prices at expiration of the contract, in which case all the assets involved in the strategy pay one unit of cash: the result is a net worth of one unit of cash.*
- *Only one of the two markets ends above the corresponding strike price: in this case only that contract delivers one unit of cash, while the other two expire worthless.*
- *Both markets end below the corresponding strike prices: in this case all the contracts expire worthless.*

Notice that the replicating portfolio is actually the dual *of the copula, and it is for this reason that the value of the product decreases when dependence increases. One could ask: what about the* co-copula? *We leave it as an exercise to check that by substituting the* put–call parity *relationship between the call option* $MDC(K_A, K_B)$ *and the corresponding put* $MDP(K_A, K_B)$, *one obtains another replicating portfolio:*

- *long one unit of the risk-free asset* $v(t,T)$,
- *short one unit of the bivariate digital put option* $MDP(K_A, K_B)$,

and this portfolio corresponds to the co-copula *above.*

Before going on it is worthwhile to invite the reader to stay focused on the problem and to address the same long/short correlation question about put options. It will turn out that this time the bivariate *put option on the maximum* (the put on the *best-of*) would be long correlation, while the put on the *worst-of* will be short. So, it is not the maximum/minimum or the best/worst feature that makes the difference, but it is the AND/OR rule that discriminates. To conclude on this topic, it is mandatory to report a couple of examples on the credit market, which is where correlation trading is most developed.

Example 2.10.2. First-to-default option, FTD. *Consider a set of exposures, that we refer to as* names, *and a contract that provides payment of a given sum (again, a* digital *contract) if at least one of the names goes into default by a given maturity. This contract is called a* first-to-default *option, or swap. Is it long or short correlation? The answer is easy. The product pays only if not all the* names *survive to maturity. Formally, let us denote by* $\mathbf{1}_j$ *the indicator function taking value 1 if the jth name defaults in the period, with* $j = 1, \ldots, n$ *and n the number of names in the basket. The price of the FTD option would then be*

$$FTD = v(t, T)(1 - \hat{C}(\mathbb{P}(1 - \mathbf{1}_1 > 0), \ldots, \mathbb{P}(1 - \mathbf{1}_n > 0))),$$

where $\hat{C}(\cdot, \cdot)$ *denotes the survival copulas of the names. It is immediate to see that an increase in dependence increases the value of the survival copula and so reduces that of the FTD option.*

By symmetry, we can construct an example of a credit product which is long correlation.

Remark 2.10.1. nth-to-default option.   In the same setting as above, that is a basket of credit exposures to *n names*, consider an option or a swap that pays a sum at the *n*th default. Of course, this means that the payment occurs only if all the *n* names have defaulted. Formally, the price will be

$$n - TD = v(t, T)C(\mathbb{P}(1 - \mathbf{1}_1 \leq 0), \ldots, \mathbb{P}(1 - \mathbf{1}_n \leq 0)),$$

and the price is clearly increasing with the dependence among defaults.

2.11 FAMILIES OF COPULAS

We are now going to illustrate the main families of copula functions, namely elliptical and Archimedean. We also give a brief account of the way they are created, and for the sake of completeness we generalize the analysis to the *n*-dimensional setting.

2.11.1 Elliptical Copulas

Going back to the way in which copula functions have been constructed in the first place, that is the probability integral transformation theorem, let us take the first multidimensional distribution that comes to mind. Define as $\Phi_n(\epsilon_1, \ldots, \epsilon_n; \mathbf{R})$ the *n*-dimension multivariate

Gaussian distribution, linking together a set of n standardized variables with correlation represented by the n-dimensional matrix $\mathbf{R}$. It is well known that the marginal distribution of each variable is also standard normal. Denote by $\Phi(x)$ the univariate standard normal distribution. You can now rewrite the joint normal distribution as

$$C(u_1, \ldots, u_n) \equiv \Phi_n \left(\Phi^{-1}(u_1), \ldots, \Phi^{-1}(u_n); \mathbf{R} \right). \tag{2.29}$$

This is known as a *Gaussian copula* function, and it is by far the most widely used in financial applications. Consider, for example, that this copula function is used to recover implied correlations in the credit market. This makes the Gaussian copula almost as famous as the Black–Scholes formula that is used to recover implied volatilities from the option market, and that is also based on the assumption of normality. In market practice, it is usual to use the Gaussian copula assuming a single correlation figure, representing a sort of "average" correlation among the assets. In this case, we will write ρ instead of $\mathbf{R}$. We may use the Gaussian copula whenever we want to preserve a Gaussian kind of dependence, even though the marginal distributions are not Gaussian. This dependence structure is radially symmetric and does not display tail dependence. So, we may use it if we want to link together variables whose distribution may well be asymmetric and "fat-tailed", but with a dependence structure that does not change with the swings of the market and such that if one market tumbles or soars, that is independent from what happens to the others.

We know that in univariate models a way to introduce fat tails is to substitute the normal distribution with the so-called *Student's t* distribution, in which a parameter v, representing the *degrees of freedom*, is responsible for generating fat tails. The same can be done with copulas. So, define as $\mathbf{T}_n(\epsilon_1, \ldots, \epsilon_n; \mathbf{R}, v)$ the joint *Student's t* distribution, and denote by $\mathbf{T}_v(x)$ the univariate *Student's t* distributions. The *Student's t copula* is then defined as

$$C(u_1, \ldots, u_n) \equiv \mathbf{T}_n \left(\mathbf{T}_v^{-1}(u_1), \ldots, \mathbf{T}_v^{-1}(u_n); \mathbf{R}, v \right). \tag{2.30}$$

Like the Gaussian copula, it is symmetric, but different from that, it has tail dependence. More precisely, the bivariate tail dependence coefficient is represented by

$$\lambda = 2\mathbf{T}_{v+1} \left(\frac{\sqrt{v+1}\sqrt{1-\rho}}{\sqrt{1+\rho}} \right). \tag{2.31}$$

So, one can actually link together assets that are reasonably normally distributed, but which we believe are likely to experience extreme upward or downward movements at the same time.

Both the Gaussian and *Student's t* copulas are part of the so-called family of *elliptical copulas*. They are obtained directly from the class of elliptical distributions. For a detailed study on elliptical distribution we refer the reader to Fang *et al.* (1990). Below we are just going to sketch the main properties needed for the sequel.

Elliptical distributions are obtained as the distributions of affine transformations of spherically distributed random variables. To be more precise, a spherically distributed random vector $\mathbf{X}$ in $\mathbb{R}^n$ can be recovered starting from n-dimensional random vectors $\mathbf{U}^{(n)}$ uniformly distributed on the unit sphere surface of $\mathbb{R}^n$ in the following way:

$$\mathbf{X} \overset{d}{=} r\mathbf{U}^{(n)},$$

where r is a random variable independent of $\mathbf{U}^{(n)}$ with support on $[0, +\infty)$. As suggested by intuition, the random vector $\mathbf{X}$ is said to have a spherically symmetric distribution. For this kind of vector (with the further assumption that $\mathbb{P}(\mathbf{X} = \mathbf{0}) = 0$), we have that

$$||\mathbf{X}||_2 = r \qquad \text{and} \qquad \frac{\mathbf{X}}{||\mathbf{X}||_2} = \mathbf{U}^{(n)}, \tag{2.32}$$

where, for $z \in \mathbb{R}^n$, $||z||_2 = \sqrt{z_1^2 + \ldots + z_n^2}$. A spherically distributed random vector $\mathbf{X}$, in general, does not necessarily possess a density. If it exists, it is of the form $g(\mathbf{X}'\mathbf{X})$ for some non-negative function $g(\cdot)$ of a scalar variable satisfying the integrability condition $\int_0^\infty z^{n/2-1} g(z) dz < +\infty$.

An n-dimensional random vector $\mathbf{Y}$ is said to have an elliptically symmetric distribution with parameters given by an n-dimensional vector μ and an $n \times n$ matrix Σ if

$$\mathbf{Y} \overset{d}{=} \mu + A'\mathbf{X},$$

with $\mathbf{X}$ spherically distributed. Matrix A is of dimension $k \times n$ and such that $\Sigma = A'A$ is symmetric with $rank(\Sigma) = k$. Obviously, any elliptically distributed random vector can be written as

$$\mathbf{Y} \overset{d}{=} \mu + r A'\mathbf{U}^{(k)}.$$

A necessary condition for an elliptical distribution to admit a density is that Σ be non-singular. In that case the density is of the form

$$\frac{1}{det(\Sigma)^{-1/2}} g((x - \mu)' \Sigma^{-1} (x - \mu)).$$

for $x \in \mathbb{R}^n$. The normal distribution is easily recovered by taking $g(t) = \frac{1}{(2\pi)^{n/2}} e^{-t/2}$ and the Student's t distribution with ν degrees of freedom by taking $g(t) = \frac{\Gamma(\frac{\nu+n}{2})}{(\pi\nu)^{n/2}\Gamma(\frac{\nu}{2})} \left(1 + \frac{t}{\nu}\right)^{-\frac{\nu+n}{2}}$.

In the case of an elliptically distributed random vector $\mathbf{Y}$ that admits a density, by Sklar's theorem, we can write the associated copula function as

$$C(u_1, \ldots, u_n) = \int_{-\infty}^{F_{Y_1}^{-1}(u)} \cdots \int_{-\infty}^{F_{Y_n}^{-1}(v)} \frac{1}{det(\Sigma)^{-1/2}} g((x - \mu)'\Sigma^{-1}(x - \mu)) \, dx_1 \ldots dx_n. \tag{2.33}$$

Copula functions of this type are symmetric and the *Spearman's* ρ and *Kendall's* τ measures are simply obtained by the correlation figure as follows:

$$\rho_S = \frac{6}{\pi} \arcsin\left(\frac{\rho}{2}\right), \; \tau = \frac{2}{\pi} \arcsin \rho. \tag{2.34}$$

Generally, elliptical distributions and copulas are used to represent symmetric structures of probability. It is worth mentioning, however, that there exist symmetric extensions of this class of distributions (see for example Azzalini and Capitanio, 1999).

2.11.2 Archimedean Copulas

The other major set of copulas often used in applications is the so-called class of *Archimedean copulas*. Differently from the elliptical copulas, it is not easy to explain how to introduce them directly from multivariate distributions. For this reason here we stick to the cookbook recipe, referring the reader to Nelson (2006) for a detailed description. Assume a function $\phi(x)$, called the generator of the Archimedean copula function. We will (shortly) below digress on the technical requirements this function must satisfy to be a generator. Now we use it to define the class of *Archimedean copulas* as follows:

$$C(u_1, \ldots, u_n) \equiv \phi^{-1}\left(\phi(u_1) + \cdots + \phi(u_n)\right). \tag{2.35}$$

To provide a general way to define the requirements that must be met to generate an Archimedean copula function, let us focus on the inverse of the generator $\psi(\cdot) \equiv \phi^{-1}(\cdot)$. In order to allow for every dimension n, this function is required to be infinitely differentiable, with derivatives of ascending order of alternate sign, and such that $\psi(0) = 1$ and $\lim_{x \to +\infty} \psi(x) = 0$. These features define a famous class of functionals, the family of *Laplace transforms*. The assumption of infinite differentiability can be lessened to the existence of the first $n - 2$ derivatives (again alternate in sign) if we are interested in defining Archimedean copulas with the given dimension n. This case is studied in McNeil and Neslehova (2009). Here the authors show that the role of the Laplace transform is played by the Williamson transform and that the class of all n-dimensional random variables $\mathbf{X}$ having an associated Archimedean copula with such a generator coincides with that of all n-dimensional $\mathbb{L}_1$-norm symmetric distributions such that

$$\mathbf{X} \stackrel{d}{=} R\mathbf{S}_n,$$

where $\mathbf{S}_n$ is an n-dimensional random vector uniformly distributed on the unit simplex $\{x \in \mathbb{R}_+^n : ||\mathbf{X}||_1 = 1\}$ (where $||x||_1 = \sum_{i=1}^n |x_i|$) and R is a non-negative random variable independent of $\mathbf{S}_n$. Similarly, as in the case of elliptical distributions (see (2.32)), if $\mathbb{P}(\mathbf{X} = \mathbf{0}) = 0$, we have that

$$||\mathbf{X}||_1 = R \qquad \text{and} \qquad \frac{\mathbf{X}}{||\mathbf{X}||_1} = \mathbf{S}_{(n)}.$$

For the sake of simplicity, here we stick to the standard *Laplace transform*. For example, $\psi(t) \equiv (1 + \theta t)^{-1/\theta}$ is the Laplace transform of the *Gamma* distribution. This yields

$$\phi(t) = \frac{1}{\theta}\left(t^{-\theta} - 1\right), \tag{2.36}$$

and equation (2.35) produces the most famous among Archimedean copulas, that is the *Clayton copula*

$$C(u_1, \ldots, u_n) = \max\left(\left(u_1^{-\theta} + \ldots + u_n^{-\theta} + n - 1\right)^{-1/\theta}, 0\right). \tag{2.37}$$

Other famous examples in the class of Archimedean copulas are represented by the *Frank copula*, obtained from the generator

$$\phi(t) = -\ln\left(\frac{\exp(-\theta t) - 1}{\exp(-t) - 1}\right). \tag{2.38}$$

The *Frank copula* is then defined as

$$C(u_1, \ldots u_n) = -\ln\left(1 + \frac{(\exp(-\theta u_1) - 1) \ldots (\exp(-\theta u_n) - 1)}{\exp(-\theta) - 1}\right). \tag{2.39}$$

The Frank copula has the same properties as the Gaussian one and is indeed very close to it in shape.

If one were to price a bivariate digital put option like that used in this chapter as an example, one would find that, assuming the same dependence figure, the Clayton copula would overprice it in case of negative association. This means that the Clayton copula is asymmetric, and shows lower tail dependence, measured by

$$\lambda_L = 2^{-1/\theta}, \tag{2.40}$$

for $\theta > 0$.

Another famous Archimedean copula is generated by $\phi(t) \equiv (-\ln t)^\alpha$, which yields the *Gumbel family*:

$$C(u_1, \ldots, u_n) = \exp\left(-\left((-\ln u_1)^\theta + \cdots + (-\ln u_n)^\theta\right)^{1/\theta}\right). \tag{2.41}$$

Differently from the families that have been reviewed up to now, which at least in the bivariate case encompass the cases of independence, perfect positive and negative dependence, the Gumbel copula is limited to positive dependence only. In the jargon of copula functions, we say it is not *comprehensive*. Moreover, with respect to the Frank copula, this time *out-of-the-money* bivariate call options will be overpriced, pointing out that the Gumbel copula has upper tail dependence, measured by

$$\lambda_U = 2 - 2^{-1/\theta}. \tag{2.42}$$

2.12 KENDALL FUNCTION

The class of Archimedean copulas is endowed with a very interesting property, that may turn out to be useful in pricing. Remember again that the integral probability transform theorem only holds for univariate marginal distributions. So, we know that $U = F_X(X)$ and $V = F_Y(Y)$ are uniformly distributed, but we cannot say what is the distribution of $C(U, V)$. Actually, this question finds an answer in the class of Archimedean copulas. The bivariate distribution in this case is represented by the so-called *Kendall function*,

$$\mathbb{P}(C(U, V) \leq t) = t - \frac{\phi(t)}{\phi'^+(t)} \equiv K_C(t), \tag{2.43}$$

where $\phi'^+(t)$ denotes the right derivative of the generator. Life is less easy in higher dimensions. However, Barbe *et al.* (1996) provide an elegant extension in this direction. Actually, one could use another plot technique known as a chi-plot for detecting dependence and tail dependence (see Boero *et al.* (2010) for an application to exchange rates).

The relevance of this finding for financial applications is immediately understood.

Example 2.12.1. *Probability distribution of a bivariate digital put option. Take the bivariate digital put options discussed in Section 2.2. We ask which is the probability distribution of its price. The Kendall function allows us to compute this distribution in closed form. So, for the Clayton copula for example we compute:* $\phi'(t) = -t^{-\theta-1}$. *Using this and equation (2.36) it is very easy to draw the probability distribution of the product. This distribution, which can be used for risk management and pricing purposes, is represented in Figure 2.4. In the solid line we report the distribution of the Clayton price corresponding to a parameter* $\theta = 2$, *corresponding to a* Kendall's τ *figure of 50%. The distribution is compared with the cases of perfect dependence (the upward-sloping straight line and the horizontal line) and independence (the curved line in the middle).*

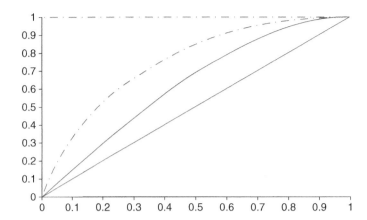

Figure 2.4 Kendall function of the bivariate digital put option in the case of a Clayton copula with dependence parameter $\theta = 2$.

Example 2.12.2. *Compound bivariate digital options.* *Let us stick to the bivariate digital put option above. Assume there is an option written on this product. This may happen if, for example, either the writer or the buyer of the option can terminate the option before the exercise time T, say at time τ, in exchange for a sum w. If this clause is with the writer, the value of the product at time τ will then be*

$$\min(w, C(DP_A(\tau), DP_B(\tau))) = C(DP_A(\tau), DP_B(\tau)) - \max(C(DP_A(\tau), DP_B(\tau)) - w, 0),$$

where we have set the risk-free rate to zero for the sake of simplicity. We see that the product now involves a call option, which in fact measures the value of the callability *clause embedded in the digital option. By the same token, if the clause were in favor of the buyer of the option, we would have*

$$\max(w, C(DP_A(\tau), DP_B(\tau))) = C(DP_A(\tau), DP_B(\tau)) + \max(w - C(DP_A(\tau), DP_B(\tau)), 0),$$

and the product would now involve a put option in favor of the buyer of the product. Under the assumption of Archimedean dependence between the assets, in both cases the price of the option could be computed by taking the integral of Kendall's function.

Example 2.12.3. *Value-at-Risk of a bivariate digital option.* *In the same example of the bivariate digital option above, the Kendall function provides a direct measure of the Value-at-Risk of the product. We remember that the Value-at-Risk defines the quantile of the price of a product over a suitably chosen unwinding period. The use of the Kendall function for the computation of Value-at-Risk follows directly from its definition.*

2.13 EXCHANGEABILITY

Let us now take a brief look at the frontier issues in the modeling of market co-movements. A major issue has to do with the so-called *exchangeability* property of all the copula functions discussed so far, and in general of the copula functions that are currently used in financial applications. The term "exchangeability" was first introduced by De Finetti in the 1930s and refers to a property of joint distributions according to which $F(x, y) = F(y, x)$. Of course, this property is particularly restrictive. First, it requires that marginal distributions have to be the same. Second, it requires the copula function to be exchangeable, that is $C(u, v) = C(v, u), \forall v, u$.

It is easy to understand that distributions of financial prices and risk factors are not exchangeable, at least on the basis of the argument that marginal distributions are different. It may, however, be interesting to address the question of what could happen if the copula functions themselves failed to be exchangeable. There is first a question of the economic meaning of *non-exchangeable copulas*. Let us go back to the case of our bivariate products. Take the bivariate digital put with marginal distributions so that the probability that market A is below the given strike price is 70% while that of market B equals 30%. Say the price of the bivariate product is about 25%. Is there any reason why the price could be different, higher or lower, if the required probability of market A falls to 30% and that of market B climbs to 70%? Say this joint probability is lower. The economic meaning is that the decrease in marginal probability in market A outweighs the effect of the increase in market B. In other words, movements in market A would play a *dominant* role over those in market B, even assuming that marginal distributions are the same.

An interesting question is how to construct a copula function with the *non-exchangeability* property. Up to now, there have been few proposals in this direction. An interesting recipe to build non-exchangeable copulas was proposed in a PhD dissertation by Khoudraji (1995). The idea is to resort to some geometric average of copulas. More precisely, select a copula function $C(u,v)$, define γ and κ in [0, 1], and set

$$\check{C}(u, v) \equiv u^\gamma v^\kappa C(u^{1-\gamma}, v^{1-\kappa}). \tag{2.44}$$

Then, the copula function $\check{C}(u, v)$ is non-exchangeable. It is interesting to ask what would happen to prices and risk measures. The empirical issue remains open. Anyway, the issue of non-exchangeability becomes unavoidable if one addresses the dependence problem in higher dimensions. In this respect the first intuitive idea that would come to mind is to arrange the variables in clusters of homogeneous dependence, and to impose a dependence structure between the clusters. This idea leads to the concept of hierarchical copulas discussed below.

2.14 HIERARCHICAL COPULAS

In every field of science, scholars have some special typical nightmare (they may actually be considered lucky if they have just one). Those in the field of copula functions have a nightmare called *compatibility*. Stating the problem is quite simple. Say we study an n-dimensional distribution and we want to build it in such a way that the $(n - 1)$-distributions have pre-specified shapes, and those in $n - 2$ have other desirable properties, all the way down to one dimension. In this chapter we have learned that the question has an easy answer at the two-dimensional level: the answer stems directly from the probability integral transformation theorem and Sklar's theorem. It is very difficult to come across viable answers for higher dimensions. The reader is referred to Joe (1997) for a discussion of some solutions available for this problem for dimension three. Here we focus on a solution that is available for the class of *Archimedean copulas*. Actually we have seen that for these copulas we have a result parallel to the integral probability transformation theorem. A major flaw of this class is that the dependence parameter is assumed to be the same across the several pairs of assets. It is easy to check that this implies the same bivariate association for all variables. In fact, it follows from the property of copula functions that

$$C(u_1, u_2, 1, 1) = \phi^{-1}(\phi(u_1) + \phi(u_2) + \phi(1) + \phi(1)) = \phi^{-1}(\phi(u_1) + \phi(u_2)),$$
$$C(1, 1, u_3, u_4) = \phi^{-1}(\phi(1) + \phi(1) + \phi(u_3) + \phi(u_4)) = \phi^{-1}(\phi(u_3) + \phi(u_4)). \quad (2.45)$$

It would be good to find a way to generalize this representation allowing for $C(u_i, u_j) \neq C(u_m, u_n), i \neq m, j \neq n$. A way to achieve this is to impose a *hierarchical* representation. An example is

$$C(u_1, u_2, u_3, u_4) \equiv \phi_3^{-1}\left(\phi_3 \circ \phi_1^{-1}(\phi_1(u_1) + \phi_1(u_2)) + \phi_3 \circ \phi_2^{-1}(\phi_2(u_3) + \phi_2(u_4))\right). \quad (2.46)$$

This representation uses different generators ϕ_1 and ϕ_2 for the couples u_1, u_2 and u_3, u_4 and a generator $\phi_3 \circ \phi_i^{-1}$ to model the dependence of the pairs u_1, u_3 and u_2, u_4. For the generator $\phi_3 \circ \phi_i^{-1}$ to be well defined, the dependence parameter must be decreasing for higher hierarchical levels. So, denoting by θ_i the dependence parameter of the generator ϕ_i, in our case it must be $\theta_3 \leq \theta_i, i = 1, 2$. Notice that this also destroys the exchangeability property of the copula. The dependence structure of variables of the same group (in a parallel with econometrics, we could call it *within dependence*) is higher than that of variables of different groups (*between dependence*). If the number of variables is limited, one could choose to arrange them in decreasing order of dependence. In the four-variable case above one could in fact choose the structure

$$C(u_1, u_2, u_3, u_4) \equiv \phi_3^{-1}\left(\phi_3(u_4) + \phi_3 \circ \phi_2^{-1}\left(\phi_2(u_3) + \phi_2 \circ \phi_1^{-1}(\phi_1(u_2) + \phi_1(u_1))\right)\right), \quad (2.47)$$

assuming the dependence parameter is increasing from ϕ_3 to ϕ_1. The reader interested in more material on hierarchical copulas is referred to Savu and Trede (2008), McNeil (2008), Hofert (2008), Durante *et al.* (2009). Finally, Mai and Scherer (2011) show that the problem of hierarchical dependence structure can also be solved in the family of Marshall–Olkin copulas.

Multivariate Digital Call on Exchange Rates

Here we price a multivariate product by computing the volume of a nested hierarchical copula function. We consider a multivariate digital call option on the exchange rates euro/yen, euro/yuan, euro/dollar, and euro/pound with strikes equal to the 25th percentile of the corresponding empirical marginal distributions.

From the historical data, we can estimate the empirical c.d.f. of the exchange rates and then the corresponding 25th percentiles. In Figure 2.5 we plot the empirical c.d.f. of the exchange rates under analysis.

To solve this pricing problem, we first examine the scatters of the historical data relative to the daily exchange rates from January 2002 till September 2010 (source: Datastream).

Figure 2.6 represents the scatter between all the possible couples of exchange rates.

In Tables 2.2–2.4 we estimate the Kendall τ and the dependence parameters of copula functions through the empirical copula. We then compute the empirical bivariate copulas of all the combinations of exchange rates and estimate the Kendall τ parameters and the parameters of a selected kind of copula function. In Table 2.2 we can see the degree of dependence of all the combinations of exchange rates.

We estimate the parameters of the Gumbel and the Clayton copula functions and the standard errors with respect to the empirical copulas. As we can check in Tables 2.3 and 2.4, the Clayton copula is the best choice in terms of minimal error.

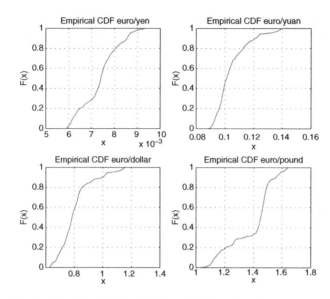

Figure 2.5 Empirical c.d.f. estimated from the historical data of exchange rates.

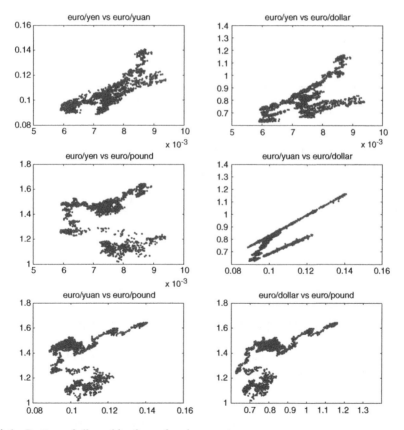

Figure 2.6 Scatters of all combinations of exchange rates.

Table 2.2 Kendall's τ estimated from the historical data

Ex-rates	EU/YEN	EU/YUAN	EU/DOLLAR	EU/POUND
EU/YEN	1	0.617	0.424	0.031
EU/YUAN	0.617	1	0.531	0.108
EU/DOLLAR	0.424	0.531	1	0.463
EU/POUND	0.031	0.108	0.463	1

Table 2.3 Gumbell parameters estimated from the historical data and the corresponding error with respect to the empirical copula

θ^G	EU/YEN	EU/YUAN	EU/DOLLAR	EU/POUND
EU/YEN		2.613	1.736	1.032
EU/YUAN	2.613		2.132	1.121
EU/DOLLAR	1.736	2.132		1.864
EU/POUND	1.032	1.121	1.864	

MSError	EU/YEN	EU/YUAN	EU/DOLLAR	EU/POUND
EU/YEN		0.003685	0.001157	0.003893
EU/YUAN	0.003685		0.002474	0.003599
EU/DOLLAR	0.001157	0.002474		0.002304
EU/POUND	0.003893	0.003599	0.002304	

Table 2.4 Clayton parameters estimated from the historical data and the corresponding error with respect to the empirical copula

θ^C	EU/YEN	EU/YUAN	EU/DOLLAR	EU/POUND
EU/YEN		3.225	1.472	0.064
EU/YUAN	3.225		2.263	0.241
EU/DOLLAR	1.472	2.263		1.728
EU/POUND	0.064	0.241	1.728	

MSError	EU/YEN	EU/YUAN	EU/DOLLAR	EU/POUND
EU/YEN		0.000552	0.000914	0.003452
EU/YUAN	0.000552		0.000411	0.002617
EU/DOLLAR	0.000914	0.000411		0.001078
EU/POUND	0.003452	0.002617	0.001078	

Finally, we rank the parameters in decreasing order and construct a nested Clayton hierarchical copula where a higher level of aggregation corresponds to a lower dependence parameter. The marginals plugged into the copula function correspond to the empirical c.d.f. at the 25th percentile.

The price of the multivariate digital call corresponds to the survival of the nested Clayton hierarchical copula described before:

$$MDC_{EU/YEN,EU/YUAN,EU/DOLLAR,EU/POUND}$$
$$= \overline{HC}(u_{EU/YEN}, u_{EU/YUAN}, u_{EU/DOLLAR}, u_{EU/POUND})$$
$$= 0.310735,$$

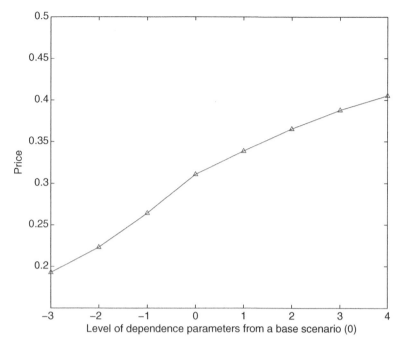

Figure 2.7   Multivariate digital call prices with parallel shift of the dependence parameters.

where u are the marginals and we assume for simplicity a null interest rate. We can say that the probability of having all the exchange rates greater than their 25th percentiles is more than 30%.

If we consider a parallel shift of the dependence parameters, the price moves in the same direction as the shift, since this product is long correlation. Figure 2.7 reports the sensitivity of the product to correlation, starting from our base scenario and shifting the dependence parameters of 50% from one scenario to the other.

2.15   CONDITIONAL PROBABILITY AND FACTOR COPULAS

In cases in which the dimension of the problem is too large, there is no hope of specifying the individual dependence relationship among the variables. The hope remains that most of the dependence structure will be explained by a limited number of *factors*. If this is the case, then one could analyze the variables in the system conditional on given scenarios of the factors and, crossing fingers, assume that once conditioned on those they be independent. In other words, one assumes that a limited number of factors suffice to explain the dependence structure of the system completely. Of course this is an approximation, but it is the best one can do for large systems.

A key concept in the approach above, called the *factor copula*, is that of *conditional probability*. Actually we will see that it is also paramount in our applications of copulas to dynamic systems that will be covered in the rest of the book. So, it turns out to be extremely useful to notice that there is a close relationship between conditional probabilities and copulas. We have already seen that writing Bayes' theorem with the language of copulas yields the conditional probability of extreme events in one market given an extreme event in another ($\mathbb{P}(X \leq x | Y < y)$). It is, however, almost as easy to come up with the conditional probability with respect to a given scenario, i.e. $Y = y$, at least in the case of continuous and strictly increasing marginal distributions:

$$
\begin{aligned}
\mathbb{P}(X \leq x | Y = y) &= \lim_{\Delta y \to 0} \mathbb{P}(X \leq x | y \leq Y \leq y + \Delta y) \\
&= \lim_{\Delta y \to 0} \frac{F(x, y + \Delta y) - F(x, y)}{F_Y(y + \Delta y) - F_Y(y)} \\
&= \lim_{\Delta y \to 0} \frac{C(F_X(x), F_Y(y + \Delta y)) - C(F_X(x), F_Y(y))}{F_Y(y + \Delta y) - F_Y(y)} \\
&= \frac{\partial C(F_X(x), v)}{\partial v} \bigg|_{v = F_Y(y)}.
\end{aligned}
\tag{2.48}
$$

Here we have also relied on the assumption that partial derivatives of copula functions exist. This is indeed true. More precisely, $\frac{\partial}{\partial u} C(u, v)$ and $\frac{\partial}{\partial v} C(u, v)$ exist for almost all u and v and for those u and v we have $0 \leq \frac{\partial}{\partial u} C(u, v) \leq 1$ and $0 \leq \frac{\partial}{\partial v} C(u, v) \leq 1$ (see Nelsen (2006), Theorem 2.2.7). Anyway, this result holds in full generality (without any assumption on the distribution functions involved) but the proof requires more technicalities (for a detailed proof see Darsow et al. (1992), Theorem 3.1).

We have just proved that conditional probability is nothing but the partial derivative of the copula with respect to the conditioning variable. Notice that this suggests an alternative way of writing a copula function, namely by integration of the conditional probability:

$$
\mathbb{P}(X \leq x, Y \leq y) = \int_0^{F_Y(y)} \frac{\partial C(F_X(x), w)}{\partial w} dw.
$$

These results will be used again and again in this book. For the time being, we show how to build a copula-based factor model. Assume the co-movement of the variables in the system is fully explained by a single factor M. Denote as $D_i C(u_1, \ldots, u_n)$ the partial derivative of the copula with respect to the ith component. Assume the common factor M is reported in the first place of the copula function. Then, for any scenario m of the risk factor, the dependence between the factor and each variable $i, i = 2, \ldots, n$, is fully represented by the conditional probability

$$
\mathbb{P}(U_i \leq u_i | M = m) = D_1 C(w, u_i),
$$

where $\mathbb{P}(M < m) = w$. Now, trusting the assumption that all co-movements are explained by the common factor and only by it, we have that upon conditioning on a specific scenario of the common factor the variables are independent of each other: this is called *conditional independence*. Under the scenario $M = m$, the joint distribution of the variables is then represented by the product copula

$$
\mathbb{P}(U_2 \leq u_2, \ldots, U_n \leq u_n | M = m) = \prod_{i=2}^{n} D_1 C(w, u_i).
$$

The dependence structure of the factor and the variables is finally obtained by integrating this conditional joint probability with respect to the scenarios:

$$C(u_1, \ldots, u_n) = \int_0^{u_1} \prod_{i=2}^{n} D_1 C(w, u_i) dw.$$

Finally, notice that, consistent with the uniform margin property of the copula function, the unconditional joint distribution of the variables is obtained by computing the copula at $u_1 = 1$, or, which is the same, by integrating out the common factor:

$$C(1, u_2, \ldots, u_n) = \int_0^1 \prod_{i=2}^{n} D_1 C(w, u_i) dw.$$

Before going on it is worthwhile to give at least one example of a conditional distribution. For the Gaussian conditional distribution we have

$$D_1 C(u_1, u_i) = \Phi\left(\frac{\Phi^{-1}(u_i) - \rho \Phi^{-1}(u_1)}{\sqrt{1 - \rho^2}}\right).$$

So, the Gaussian copula can be written in two possible ways:

$$C(u_1, u_i) = \Phi_2(\Phi^{-1}(u_1), \Phi^{-1}(u_i); \rho) = \int_0^{u_1} \Phi\left(\frac{\Phi^{-1}(u_i) - \rho \Phi^{-1}(w)}{\sqrt{1 - \rho^2}}\right) dw.$$

A final point on the conditional distribution is the extension to more factors. Assume that in the n-dimensional system above the first two variables are the factors that explain all the co-movements in the system. How do we condition with respect to these two factors? We first lay out the rule and then discuss it. We have:

$$\mathbb{P}(U_3 \leq u_3, \ldots, U_n \leq u_n | U_1 = u_1, U_2 = u_2) = \frac{\frac{\partial^2 C(u_1, u_2, u_3, \ldots, u_n)}{\partial u_1 \partial u_2}}{\frac{\partial^2 C(u_1, u_2, 1, \ldots, 1)}{\partial u_1 \partial u_2}}.$$

Here we propose a sketch of the proof for the two-factor conditional distribution case:

$$\mathbb{P}(U_3 \leq u_3, \ldots, U_n \leq u_n | U_1 = u_1, U_2 = u_2)$$

$$= \lim_{\substack{\Delta x \to 0 \\ \Delta y \to 0}} \mathbb{P}(U_3 \leq u_3, \ldots, U_n \leq u_n | u_1 \leq U_1 \leq u_1 + \delta x, u_2 \leq U_2 \leq u_2 + \delta y)$$

$$= \lim_{\substack{\Delta x \to 0 \\ \Delta y \to 0}} \frac{\frac{C(u_1+\delta x, u_2+\delta y, u_3, \ldots, u_n) - C(u_1, u_2+\delta y, u_3, \ldots, u_n) + C(u_1, \ldots, u_n) - C(u_1+\delta x, u_2, \ldots, u_n)}{\delta x \delta y}}{\frac{C(u_1+\delta x, u_2+\delta y, 1, \ldots, 1) - C(u_1, u_2+\delta y, 1, \ldots, 1) + C(u_1, u_2, 1, \ldots, 1) - C(u_1+\delta x, u_2, 1, \ldots, 1)}{\delta x \delta y}}$$

$$= \frac{\frac{\partial^2 C(u, v, u_3, \ldots, u_n)}{\partial u \partial v}}{\frac{\partial^2 C(u, v, 1, \ldots, 1)}{\partial u \partial v}} \Bigg|_{u=u_1, v=u_2}. \tag{2.49}$$

These representations, where we condition with respect to two factors, may be extended to more factors. For a general proof we refer to Schmitz (2003). In the end, let us define

$$c(u_1, u_2) \equiv \frac{\partial^2 C(u_1, u_2, 1, \ldots, 1)}{\partial u_1 \partial u_2}.$$

and the copula function of the whole system can be written

$$C(u_1, \ldots, u_n) = \int_0^{u_1} \int_0^{u_2} \mathbb{P}(U_3 \leq u_3, \ldots, U_n \leq u_n | U_1 = u_1, U_2 = u_2) c(w, \xi) dw d\xi.$$

It is easy to guess what function $c(u_1, u_2)$ is. It is actually the joint density of the uniformly distributed variables U_1 and U_2, which have joint distribution given by the copula $C(u_1, u_2, 1, \ldots, 1)$. In a word, it is the density of the copula linking the factors. So, we have discovered another fundamental result of copula function theory, namely that the cross-derivative of a copula is its density. We will see some applications below and we will also make quite frequent use of the tool, even though it is subject to some restrictions, as we will soon see.

2.16 COPULA DENSITY AND VINE COPULAS

Let us now see how to represent joint distributions in terms of the density of copula functions, instead of copula functions themselves, and recent methodologies proposed to provide workable representations of multivariate densities. Let us take again a system of variables X_i, $i = 1, 2, .., n$ and recall that, in some cases, we can write the joint distribution as

$$\mathbb{P}(X_1 \leq x_1, X_2 \leq x_2, \ldots, X_n \leq x_n) = F(x_1, \ldots, x_n)$$
$$= \int_\infty^{x_1} \int_\infty^{x_2} \ldots \int_\infty^{x_n} f(w_1, w_2, \ldots, w_n) dw_1 dw_2 \ldots dw_n,$$

where $f(x_1, x_2, \ldots, x_n)$ is the joint density function. Extending our previous analysis for the two-factor model, and taking into account that a copula function is nothing else than a joint distribution function with uniformly distributed marginals, let us introduce (assuming it exists) the cross-derivative

$$c(u_1, u_2, \ldots, u_n) = \frac{\partial^n C(u_1, u_2, \ldots, u_n)}{\partial u_1 \partial u_2 \ldots \partial u_n}.$$

This is the density of the copula function linking the variables, or the *copula density*, for short. The joint density function can then be decomposed as

$$f(x_1, x_2, \ldots, x_n) = c(u_1, u_2, \ldots, u_n) f_{X_1}(x_1) f_{X_2}(x_2) \ldots f_{X_n}(x_n),$$

where $u_i = F_{X_i}(x_i)$ and $f_{X_i}(x_i)$ is the marginal density function of the ith variable. Just like the copula function, one could then refer to the copula density to represent the dependence structure of the system. This representation is also called the *canonical representation* of a copula function.

The approach of representing copulas in terms of densities, which is used particularly in econometric applications, is subject to an important warning. Indeed, not all copulas can be completely represented in terms of density, and the reason is that some of them may have positive mass in some region with zero Lebesgue measure of the unit square. To be more precise on this point, copulas are functions of uniform variables, and as such they cannot have *atoms*, that is points with positive mass, but they may have lines or regions (with zero Lebesgue measure) endowed with positive mass. As the simplest example, take the maximum copula $\min(u, v)$. In this case, all the probability mass is concentrated on the main diagonal, which is also the only set in which the density is not equal to zero. So, it is not feasible to represent the copula by means of its density, computing

$$\int_0^u \int_0^v \frac{\partial^2 C(s,t)}{\partial s \partial t} ds dt.$$

In this case the value of this double integral is in fact equal to zero almost everywhere, and the copula is said to be *singular*. In the opposite case, that is when

$$C(u,v) = \int_0^u \int_0^v \frac{\partial^2 C(s,t)}{\partial s \partial t} ds dt$$

holds for all u and v in the unit interval, we say that the copula is *absolutely continuous*. In general we may decompose the copula function into two separate components: the absolutely continuous component

$$A_C(u,v) = \int_0^u \int_0^v \frac{\partial^2 C(s,t)}{\partial s \partial t} ds dt$$

and the singular component

$$S_C(u,v) = C(u,v) - A_C(u,v).$$

The measure of the singular component is given by $S_C(1,1)$. In the case of the maximum copulas, it is easy to check that the singular component has measure 1. There are cases in which the copulas have both an absolutely continuous and a singular component. As an example, the reader is referred to the Marshall–Olkin copula (Nelsen (2006), paragraph 3.1).

For these reasons, purists of copula function theory refrain from the canonical representation whenever they can. It is somewhat like what happens with people who use characteristic functions instead of densities in univariate models: they do not have to care about the existence of densities, because characteristic functions always exist.

Having pointed this out, one must also admit that the restriction to cases in which the copula density is defined almost surely grants the possibility of exploiting the strongest and most popular estimation technique available in statistics, that is *Maximum Likelihood Estimation* (MLE). We remember in fact that the likelihood is written in terms of density. For our set it would be

$$L = \prod_{j=1}^m f(x_1^j, \ldots, x_n^j) = \prod_{j=1}^m c(u_1^j, u_2^j, \ldots, u_n^j) f_{X_1}(x_1^j) f_{X_2}(x_2^j) \ldots f_{X_n}(x_n^j).$$

where $u_i^j = F_{X_i}(x_i^j)$.

For estimation purposes we maximize the logarithm of the likelihood, so that we have

$$\log L = \sum_{j=1}^m \log(c(u_1^j, u_2^j, \ldots, u_n^j)) + \sum_{j=1}^m \sum_{i=1}^n \log(f_{X_i}(x_i^j)).$$

We will devote a whole chapter to estimation issues, so in this introduction we only notice that the log-likelihood can be decoupled, and estimation of the marginal densities and the copula density can be carried out separately. In one approach, called *inference for margins*, the marginal densities are estimated in the first step and the copula density in the second one. In the canonical estimation method, the likelihood is estimated directly using non-parametric representations of the marginal distributions in the copula density. This makes estimation quite easy, except for a problem: the computation of the copula density. A recent approach, known as *vine copulas*, aims to break the copula representation into a combination of representations of lower dimensions, for example bivariate densities. Here we provide a brief sketch of how the approach works, following Aas *et al.* (2009), to which we refer for deeper analysis of the topic.

The key point is that the joint density function can be factorized using conditional densities as

$$f(x_1, x_2, \ldots, x_n) = f(x_1)f_{2|1}(x_2|x_1)f_{3|12}(x_3|x_1, x_2) \ldots f_{n|12\ldots n-1}(x_n|x_1, \ldots, x_{n-1}),$$

where $f_{i|j}$ denotes the conditional density of variable i with respect to variable j. Let us take the case $n = 3$ and see how to use conditional densities. Let us take the density between the first and the second variable:

$$f(x_1, x_2) = c_{12}(F_{X_1}(x_1), F_{X_2}(x_2))f(x_1)f(x_2).$$

The conditional density of the second variable with respect to the first yields

$$f_{2|1}(x_2|x_1) = c_{12}(F_{X_1}(x_1), F_{X_2}(x_2))f(x_2).$$

Now let us compute

$$f_{3|12}(x_3|x_1, x_2) = \frac{f_{23|1}(x_2, x_3|x_1)}{f_{2|1}(x_2|x_1)} = c_{23|1}(F_{X_2|X_1}(x_2|x_1), F_{X_3|X_1}(x_3|x_1))f_{3|1}(x_3|x_1).$$

Now apply to $f_{3|1}$ the same treatment as that applied to $f_{2|1}$ and substitute everything inside the joint density factorization to yield

$$f(x_1, x_2, x_2) = f(x_1)f(x_2)f(x_3)c_{12}(F_{X_1}(x_1), F_{X_2}(x_2))c_{13}(F_{X_1}(x_1), F_{X_3}(x_3))$$
$$\times c_{23|1}(F_{X_2|X_1}(x_2|x_1), F_{X_3|X_1}(x_3|x_1)).$$

What remains to be computed are the conditional distributions $F_{X_2|X_1}$ and $F_{X_3|X_1}$, but we know how to do this (remember that conditional distributions are partial derivatives of copulas). Notice that there are six possible permutations of the three-variable models, even though only three give different decompositions. So, even in the three-variable case, there is no unique way to perform the decomposition. The problem becomes more dramatic for higher-dimension systems, even though this dimension remains modest. In the five-variable case, we have 240 different constructions for the same density. In order to give some order to these factorizations, it is common to refer to the concept of *vine*. The concept was introduced by Bedford and Cooke (2002) and is defined as a set of trees embedded one in the other. If the trees have at least one node in common, the vine is called *regular*. Further restrictions allow us to distinguish two different categories of regular vines:

- D-vine: in each tree, no node is connected to more than two edges.
- C-vine (canonical vine): the node of the T_j tree in the vine is linked to $n-j$ nodes.

Even after this classification, there is still plenty of room for choice. Sticking again to the five-variable case, Aas *et al.* (2009) compute that of the 240 possible choices, 60 are vines of the canonical family, 60 are D-vines, and the remaining 120 are regular vines that are not part of either class.

2.17 DYNAMIC COPULAS

An important issue is the change of dependence and market co-movement in different periods of the market. Early work has documented this issue for equity markets (see e.g. Erb *et al.* (1994), Longin and Solnik (2001), Ang and Chen (2002)), along with evidence that changes in correlation caused deviations from multivariate normality characterized by asymmetric dependence. Intuitively, this calls for the specification of a *dynamic copula*, that is a copula whose shape and parameters change with time. Models in which the copula function representing the dependence of financial time series change across time were proposed by Rockinger and Jondeau (2001), and applied to the evaluation of options in van der Goorberg *et al.* (2005), along with other contributions. Here we focus more in detail on the two most structured proposals on the topic, due to the work by Patton (2006a,b) and Fermanian and Wegkamp (2004), for the reason that we explain below.

It is important to notice that the problem of providing dynamical models of copula functions is different from that of using copulas to provide a representation of the dynamics of a system, and the latter is the main goal of this book. Nevertheless, the main approaches that were proposed to make the copula dynamic faced a key problem which is common to the use of copulas for the representation of market dynamics. The point is the definition of the set of information that is supposed to drive these changes. Must it be the same for marginal distributions and the copula, or can it be different? We will focus on a similar question in Chapter 3. As for dynamic copulas, this question was addressed in the two main approaches, namely in the concept of a *conditional copula* and the more general concept of a *pseudo-copula*.

2.17.1 Conditional Copulas

Patton (2001, 2006a,b) introduces the concept of a conditional copula, which can be used in the analysis of time-varying conditional dependence and for multivariate density modeling. The theoretical framework is the following: X and Y are the variables of interest and Z is the conditioning variable. Let F_{XYZ} be the distribution of the random vector (X, Y, Z), $F_{XY|Z}$ be the conditional distribution of (X, Y) given Z, and $F_{X|Z}$ and $F_{Y|Z}$ be the conditional marginal distributions of X given Z, and of Y given Z, respectively. The conditional bivariate distribution of (X, Y) given Z can be derived from the unconditional joint distribution of (X, Y, Z) as follows:

$$F_{XY|Z}(x, y|z) = f_Z(z)^{-1} \frac{\partial F_{XYZ}(x, y, z)}{\partial z},$$

where f_Z is the unconditional density of Z. The conditional copula of (X, Y) given Z cannot be derived from the unconditional copula of (X, Y, Z): further information is required. Patton defines the conditional copula as follows:

Definition 2.17.1. The conditional copula of (X, Y) given $Z = z$, where X given $Z = z$ has distribution $F_{X|Z=z}(\cdot|z)$ and Y given $Z = z$ has distribution $F_{Y|Z=z}(\cdot|z)$, is the conditional distribution function of $U \equiv F_{X|Z}(X|z)$ and $V \equiv F_{Y|Z}(Y|z)$ given $Z = z$.

Let $\mathcal{Z}$ be the support of Z. So, an extension of Sklar's theorem for conditional distributions shows that a conditional copula has the properties of an unconditional copula for each $z \in \mathcal{Z}$.

Theorem 2.17.1 *(Patton, 2001).* *Let $F_{X|Z}(\cdot|z)$ be the conditional distribution of X given $Z = z$, $F_{Y|Z}(\cdot|z)$ be the conditional distribution of $Y|Z = z$, $F_{XY|Z}(\cdot, \cdot|z)$ be the joint conditional distribution of (X, Y) given $Z = z$, and $\mathcal{Z}$ be the support of Z. Assume that $F_{X|Z}(\cdot|z)$ and $F_{Y|Z}(\cdot|z)$ are continuous in x and y for all $z \in \mathcal{Z}$. Then there exists a unique conditional copula $C(\cdot|z)$ such that*

$$F_{XY|Z}(x, y|z) = C(F_{X|Z}(x|z), F_{Y|Z}(y|z)|z), \quad \forall(x, y) \in \bar{\mathbb{R}} \times \bar{\mathbb{R}}, \ \forall z \in \mathcal{Z}. \tag{2.50}$$

Conversely, if we let $F_{X|Z}(\cdot|z)$ be the conditional distribution of X given $Z = z$, $F_{Y|Z}(\cdot|z)$ be the conditional distribution of Y given $Z = z$, and $\{C(\cdot, \cdot|z)\}$ be a family of conditional copulas that is measurable in z, then the function $F_{XY|Z}(\cdot, \cdot|z)$ defined by equation (2.50) is a conditional bivariate distribution function with conditional marginal distributions $F_{X|Z}(\cdot|z)$ and $F_{Y|Z}(\cdot|z)$.

The key point in this "conditional" version of Sklar's theorem is that the conditioning variable Z must be the same for both marginal distributions and the copula, and this is a fundamental condition. For example, suppose that we condition X on Z_1 and Y on Z_2 and the copula on (Z_1, Z_2). We specify $\tilde{F}_{XY|Z_1, Z_2}(x, y|z_1, z_2) = C(F_{X|Z_1}(x|z_1), F_{Y|Z_2}(y|z_2)|z_1, z_2)$. Then

$$\tilde{F}_{XY|Z_1, Z_2}(x, \infty|z_1, z_2) = C(F_{X|Z_1}(x|z_1), F_{Y|Z_2}(\infty|z_2)|z_1, z_2)$$
$$= C(F_{X|Z_1}(x|z_1), 1|z_1, z_2) = F_{X|Z_1}(x|z_1),$$

which is the conditional marginal distribution of X given (Z_1, Z_2), and

$$\tilde{F}_{XY|Z_1, Z_2}(\infty, y|z_1, z_2) = C(F_{X|Z_1}(\infty|z_1), F_{Y|Z_2}(y|z_2)|z_1, z_2)$$
$$= C(1, F_{Y|Z_2}(y|z_2)|z_1, z_2) = F_{Y|Z_2}(y|z_2),$$

which is the conditional marginal distribution of Y given (Z_1, Z_2). This implies that $\tilde{F}_{XY|Z_1, Z_2}$ will not be the joint distribution of $(X, Y)|(Z_1, Z_2)$ in general. A special case where this happens is when $F_{X|Z_1}(x|z_1) = F_{X|Z_1, Z_2}(x|z_1, z_2)$ for all $(x, z_1, z_2) \in \mathbb{R} \times \mathcal{Z}_1 \times \mathcal{Z}_2$ and $F_{Y|Z_2}(y|z_2) = F_{Y|Z_1, Z_2}(y|z_1, z_2)$ for all $(y, z_1, z_2) \in \mathbb{R} \times \mathcal{Z}_1 \times \mathcal{Z}_2$.

2.17.2 Pseudo-copulas

Fermanian and Wegkamp (2004) extend the definition of a conditional copula to cover a much larger scope of situations, introducing the concept of a *pseudo-copula*. We give the definition in the d-dimensional case (here $\mathbf{u} = (u_1, \dots, u_d)$ and $\mathbf{v} = (v_1, \dots, v_d)$).

Definition 2.17.2. 1. A d-dimensional pseudo-copula is a function $C : [0, 1]^d \to [0, 1]$ such that:

(a) for every $\mathbf{u} \in [0, 1]^d$, $C(u) = 0$ when at least one coordinate of $\mathbf{u}$ is zero;
(b) $C(1, \dots, 1) = 1$;
(c) for every $\mathbf{u}$ and $\mathbf{v}$ in $[0, 1]^d$ such that $\mathbf{u} \leq \mathbf{v}$, the C-volume of $[u, v]$ is positive.

Therefore, a pseudo-copula satisfies all the properties of a copula except that $C(\mathbf{u})$ is not necessarily u_k when all coordinates of $\mathbf{u}$ except u_k are equal to 1. Nonetheless, it satisfies a similar version of the Sklar theorem.

Theorem 2.17.2 (*Fermanian and Wegkamp, 2004*). *Let H be a c.d.f. on $\mathbb{R}^d$, and let $F_1, \ldots, F_d$ be d univariate c.d.f.s on $\mathbb{R}$. Assume that for every $x = (x_1, \ldots, x_d)$, $\tilde{x} = (\tilde{x}_1, \ldots, \tilde{x}_d) \in \mathbb{R}^d$, $F_j(x_j) = F_j(\tilde{x}_j)$ for all $1 \le j \le d \Rightarrow H(x) = H(\tilde{x})$. Then there exists a pseudo-copula C such that $H(x) = C(F_1(x_1), \ldots, F_d(x_d))$, for every $x = (x_1, \ldots, x_d) \in \mathbb{R}^d$. C is uniquely defined on $\text{Ran} F_1 \times \cdots \times \text{Ran} F_d$, the product of the values taken by the F_j. Conversely, if C is a pseudo-copula and if $F_1, \ldots, F_d$ are some univariate c.d.f.s, then the function H defined above is a d-dimensional c.d.f.*

Note that the pseudo-copula C is a (true) copula if and only if $H(\infty, \ldots, x_j, \ldots, \infty) = F_j(x_j)$, for every $j = 1, \ldots, d$ and $x = (x_1, \ldots, x_d) \in \mathbb{R}^d$.

Let $\mathbf{X}$ be a d-dimensional random vector from $(\Omega, \mathcal{F}, \mathbb{P})$ to $\mathbb{R}^d$. Consider some arbitrary sub-σ-algebras $\mathcal{F}_1, \ldots, \mathcal{F}_d$ and $\mathcal{B}$ such that for all $x = (x_1, \ldots, x_d)$ and $\tilde{x} = (\tilde{x}_1, \ldots, \tilde{x}_d) \in \mathbb{R}^d$ and for almost every $\omega \in \Omega$, $\mathbb{P}(X_j \le x_j | \mathcal{F}_j)(\omega) = \mathbb{P}(X_j \le \tilde{x}_j | \mathcal{F}_j)(\omega)$ for every $j = 1, \ldots, d$ implies $\mathbb{P}(\mathbf{X} \le \mathbf{x} | \mathcal{B})(\omega) = \mathbb{P}(\mathbf{X} \le \tilde{\mathbf{x}} | \mathcal{B})(\omega)$. Two simple cases in which this condition is satisfied are if the conditional c.d.f.s of $X_1, \ldots, X_d$ are strictly increasing and also if $\mathcal{F}_1 = \ldots = \mathcal{F}_d = \mathcal{B}$. If we denote $\mathcal{F} = (\mathcal{F}_1, \ldots, \mathcal{F}_d)$ and set $F_j(x_j | \mathcal{F}_j) = \mathbb{P}(X_j \le x_j | \mathcal{F}_j)$, the following result characterizes a pseudo-copula:

Theorem 2.17.3 (*Fermanian and Wegkamp, 2004*). *There exists a random function C : $[0, 1]^d \times \Omega \to [0, 1]$ such that*

$$\mathbb{P}(X \le x | \mathcal{B})(\omega) = C(\mathbb{P}(X_1 \le x_1 | \mathcal{F}_1)(\omega), \ldots, \mathbb{P}(X_d \le x_d | \mathcal{F}_d)(\omega), \omega)$$
$$\equiv C(\mathbb{P}(X_1 \le x_1 | \mathcal{F}_1), \ldots, \mathbb{P}(X_d \le x_d | \mathcal{F}_d)(\omega),$$

for every $x = (x_1, \ldots, x_d) \in \mathbb{R}^d$ and almost every $\omega \in \Omega$. This function C is $\mathcal{B}([0, 1]^d) \otimes (\mathcal{F}, \mathcal{B})$-measurable. For almost every $\omega \in \Omega$, $C(\cdot, \omega)$ is a pseudo-copula and is uniquely defined on the product of the values taken by $x_j \mapsto \mathbb{P}(X_j \le x_j | \mathcal{F}_j(\omega))$, $j = 1, \ldots, d$.

If C is unique, Fermanian and Wegkamp refer to it as the conditional $(\mathcal{F}, \mathcal{B})$-pseudo copula associated with X and denoted by $C(\cdot | \mathcal{F}, \mathcal{B})$. Note that in general $C(\cdot | \mathcal{F}, \mathcal{B})$ are not copulas, but by the above condition $H(\infty, \ldots, x_j, \ldots, \infty) = F_j(x_j)$, for every $j = 1, \ldots, d$, we see that $C(\cdot | \mathcal{F}, \mathcal{B})$ is a true copula if and only if $\mathbb{P}(X_j \le x_j | \mathcal{B}) = \mathbb{P}(X_j \le x_j | \mathcal{F}_j)$, a.e. for all $j = 1, \ldots, d$ and $x = (x_1, \ldots, x_d) \in \mathbb{R}^d$. This last equality has the following intuitive consequence: the sub-σ-algebra $\mathcal{B}$ cannot provide more information about X_j than the sub-σ-algebra $\mathcal{F}_j$, for every j. Patton's conditional copula corresponds to the particular case $\mathcal{B} = \mathcal{F}_1 = \ldots = \mathcal{F}_d$. Typically, suppose that $(\mathbf{X}_n)_{n \in \mathbb{N}}$ is a d-dimensional process. Then the previous considered sub-σ-algebras are those generated by each component and by the whole process and so they depend on the past values of the process. For instance, we may set $\mathcal{F}_{i,n} = \sigma(X_{i,n-1}, X_{i,n-2}, \ldots)$ the filtration generated by the ith component and $\mathcal{B}_n = \sigma(\mathbf{X}_{n-1}, \ldots)$ the general filtration.

3

Copula Functions and Asset Price Dynamics

Almost all the applications of copula functions to finance, both in asset pricing and risk management, have been devoted to modeling the dependence across markets, financial products, and risk factors. This kind of application would be called spatial correlation by statisticians or cross-sectional dependence by econometricians. Actually, the study of dependence might find applications in a different perspective. Consider what a stochastic process is: it is a collection of variables indexed by time. Then, one could study the dependence structure of these variables, which is actually known as the autocorrelation structure of the process. So, a univariate process is also a multivariate problem, and copula functions are candidate tools for their flexible representation. This is called temporal correlation, and it has been the focus of most of the copula applications to econometrics, which we will cover in the next chapter. Surprisingly, the perspective of temporal dependence is almost unknown in financial applications, and this has left room for criticism about the use of copulas. Actually, the application of copulas should allow for both spatial and temporal dependence, in what econometricians call *panel data* models.

Moreover, the economics of financial markets and mathematical finance impose relevant restrictions on the dynamics of prices in speculative markets, that have to be taken into account particularly in pricing applications. The Markov property and the martingale requirement are typical conditions that have to be imposed on the dynamic asset prices to make the results consistent with theory.

Luckily, copula functions have been applied to Markov processes since the pioneering work by Darsow *et al.* (1992). This work was recently extended to kth order Markov processes, and so implicitly to k-variate Markov processes as well, by Ibragimov (2005, 2009). The contribution from us, the Bologna school, has been to disentangle processes with independent and dependent increments, and to show how to impose the martingale restriction on a multivariate model of asset prices. The aim of this chapter is to review this literature.

3.1 THE DYNAMICS OF SPECULATIVE PRICES

The study of the dynamics of prices in competitive markets represents one of the first laboratories for the creation and application of the theory of stochastic processes, at the beginning of the last century. The other, maybe more famous, application was to particle dynamics. In a nutshell, the idea, first laid out by Bachelier in his thesis under the supervision of Poincaré at Sorbonne University in Paris, was that the movements of stock prices traded in the market should be unpredictable, and the prices should follow a process called a *random walk*. A random walk is given by a sequence of independent increments with zero mean. If one assumes that these increments are independent and have finite variance, it turns out that the variance of the price differences should increase linearly with the length of time on which the first difference is measured. While in the original Bachelier model the first difference was computed on prices, with the flaw of allowing negative prices, in the 1960s the approach was re-elaborated by Eugene Fama and Paul Samuelson in what is known as the *Efficient Market Hypothesis* (EMH). The focus of the new approach was the logarithm of prices, instead of the price itself. Given the price of the asset S_t, the variable whose dynamic is studied is X_t, defined from the relationship

$$S_t = e^{X_t}.$$

The idea is to model percentage changes of the prices instead of changes in value. So, changes are continuously compounded in an investment that is done at some time t. So, holding the investment until time T would give a return

$$\frac{S_T}{S_t} - 1 = e^{X_T - X_t} - 1.$$

The key point is that, given the set of information available at time t, the return $X_T - X_t$ must be unpredictable. For these reasons, in most of the financial literature the process X_t is studied in continuous time, using processes with independent increments. All the models used may be decomposed into three components: a diffusive part, jumps with finite activity and with infinite activity. This way, the dynamics of the process can be collected in a unique equation, called the *Lévy–Kintchine equation*, which characterizes the process. Allowing for independent increments also enables us to use Fourier transforms to study the dynamics of the process and evaluate derivatives.

Our approach is in discrete time, with the purpose of being more general and flexible than those in continuous time. So, we model

$$X_i = X_{i-1} + Y_i, \qquad (3.1)$$

where Y_i is called the *innovation*, or the *shock* reaching the price at time t_i. Under EMH, the price should embed all information available to people participating in the market. The first implication of this is the *Markov property* of process X_i, meaning that the past history of the process does not carry more information than the value of the process itself at the date of observation t_i. Formally, we have

$$\mathbb{P}(X_i \leq x | \mathcal{F}_j) = \mathbb{P}(X_i \leq x | X_j), \qquad \forall i > j, \qquad (3.2)$$

where $\mathcal{F}_j$ represents information and is called the information set in the econometrics literature and filtration in probability theory. The second implication of EMH refers to the distribution of innovations: every such innovation should be unpredictable. Innovations must then be endowed with the martingale property with respect to all information available at time t_j, namely

$$\mathbb{E}[Y_i | \mathcal{F}_j] = 0, \qquad \forall i > j. \qquad (3.3)$$

In plain words, if markets are efficient, then it should not be possible to exploit any piece of information to earn expected returns greater than zero over any investment horizon in the future.

In the sections that follow, we will show how to use copulas to design discrete time processes endowed with the properties required by the EMH and discussed above: the Markov property and the martingale property.

3.2 COPULAS AND MARKOV PROCESSES: THE DNO APPROACH

We start by illustrating how to apply copulas to represent the dynamics of a Markov process. This relationship was first analyzed by Darsow *et al.* (1992) in a path-breaking paper. The idea is that every Markov process can be represented in terms of a chain of copulas linked to each other by a particular operator, that we will describe below, which yields the copula of the Markov process as an outcome. On the contrary, given every Markov process one can recover a sequence of copulas that link subsequent values of the process.

Before trying to understand this connection by intuition, let us introduce the formal definition of a Markov process.

Definition 3.2.1 (Markov process). Let $(\Omega, \mathcal{F}, (\mathcal{F}_t)_{t \in T}, \mathbb{P})$, where T is some set of real numbers, be a filtered probability space and $X = (X_t)_{t \in T}$ be an adapted stochastic process. Then, X is a Markov process if and only if for all finite index sets $t_1 < t_2 < \ldots < t_n < t$

$$\mathbb{P}\left(X_t \leq x | X_{t_1}, \ldots, X_{t_n}\right) = \mathbb{P}\left(X_t \leq x | X_{t_n}\right) \quad \text{(Markov property)}.$$

In plain words, the Markov property states that all information available to forecast future values of the process is embedded in its current value. From this it follows that, upon conditioning on such information, any other changes of future values are independent of each other, and would be represented by the product copula. We should also recall that conditional probabilities are actually partial derivatives of copulas. Collecting everything together, we have that for times $s < \tau$

$$\mathbb{P}(X_s \leq x_s \mid X_\tau = x_\tau) = D_2 C_{s\tau}(F_s(x_s), F_\tau(x_\tau)),$$

where $C_{s\tau}$ is the copula function associated with the random vector (X_s, X_τ), $F_s(\cdot)$ and $F_\tau(\cdot)$ the corresponding marginal distributions and $D_i(\cdot, \cdot)$ denotes the copula partial derivative with respect to the ith argument. The same holds for the time $t > \tau$

$$\mathbb{P}(X_t \leq x_t \mid X_\tau = x_\tau) = D_1 C_{\tau t}(F_\tau(x_\tau), F_t(x_t)),$$

where the notation is consistent with that introduced above.

If we now ask what is the joint probability of $X_t \leq x_t$ and $X_s \leq x_s$, we have to multiply the two conditional probabilities above (by conditional independence) for each scenario $X_\tau = x_\tau$ and then integrate over all the scenarios. If we now apply the integral probability transform to work with uniformly distributed variables, $F_s(X_s) = U_s$, $F_\tau(X_\tau) = U_\tau$ and $F_t(X_t) = U_t$, we may collect everything together and get

$$\mathbb{P}(U_t \leq u_t, U_s \leq u_s) = \int_0^1 D_2 C_{s\tau}(u_s, u_\tau) D_1 C_{\tau t}(u_\tau, u_t) \, du_\tau.$$

What we have obtained is the copula version of the famous Chapman–Kolmogorov equation. If we think of it more carefully, we recognize that what we have actually faced is a compatibility problem. We have asked the question: given that $\{U_s, U_\tau, U_t\}$ is a Markov process what should the joint distribution of $\{U_s, U_t\}$ be? The answer is obtained by integrating out U_τ: $\mathbb{P}(U_s \leq u_s, U_\tau \leq 1, U_t \leq u_t)$. If we now ask what is the joint probability of the three variables $\mathbb{P}(U_s \leq u_s, U_\tau \leq u_\tau, U_t \leq u_t)$, it is easy to guess what the answer is:

$$\mathbb{P}(U_s \leq u_s, U_\tau \leq u_\tau, U_t \leq u_t) = \int_0^{u_\tau} D_2 C_{s\tau}(u_s, u_\tau) D_1 C_{\tau t}(u_\tau, u_t) \, du_\tau.$$

This is the intuition of the remarkable result that characterizes the link between copula functions and Markov processes, as first discovered and analyzed by Darsow *et al.* (1992), and that we label DNO throughout this book. Having laid out the intuition of the idea, we now go into the details of the approach.

3.2.1 The ∗ and ⋆ Product Operators

We are now going to formally define the operations that we applied to copulas in the previous section. Then, we will have to explore in more detail the relationship between these operators and Markov processes. Let $\mathcal{C}$ denote the set of all bivariate copulas on $[0, 1]^2$, satisfying the usual boundary and monotonicity conditions.

The ∗ product operation on copulas.   Let $A, B \in \mathcal{C}$. For $u, v \in [0, 1]$, the ∗ product operation is defined as follows:

$$(A * B)(u, v) = \int_0^1 D_2 A(u, t) D_1 B(t, v)\, dt, \tag{3.4}$$

where we remember that $D_i C(\cdot, \cdot)$, $i = 1, 2$, denotes the partial derivative of copula $C(\cdot, \cdot)$ with respect to the ith argument.

We can observe that, since A and B are absolutely continuous in each place by the monotonicity condition, the integral in (3.4) exists.

It is very easy to prove that the product operator returns a copula. As for the boundary conditions, we have

$$(A * B)(0, v) = \int_0^1 D_2 A(0, t) D_1 B(t, v)\, dt = 0 \tag{3.5}$$

and the same holds for $(A * B)(u, 0)$. Also,

$$(A * B)(1, v) = \int_0^1 D_2 A(1, t) D_1 B(t, v)\, dt = \int_0^1 D_1 B(t, v)\, dt = B(1, v) = v \tag{3.6}$$

and the same obtains for $(A * B)(u, 1)$. As for the monotonicity condition, i.e. non-negative volume, it is easy to see that it may be written as

$$V_{(A*B)}([u_1, u_2], [v_1, v_2]) = \int_0^1 D_2[A(u_2, t) - A(u_1, t)] D_1[B(t, v_2) - B(t, v_1)]dt, \tag{3.7}$$

and the result follows from the standard finding that both $D_2[A(u_2, t) - A(u_1, t)]$ and $D_1[B(t, v_2) - B(t, v_1)]$ are non-negative (see Lemma 2.1.3 in Nelsen (2006)).

Let us now come to the main properties of the product operator, leaving the others to the applications in which they will be applied. We start with the associativity property, which is crucial for the DNO result:

$$A * (B * C) = (A * B) * C, \qquad \text{for all } A, B, C \in \mathcal{C}.$$

Here we just sketch the proof, assuming that all required copula densities exist and all integrals are well defined, referring to the original DNO paper for the proof in full generality. Fix u and

v and set $f(t) = A(u, t)$, $g(s) = C(s, v)$, and $b(t, s) = \frac{\partial^2 B}{\partial t \partial s}(t, s)$. Then, by Fubini's theorem

$$[A * (B * C)](u, v) = \int_0^1 f'(t) \frac{d}{dt} \left(\int_0^1 D_2 B(t, s) g'(s) \, ds \right) dt$$

$$= \int_0^1 \int_0^1 f'(t) b(t, s) g'(s) \, ds \, dt$$

$$= \int_0^1 g'(s) \frac{d}{ds} \left(\int_0^1 f'(t) D_1 B(t, s) \, dt \right) ds$$

$$= [(A * B) * C](u, v).$$

Now, we may ask whether there is a copula that plays the role of zero (so that every copula multiplied times that element returns that element) and an identity element (so that every element times the identity returns the element itself). As for the first, it is easy to check that

$$C * \Pi(u, v) = \int_0^1 D_2 C(u, t) D_1(tv) \, dt = \int_0^1 D_2 C(u, t) v \, dt = v \int_0^1 D_2 C(u, t) \, dt = uv = \Pi,$$
(3.8)

and it is easy to check that the same result is obtained for $\Pi * C$. As for the identity element, denote by $M = \min(u, v)$ the maximum copula. Then, $D_1(\min(u, v))$ is equal to 1 if $u < v$ and zero for $u > v$, and we have

$$C * M(u, v) = \int_0^1 D_2 C(u, t) D_1(\min(t, v)) \, dt = \int_0^v D_2 C(u, t) \, dt = C(u, v), \quad (3.9)$$

and the same for $M * C(u, v)$. So, we have found the identity element.

These results lead to the natural question of what happens if one element of the product is the minimum copula $W = \max(u + v - 1, 0)$. Since this copula corresponds to perfect negative association, one expects that this element plays the role of -1, turning dependence between X and Y into that between X and $-Y$. It is easy to prove that this is in fact the case. Notice that $D_1(\max(u + v - 1, 0))$ is equal to 1 if $u > 1 - v$, and zero for $u < 1 - v$. Then, we obtain

$$C * W(u, v) = \int_0^1 D_2 C(u, t) D_1(\max(t + v - 1, 0)) \, dt \qquad (3.10)$$

$$= \int_{1-v}^1 D_2 C(u, t) \, dt = u - C(u, 1 - v). \qquad (3.11)$$

However, this time we have

$$W * C(u, v) = v - C(1 - u, v), \qquad (3.12)$$

and the product operator is not endowed with the commutative property. Anyway, to confirm that the element W corresponds to -1, it is easy to show that the product of the element times itself returns the identity element M:

$$W * W(u, v) = v - W(1 - u, v) = v - \max(v - u, 0) = \min(u, v) = M(u, v). \quad (3.13)$$

It is also no surprise that if we pre- and post-multiply a copula $C(u,v)$ representing the dependence between X and Y by the element W we should obtain the dependence between

$-X$ and $-Y$, and so the survival copula. By using the associativity property, we have in fact

$$W * C * W(u, v) = W * (u - C(u, 1 - v)) = v - (1 - u - C(1 - u, 1 - v))$$
$$= v + u - 1 + C(1 - u, 1 - v) = \hat{C}(u, v). \tag{3.14}$$

There are a lot of properties of the $*$ product that could be investigated. Let A and B be two 2-copulas. For every copula C we set (see (2.7))

$$C^u(u, v) = v - C(1 - u, v), \tag{3.15}$$

and (see (2.6))

$$C^v(u, v) = u - C(u, 1 - v). \tag{3.16}$$

Then

$$A^v * B^u = A * B. \tag{3.17}$$

In fact

$$A^v * B^u(u, v) = \int_0^1 D_2 A^v(u, t) D_1 B^u(t, v) \, dt$$

$$= \int_0^1 D_2 A(u, 1 - t) D_1 B(1 - t, v) \, dt$$

$$= \int_0^1 D_2 A(u, s) D_1 B(s, v) \, ds$$

$$= A * B(u, v).$$

Similarly

$$A * B^v = (A * B)^v. \tag{3.18}$$

In fact

$$A * B^v(u, v) = \int_0^1 D_2 A(u, t) D_1 B^v(t, v) \, dt$$

$$= \int_0^1 D_2 A(u, t)(1 - D_1 B(t, 1 - v)) \, dt$$

$$= \int_0^1 D_2 A(u, t) \, dt - \int_0^1 D_2 A(u, t) D_1 B(t, 1 - v) \, dt$$

$$= u - A * B(u, 1 - v)$$

$$= (A * B)^v(u, v).$$

Following the same lines, it is trivial to show that

$$A^u * B = (A * B)^u. \tag{3.19}$$

Before we proceed, it is useful to provide a more general definition of the product, which includes the $*$ product as a particular case, and that will be of great help in the following section.

The ⋆ product operation on copulas. Let A be an m-copula and B an n-copula. The ⋆ product operation is defined as follows:

$$(A \star B)(u_1, \ldots, u_{m+n-1}) = \int_0^{u_m} D_m A(u_1, \ldots, u_{m-1}, \xi) D_1 B(\xi, u_{m+1}, \ldots, u_{m+n-1}) d\xi,$$

where we remember that $D_i C$ denotes the partial derivative of copula C with respect to the ith argument.

By arguments similar to those used for the ∗ product, it is easily verified that $A \star B$ is an $(n + m - 1)$-copula and satisfies the same properties listed above. Note that the fact that the ∗ product is a special case of the ⋆ product is immediately evident by inspection of the definitions: $A * B(x, y) = A \star B(x, 1, y)$.

3.2.2   Product Operators and Markov Processes

We now show how to use the operators defined above to provide the representation of Markov processes in terms of copula functions. We go through the proofs reported in the original DNO paper. We first provide a rigorous proof of how to write the Chapman–Kolmogorov equation in terms of copulas. We stick to the case of continuous marginals, even though the proof can readily be extended to the general case along the lines proposed in the original paper. Before approaching the result, let us introduce some notation and preliminaries. If, for fixed y, we set $f(x) = g(x, y)$, we use the notation

$$\int_a^b h(x) g(dx, y)$$

for the Stieltjes integral

$$\int_a^b h(x) \, df(x).$$

In the proof of the theorem below we will need the following relation among two given copula functions A and B:

$$\frac{\partial}{\partial x} \int_0^1 D_1 A(t, y) D_2 B(x, t) dt = \int_0^1 D_1 A(t, y) D_1 B(x, dt). \tag{3.20}$$

For a detailed proof of (3.20) we address the reader to the DNO paper.

Theorem 3.2.1 (DNO, 1992). *Let $X_t, t \in T$ be a real stochastic process, and for each $s, t \in T$ let C_{st} denote the copula of the random variables X_s and X_t. The following are equivalent:*

(i) The transition probabilities $P(s, x, t, A) = P(X_t \in A | X_s = x)$ of the process satisfy the Chapman–Kolmogorov equations

$$P(s, x, t, A) = \int_{-\infty}^{+\infty} P(u, \xi, t, A) P(s, x, u, d\xi),$$

for all Borel sets A, for all $s < t$ in T, for all $u \in (s, t) \cap T$, and for almost all $x \in \mathbb{R}$.
(ii) For all $s, u, t \in T$ satisfying $s < u < t$,

$$C_{st} = C_{su} * C_{ut}. \tag{3.21}$$

Proof. (ii)→(i) We observe that $A \to P(s, x, t, A)$ is a probability measure for all $s < t$ and almost all x: this implies that it is sufficient to verify the Chapman–Kolmogorov equation for only Borel sets of type $A = (-\infty, a]$ and that the transition probabilities are given in terms of copulas by

$$P(s, x, t, A) = D_1 C_{st}(F_s(x), F_t(a)), \quad \text{a.s.,} \tag{3.22}$$

by the conditional probability property (2.48). For the set A of this form, for almost all x we have

$$\int_{-\infty}^{\infty} P(u, \xi, t, A) P(s, x, u, d\xi) = \int_{-\infty}^{\infty} D_1 C_{ut}(F_u(\xi), F_t(a))) D_1 C_{su}(F_s(x), F_u(d\xi))$$

$$= \int_0^1 D_1 C_{ut}(\eta, F_t(a)) D_1 C_{su}(F_s(x), d\eta)$$

$$= \frac{\partial}{\partial \zeta} \int_0^1 D_1 C_{ut}(\eta, F_t(a)) D_2 C_{su}(\zeta, \eta) d\eta \Big|_{\zeta = F_s(x)}$$

$$= D_1 (C_{su} * C_{ut})(F_s(x), F_t(a))$$

$$= D_1 C_{st}(F_s(x), F_t(a))$$

$$= P(s, x, t, A), \tag{3.23}$$

where (3.20) was used in the third step. (i)→(ii) If the Chapman–Kolmogorov equations hold, (3.23) is equal to $D_1 C_{st}(F_s(x), F_t(a))$ for almost all x by (3.22), so that (ii) holds. □

Now, we know that satisfying the Chapman–Kolmogorov equation is a necessary but not sufficient condition for a Markov process. To state a sufficient condition in terms of copula functions we need a generalization of the $*$ product for copulas of dimension greater than two. From a glance at the previous paragraph it is not hard to guess what this generalization is going to be. Here we report the formal statement and a proof of the result.

Theorem 3.2.2. *A real-valued stochastic process $X_t, t \in T$ is a Markov process if and only if for all positive integers n and for all $t_1, \ldots, t_n \in T$ satisfying $t_k < t_{k+1}, k = 1, \ldots, n - 1$,*

$$C_{t_1 \ldots t_n} = C_{t_1 t_2} \star C_{t_2 t_3} \star \ldots \star C_{t_{n-1} t_n}, \tag{3.24}$$

where $C_{t_1 \ldots t_n}$ is the copula of $X_{t_1}, \ldots, X_{t_n}$ and $C_{t_k t_{k+1}}$ is the copula of X_{t_k} and $X_{t_{k+1}}$.

Proof. We start by recalling that a process $X_t, t \in T$ is a Markov process if for all finite index sets $t_1, \ldots, t_n$ and $t \in T$ satisfying $t_1 < t_2 < \ldots < t_n < t$,

$$\mathbb{P}(X_t < \lambda | X_{t_1}, \ldots, X_{t_n}) = \mathbb{P}(X_t < \lambda | X_{t_n}). \tag{3.25}$$

We first observe that for $t_1, t_2, t_3 \in T$ such that $t_1 < t_2 < t_3$, the conditional independence property (3.25) for $n = 2$ holds if and only if

$$\mathbb{P}(X_{t_1} < \lambda_1, X_{t_3} < \lambda_3 | X_{t_2}) = \mathbb{P}(X_{t_1} < \lambda_1 | X_{t_2}) \mathbb{P}(X_{t_3} < \lambda_3 | X_{t_2}), \quad a.s. \tag{3.26}$$

To see that (3.25) implies (3.26), observe that

$$\mathbb{P}(X_{t_1} < \lambda_1, X_{t_3} < \lambda_3 | X_{t_2} = x) = \int_{-\infty}^{\lambda_1} \mathbb{P}(X_{t_3} < \lambda_3 | X_{t_2} = x, X_{t_1} = z) \, dF_{X_{t_1}|X_{t_2}}(z|x)$$

$$= \int_{-\infty}^{\lambda_1} \mathbb{P}(X_{t_3} < \lambda_3 | X_{t_2} = x) \, dF_{X_{t_1}|X_{t_2}}(z|x)$$

$$= \mathbb{P}(X_{t_3} < \lambda_3 | X_{t_2} = x) \int_{-\infty}^{\lambda_1} dF_{X_{t_1}|X_{t_2}}(z|x)$$

$$= \mathbb{P}(X_{t_3} < \lambda_3 | X_{t_2} = x) \mathbb{P}(X_{t_1} < \lambda_1 | X_{t_2} = x).$$

The converse is proved similarly.

Integrating (3.26) over $(-\infty, \lambda_2]$ yields:

$$F_{t_1 t_2 t_3}(\lambda_1, \lambda_2, \lambda_3) = \int_{-\infty}^{\lambda_2} D_2 C_{t_1 t_2}(F_{t_1}(\lambda_1), F_{t_2}(\xi)) D_1 C_{t_2 t_3}(F_{t_2}(\xi), F_{t_3}(\lambda_3)) \, dF_{t_2}(\xi)$$

$$= \int_0^{F_{t_2}(\lambda_2)} D_2 C_{t_1 t_2}(F_{t_1}(\lambda_1), \eta) D_1 C_{t_2 t_3}(\eta, F_{t_3}(\lambda_3)) d\eta$$

$$= C_{t_1 t_2} \star C_{t_2 t_3}(F_{t_1}(\lambda_1), F_{t_2}(\lambda_2), F_{t_3}(\lambda_3)),$$

which yields the thesis for $n = 3$.

Conversely, we assume the thesis holds for $n = 3$ and show that (3.26) is satisfied:

$$\int_{-\infty}^{\lambda_2} \mathbb{P}(X_{t_1} < \lambda_1, X_{t_3} < \lambda_3 | X_{t_2} = z) \, dF_{t_2}(z) = C_{t_1 t_2 t_3}(F_{t_1}(\lambda_1), F_{t_2}(\lambda_2), F_{t_3}(\lambda_3))$$

$$= \int_0^{F_{t_2}(\lambda_2)} D_2 C_{t_1 t_2}(F_{t_1}(\lambda_1), \xi) D_1 C_{t_2 t_3}(\xi, F_{t_3}(\lambda_3)) d\xi$$

$$= \int_{-\infty}^{\lambda_2} \mathbb{P}(X_{t_1} < \lambda_1 | X_{t_2} = z) \mathbb{P}(X_{t_3} < \lambda_3 | X_{t_2} = z) \, dF_{t_2}(z).$$

The argument for $n > 3$ proceeds by similar reasoning and by induction. □

This representation of Markov processes based on copulas leads to a construction technique which is quite different from the conventional one; once a family of bivariate copulas have been specified satisfying the $*$ product closure condition (3.21), the marginal distributions can be specified as well, subject to a continuity condition. Holding the copulas of the process fixed and varying the initial distribution does not affect any other marginal distributions.

In contrast, in the conventional approach one must specify the initial distribution and a family of transition probabilities satisfying the Chapman–Kolmogorov equations; this implies that, holding the transition probabilities of a Markov process fixed and varying the initial distribution, all of the marginal distributions are required to vary as well. We will see that this is a key point in pricing application, in order to grant consistent prices for the same product at different maturity dates.

We give here a general description of this technique to construct Markov processes, before introducing the most standard specifications. Simply take a set of dates $\{t_1, t_2, \ldots, t_m\}$, and a copula function C defining the dependence between the state of the process at time t_i and time t_{i+1}. Now define the k-fold $*$ product of C with itself as $C^k = C * \cdots * C$ (k-times). Then we will have defined the copula $C_{i,i+k}$ between time t_i and t_{i+k} as $C_{i,i+k} = C^k$. Then, given that for example each time length between t_i and t_{i+1} refers to a day, we could aggregate the movements in weeks taking for example $k = 5$, and then we could again describe the Markov process at different frequencies. While this construction provides a fully flexible way to represent Markov processes, such flexibility may even turn out to be too much, making the approach difficult to use. This is the case when the copula family chosen is not closed under the $*$ product operator. To put it in plain words, in cases in which there is no hope that C and $C_{i,i+k}$ are part of the same family of copulas. In this case, we have different shapes of copulas at different frequencies, with little hope, to be optimistic, of preserving a close-form solution for them. So, we would have different copulas for daily dependence with respect to weekly or monthly dependence. It would then be very difficult to characterize $C_{i,i+k}$ for all k given a specification of C. And it would be impossible to recover C given a specific choice of $C_{i,i+k}$. The task is easier, obviously at the cost of less flexibility, if one restricts search to the cases in which C and $C_{i,i+k}$ are part of the same class of copulas, and possibly differ from each other only by the value of the parameter. In these cases, the Markov system that we represent is endowed with a sort of *self-similarity* that may help. Below, we present the most famous and useful examples of these copula functions. We will see that the DNO approach is a mandatory requirement to represent temporal consistency among prices in finance.

3.2.3   Self-similar Copulas

Using the approach described above we can use copulas to represent the dependence structure between the values of the process at dates separated by the unit period of time, as well as the dependence between values separated by k periods of time. We called C the first copula and $C_{i,i+k}$ the second one. The two copulas are linked to each other by the $*$ product. If one wants to characterize the process, two questions arise quite naturally. The first is the behavior of $C_{i,i+k}$ as k tends to infinity. The second is convergence of $C_{i,i+1/k}$ as k tends to infinity. The answer to the first question denotes the asymptotic behavior of the process. Namely, if one finds

$$\lim_{k \to \infty} C_{i,i+k} = \Pi, \tag{3.27}$$

where we remember that Π is the product copula, we may say that the process is compliant with some *mixing* condition, suitably defined. In plain terms, it means that the behavior of the process very far in the future is independent of the initial conditions. We will come back to this with more details in Chapter 4.

At the other extreme, if one finds

$$\lim_{1/k \to \infty} C_{i,i+k} = M, \tag{3.28}$$

where M is the maximum copula, one may conjecture that this means convergence to a right continuous process in continuous time. This topic is actually an open problem and is left for future research.

Now, as we pointed out in the previous paragraph, it is very hard to address this problem in full generality, and to exploit the flexibility of copulas completely. The analysis may be simplified if we stick to the subset of cases in which the copulas C and $C_{i,t+k}$ are of the same class. We may define them as *self-similar* copulas because the shape of the copula is preserved when the process is analyzed at different frequencies. Say for example that at all frequencies the copulas are $C_{i,i+k} = C_{\theta(k)}$, where $\theta(k)$ denotes the dependence structure at the frequency of k units of time. In what follows, we will give three standard examples of such processes.

Brownian Dynamics

One way of constructing a *self-similar* copula is to recover it from processes that we know are endowed with the same property. The first case that comes to mind is *Brownian motion*. It is well known that in this case the conditional distribution of the process X at time t given X_s is normal with zero mean and variance equal to $t-s$. Then,

$$\mathbb{P}(X_t \leq x_t | X_s = x_s) = D_1 C_{st}(\mathbb{P}(X_s \leq x_s), \mathbb{P}(X_t \leq x_t)) = \Phi\left(\frac{x_t - x_s}{\sqrt{t-s}}\right), \qquad (3.29)$$

where $\Phi(x)$ is the standard normal distribution. Furthermore, if we sit at time 0 and observe $X_0 = 0$, it follows that $\mathbb{P}(X_t \leq x_t) = \Phi(x_t/\sqrt{t})$ and $\mathbb{P}(X_s \leq x_s) = \Phi(x_s/\sqrt{s})$. Then, we have

$$D_1 C_{st}(u_s, u_t) = \Phi\left(\frac{\sqrt{t}\Phi^{-1}(u_t) - \sqrt{s}\Phi^{-1}(u_s)}{\sqrt{t-s}}\right) \qquad (3.30)$$

and

$$C_{st}(u_s, u_t) = \int_0^{u_s} \Phi\left(\frac{\sqrt{t}\Phi^{-1}(u_t) - \sqrt{s}\Phi^{-1}(w)}{\sqrt{t-s}}\right) dw.$$

Notice that the Brownian copula is actually a Gaussian copula with a specific correlation parameter. In fact, we may rewrite it as

$$C_{st}(u_s, u_t) = \int_0^{u_s} \Phi\left(\frac{\Phi^{-1}(u_t) - \sqrt{s/t}\Phi^{-1}(w)}{\sqrt{1-s/t}}\right) dw \qquad (3.31)$$

and we obtain a Gaussian copula with correlation parameter $\rho(s, t) = \sqrt{s/t}$. Since the Brownian motion is a Markov process, necessarily

$$C_{st} = C_{s\tau} * C_{\tau t},$$

and it follows that $\rho(s, t) = \rho(s, \tau)\rho(\tau, t)$.

Gaussian Copula-based Dynamics

In other cases, one can check whether some copula functions in a class are endowed with the *self-similarity* property. We start by considering a generalization of the case of Brownian copulas. Let us consider a general Gaussian copula

$$C^\rho(u_s, u_t) = \int_0^{u_s} \Phi\left(\frac{\Phi^{-1}(u_t) - \rho\Phi^{-1}(w)}{\sqrt{1-\rho^2}}\right) dw \qquad (3.32)$$

with $\rho \in (-1, 1)$.

Let $\rho_1, \rho_2 \in (0, 1)$. Consider a Brownian motion W_t at times $s = \rho_1^2$, $\tau = 1$, $t = \frac{1}{\rho_2^2}$. Then

$$C_{\rho_1^2 1} * C_{1\frac{1}{\rho_2^2}} = C_{\rho_1^2 \frac{1}{\rho_2^2}}$$

and $\rho(\rho_1^2, \frac{1}{\rho_2^2}) = \rho(\rho_1^2, 1)\rho(1, \frac{1}{\rho_2^2}) = \rho_1 \cdot 1/\frac{1}{\rho_2} = \rho_1\rho_2$. Then

$$C^{\rho_1} * C^{\rho_2} = C^{\rho_1\rho_2}. \tag{3.33}$$

If $\rho_1, \rho_2 \in (-1, 0)$ then, for example, $C^{\rho_1} = \left(C^{|\rho_1|}\right)^v$ (see (3.16)) and $C^{\rho_2} = \left(C^{|\rho_2|}\right)^u$ (see (3.15)). Hence, by (3.17) and (3.33)

$$C^{\rho_1} * C^{\rho_2} = \left(C^{|\rho_1|}\right)^v * \left(C^{|\rho_2|}\right)^u = C^{|\rho_1|} * C^{|\rho_2|} = C^{\rho_1\rho_2}.$$

If $\rho_1 \in (0, 1)$ and $\rho_2 \in (-1, 0)$ then, for example, $C^{\rho_2} = \left(C^{|\rho_2|}\right)^v$ (see (3.16)). Hence, by (3.18) and (3.33)

$$C^{\rho_1} * C^{\rho_2} = C^{\rho_1} * \left(C^{|\rho_2|}\right)^v = \left(C^{\rho_1} * C^{|\rho_2|}\right)^v = \left(C^{\rho_1|\rho_2|}\right)^v = C^{\rho_1\rho_2}.$$

Again, if $\rho_1 \in (-1, 0)$ and $\rho_2 \in (0, 1)$ then, for example, $C^{\rho_1} = \left(C^{|\rho_1|}\right)^u$ (see (3.15)). Hence, by (3.19) and (3.33), we get

$$C^{\rho_1} * C^{\rho_2} = \left(C^{|\rho_1|}\right)^u * C^{\rho_2} = \left(C^{|\rho_1|} * C^{\rho_2}\right)^u = \left(C^{|\rho_1|\rho_2}\right)^u = C^{\rho_1\rho_2}.$$

Finally, if one of the parameters is zero, for example $\rho_1 = 0$, then $C^{\rho_1} = \Pi$ and

$$C^{\rho_1} * C^{\rho_2} = \Pi = C^0 = C^{\rho_1\rho_2}.$$

Concluding, (3.33) holds for all $\rho_1, \rho_2 \in (-1, 1)$.

Let us now consider a stochastic process X_t whose 2-level copulas are of the Gaussian type. Let ρ_{st} be the parameter associated with the copula C_{st}. The required closure property implies that, for $s < \tau < t$

$$\rho_{st} = \rho_{s\tau}\rho_{\tau t}.$$

Remark 3.2.1. Assume we model the dynamics at some frequency, say the daily frequency, by the Gaussian copula with dependence parameter ρ. Then, the dependence at lower frequencies, say that of n days, is given by ρ^n. As n tends to infinity, the corresponding copula tends to the product one.

Conversely, consider higher frequencies corresponding to intervals of length $\frac{1}{n}$. Then (assuming a constant dependent structure among subsequent intervals) the parameter is $\rho^{1/n}$ that, if $\rho \neq 0$ converges to 1, allows for perfect positive dependence.

A further self-similar generalization of the Brownian copula, but in a different direction, is given by the *Stable copula* that will be introduced in Section 3.4.2, Example 3.4.6.

Farlie–Gumbel–Morgenstern Dynamics

In order to introduce another self-similar copula family which generates a model that is widely used, we follow the analysis in Nelsen (2006) concerning symmetric copulas with quadratic sections. Since they must have quadratic sections in u, they can be written as

$$C(u, v) = a(v)u^2 + b(v)u + c(v). \tag{3.34}$$

Obviously, the grounded property requires $c(v) = 0$ and the uniform marginal property implies $a(v) + b(v) = v$. Let us define $\theta(v) = -a(v)$. Then, all quadratic section copulas can also be written as

$$C(u, v) = uv + \theta(v)u(1 - u). \tag{3.35}$$

Since C is symmetric,

$$C(u, v) = uv + \theta(u)v(1 - v), \tag{3.36}$$

consequently $\theta(v) = \theta v(1 - v)$ for some parameter θ, so that

$$C_\theta(u, v) = uv + \theta uv(1 - u)(1 - v). \tag{3.37}$$

The C_θ-volume of a rectangle $[u_1, u_2] \times [v_1, v_2]$ is given by

$$V_{C_\theta}([u_1, u_2] \times [v_1, v_2]) = (u_2 - u_1)[1 + \theta(1 - u_1 - u_2)(1 - v_1 - v_2)].$$

Since $(1 - u_1 - u_2)(1 - v_1 - v_2)$ is in $[-1, 1]$ it follows that C_θ is 2-increasing if and only if $\theta \in [-1, 1]$. This family is known as the *Farlie–Gumbel–Morgenstern (FGM) copula*.

Let us now show that this family is endowed with the self-similarity property. Take $s < \tau < t$, consider

$$C_{s\tau}(u_s, u_\tau) = u_s u_\tau + \theta_{s\tau} u_s u_\tau (1 - u_s)(1 - u_\tau)$$

and

$$C_{\tau t}(u_\tau, u_t) = u_\tau u_t + \theta_{\tau t} u_\tau u_\tau (1 - u_t)(1 - u_\tau).$$

Then

$$D_2 C_{s\tau}(u_s, u_\tau) = u_s + \theta_{s\tau} u_s (1 - u_s)(1 - 2u_\tau)$$

and

$$D_1 C_{\tau t}(u_\tau, u_t) = u_t + \theta_{\tau t} u_t (1 - u_t)(1 - 2u_\tau).$$

Now some algebra shows that

$$\int_0^1 (u_s + \theta_{s\tau} u_s (1 - u_s)(1 - 2u_\tau))(u_t + \theta_{\tau t} u_t (1 - u_t)(1 - 2u_\tau)) \, du_\tau$$
$$= u_s u_t + \frac{\theta_{s\tau} \theta_{\tau t}}{3}(1 - u_s)(1 - u_t). \tag{3.38}$$

Hence, we obtain an FGM copula function with

$$\theta_{st} = \frac{\theta_{s\tau} \theta_{\tau t}}{3}. \tag{3.39}$$

and the FGM family is closed under the $*$ product operator. The copula that has quadratic section and is closed under the $*$ product operator is

$$C_{st}(u_s, u_t) = u_s u_t + \theta_{st} u_s u_t (1 - u_s)(1 - u_t). \tag{3.40}$$

The *FGM* copula has further very interesting properties, namely it is radially symmetric, and closed with respect to mixture, that is

$$\kappa C_\alpha + (1 - \kappa)C_\beta = C_{\kappa\alpha+(1-\kappa)\beta},$$

where C_θ denotes the FGM copula with parameter θ. The parameter ranges from -1 to 1 with $\theta = 0$ corresponding to independence. A flaw of this kind of copula is that it is not comprehensive, so that even for extreme values of the dependence parameter it is not possible to reach perfect association. Moreover, the maximum level of association that can be reached is quite low.

Remark 3.2.2. Assume we model the dynamics at some frequency, say the daily frequency, by the FGM with dependence parameter θ. Then, the dependence at lower frequencies, say that obtained by the product of n copulas, is given by

$$\theta_n = 3 \left(\frac{\theta}{3} \right)^n.$$

It is easy to see that the dependence decreases with n, and tends to the product copula as n tends to infinity. Also notice that the same works if we try to recover the dependence structure of higher and higher frequencies:

$$\theta_{1/n} = \min \left(1, 3 \left(\frac{\theta}{3} \right)^{1/n} \right).$$

Anyway, remember that the FGM copula is not comprehensive and cannot reach perfect dependence at any frequency.

Fréchet Dynamics

Another important family of copulas which is closed with respect to the product operator is the Fréchet family. The result follows directly from the linearity of the integral and the properties of the perfect association copulas and the product copula. For the sake of simplicity, we show the result for the *mixture* copula $C_\rho = \rho M + (1 - \rho)\Pi$:

$$C_{\rho_{st}}(u_s, u_t) = C_{\rho_{s\tau}}(u_s, u_\tau) * C_{\rho_{\tau t}}(u_\tau, u_t) \tag{3.41}$$
$$= \rho_{s\tau}\rho_{\tau t} M(u_s, u_t) + (1 - \rho_{s\tau}\rho_{\tau t})\Pi(u_s, u_t) = C_{\rho_{s\tau}\rho_{\tau t}}(u_s, u_t).$$

Notice also that mixture copulas generate asymptotic independence.

3.2.4 Simulating Markov Chains with Copulas

Since Markov processes involve sequences of bivariate copulas, it is very easy to generate sequences of pairs of variables with a given associated copula function that represent Markov chains. The ideal technique is conditional sampling. Formally, if we define

$$h(u_t|u_s) \equiv D_1 C_{st}(u_s, u_t), \tag{3.42}$$

conditional sampling is carried out by generating a random variable with uniform distribution ζ and computing

$$u_t = h^{-1}(\zeta|u_s), \tag{3.43}$$

given the value u_s of a random variable previously simulated with uniform distribution and independent of ζ and where $h^{-1}(\cdot|u_s)$ is the generalized inverse defined in (2.1).

Conditional sampling can only be iterated for Markov systems generated by sets of copulas that are not closed under the product operator. For self-similar copulas, the algorithm could also be applied by generating scenarios for different horizon conditionals on the same starting point. This is another advantage of self-similar copulas: one could directly generate one scenario for the possible dynamics a year from the starting date in a single shoot, with no need to generate and iterate daily dynamics.

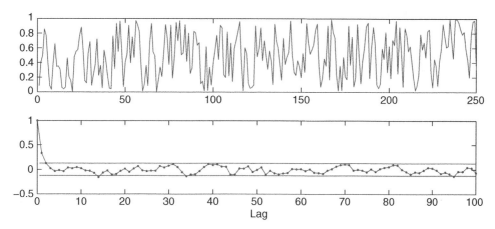

Figure 3.1 Example of the dynamics of a Markov process generated by an FGM copula and the relative autocorrelation function.

Example 3.2.1. *In this example we generate a Markov process using the simulation procedure described above. The copula function C_{st} is an FGM copula with dependence parameter equal to 1 on a weekly basis. In Figure 3.1 we report the dynamics and the relative autocorrelation function of the simulated trajectory. Here we derive the conditional sampling formulas for this kind of copula:*

$$D_1 C_\theta(u, v) = h(v|u) = v + \theta(1 - 2u)v(1 - v) = av - (a - 1)v^2,$$

with $a \equiv 1 + \theta(1 - 2u)$. We then obtain

$$v = \frac{a - \sqrt{a^2 - 4(a - 1)h(v|u)}}{2(a - 1)}.$$

3.3   TIME-CHANGED BROWNIAN COPULAS

We now address the problem of how to make the approach above more general. One first route is to apply the *time change* technique, that is usual in the theory of stochastic processes and largely used in finance.

A stochastic clock (or time change) is modeled through an increasing stochastic process $t \to T_t$ with $T_0 = 0$. In order to understand how the time change works, we assume a stochastic process X_t. As the simplest case, we start by considering the case in which $T_t(\omega)$ has continuous strictly increasing trajectories with $T_0(\omega) = 0$ and $T_{+\infty}(\omega) = +\infty$. In this case, the time-changed process $Z_t = X_{T_t}$ visits the same states as X in the same order and performing the same jumps as X, but at a different speed. If the trajectory $T_t(\omega)$ is not strictly increasing, that is there exists an interval $[t_1, t_2)$ on which it is constant, then $Z_t(\omega) = X_{T_t(\omega)}(\omega) = X_{T_{t_1}(\omega)}(\omega)$ for all $t \in [t_1, t_2)$. If $T_t(\omega)$ admits (upward) jumps, then $Z_t(\omega) = X_{T_t(\omega)}(\omega)$ does not evaluate X everywhere.

The importance of the time change technique relies on the fact that it allows us to represent general processes by a suitable change of time. In its most general form it is due to Monroe (1978).

Theorem 3.3.1 (Monroe).   *Every semi-martingale is equivalent to a time-changed Brownian motion.*

This powerful result provides a generalization of earlier findings by Dambis (1965) and Dubins and Schwarz (1965), that were limited to martingales. In this case the time change is performed by using the quadratic variation of the process.

The other way around, one can generate a non-Gaussian process by sampling a Brownian motion at prespecified random times. The stochastic clock can either be modeled using increasing Lévy processes, called *subordinators*, or other increasing processes, such as an exponentially cumulated process representing the dynamics of volatility or of the intensity of jumps. Here we show how to apply this technique to the copula-based representation of Markov processes.

To the best of our knowledge, the idea of a time-changed copula has been applied for the first time by Schmitz (2003), limited to cases in which the change of time is deterministic. We provide here a general representation that rephrases the literature of the stochastic clock in the language of copula functions.

Time-Changed Copula (Schmitz, 2003)   Consider a Brownian motion W_t and an independent stochastic clock $T_t(\omega)$ and the associated time-changed stochastic process $Z_t = W_{T_t}$. Then for each trajectory $T_t(\omega)$, the copula function

$$C_{st}(u_s, u_t; \omega) = \int_0^{u_s} \Phi \left(\frac{\sqrt{T_t(\omega)}\Phi^{-1}(u_t) - \sqrt{T_s(\omega)}\Phi^{-1}(r)}{\sqrt{T_t(\omega) - T_s(\omega)}} \right) dr \qquad (3.44)$$

is called a **time-changed copula**.

Notice that if $T_t(\omega)$ is very close to $T_s(\omega)$, so that the clock moves very slowly, we have the limit $C_{\tau t}(u_\tau, u_t) = \min(u_\tau, u_t)$, which implies $C_{s\tau}(u_s, u_\tau) * \min(u_\tau, u_t) = C_{st}(u_s, u_t)$, and the dependence structure does not change. On the other hand, a huge increase of time much further from $T_s(\omega)$ will cause $C_{\tau t}(u_\tau, u_t)$ to approach the product copula $u_\tau u_t$ in the limit, with the result $C_{s\tau}(u_s, u_\tau) * u_\tau u_t = u_s u_t$ and time series dependence will decrease towards independence.

Below we report two typical examples of stochastic clocks, commonly used in financial applications.

3.3.1   CEV Clock Brownian Copulas

The first example is inspired by the stochastic clock proposed by Carr and Wu (2004). The idea is to introduce a stochastic process described by *constant elasticity of variance* to model the stochastic clock. For econometric applications, the reader is referred to Yu and Phillips (2001). A particular case would be a non-stochastic clock, generated by an Ornstein–Uhlenbeck process, leading to the special case in Schmitz (2003).

Consider a stochastic process with dynamics

$$dV_t = -\alpha V_t + \sigma V_t^\gamma d\omega_t, \qquad (3.45)$$

where α denotes the mean reversion parameter, σ the diffusion parameter, and ω_t a *Wiener process*. The parameter γ enables us to embed different conditional distributions. Among the most renowned cases one has $\gamma = 0.5$, which is called the *square root* process, and for which the conditional distribution is *non-central chi-square*. The other case is $\gamma = 0$, which yields the Ornstein–Uhlenbeck process, and for which the conditional distribution is Gaussian. The solution of the stochastic differential equation is

$$V_t = V_0 e^{-\alpha t} + \sigma \int_0^t e^{-\alpha(t-s)} V_s^\gamma d\omega_s. \tag{3.46}$$

We change the dynamics to a martingale by setting

$$X_t \equiv V_t e^{\alpha t}. \tag{3.47}$$

The quadratic variation of this martingale is computed as

$$\mathcal{T}_t \equiv \langle X_t - X_0 \rangle = \sigma^2 \int_0^t e^{2\alpha(t-s)} V_s^{2\gamma} ds. \tag{3.48}$$

In the case of the Ornstein–Uhlenbeck process we compute

$$\mathcal{T}_t = \frac{\sigma^2}{2\alpha}(e^{2\alpha t} - 1). \tag{3.49}$$

This leads to a temporal dependence structure which is Gaussian with time-changed correlation. Namely, we have

$$C(u, v) = \int_0^u \Phi\left(\frac{\Phi^{-1}(v) - \rho(s, t)\Phi^{-1}(w)}{\sqrt{1 - \rho(s, t)^2}}\right) dw, \tag{3.50}$$

with

$$\rho(s, t) \equiv \sqrt{\frac{\mathcal{T}_s}{\mathcal{T}_t}} = \sqrt{\frac{e^{2\alpha s} - 1}{e^{2\alpha t} - 1}}. \tag{3.51}$$

The dependence figure ρ is a function of the mean reversion parameter only (and possibly of the state variable in the more general CEV model) and does not depend on the scale of the diffusion parameter σ. Furthermore, it is easy to check that

$$\lim_{\alpha \to 0} \rho(s, t) = \sqrt{\frac{s}{t}}, \tag{3.52}$$

$$\lim_{\alpha \to \infty} \rho(s, t) = 0. \tag{3.53}$$

We can observe in Figure 3.2 that the surface of $\rho(s, t)$ moves up for lower values of the parameter α.

3.3.2 VG Clock Brownian Copulas

This is the case for a time change realized through a subordinator. In this case the stochastic clock is in fact given by a Gamma process which is a Lévy process with Gamma distributed increments. More precisely, we assume $\mathcal{T}_t \overset{d}{=} \Gamma(\alpha t, \alpha)$.

The variance Gamma process can be defined by time changing a Brownian motion with drift a and volatility σ by an independent Gamma process. Then the variance Gamma process X_t may be written as

$$X_t = a\mathcal{T}_t + \sigma W_{\mathcal{T}_t},$$

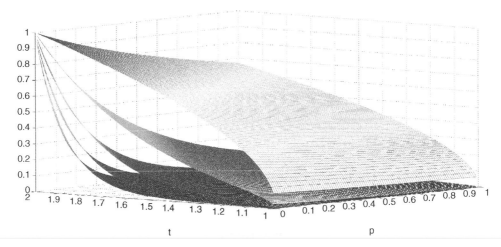

Figure 3.2   Surfaces of $\rho(s,t)$ with respect to s and t for different values of α.

where W is an independent Brownian motion. This leads to a temporal dependence structure which is Gaussian with time-changed correlation. Namely, we have

$$C(u,v) = \int_0^u \Phi \left(\frac{\Phi^{-1}(v) - \rho(s,t)\Phi^{-1}(w)}{\sqrt{1 - \rho(s,t)^2}} \right) dw, \tag{3.54}$$

with

$$\rho(s,t) \equiv \sqrt{\frac{\mathcal{T}_s}{\mathcal{T}_t}}. \tag{3.55}$$

3.4   COPULAS AND MARTINGALE PROCESSES

We are now going to address the problem of how to impose the martingale condition in a copula-based representation of a Markov process. We remember that we need this to ensure that the innovation process is unpredictable, as required by the efficient market hypothesis. Notice that up to this point we have worked with the levels of a process, rather than first differences, and this makes quite involved the task of imposing the martingale condition on it. Actually, we may follow two strategies. The first is to restrict the analysis to cases in which the actual process can be transformed into a Browian motion under a suitable change of time. In this case, all non-Gaussian behavior is moved to the *stochastic clock* and all we have to do is impose the martingale condition on the marginal distributions, setting the mean of each of them on the current value of the process X_t. This way, Brownian motion ensures that the martingale property is satisfied. The shortcoming of this choice is that we give away our task of adopting copula-based representation techniques of the price process. Actually, the only copula that we would use in this case is the Brownian copula. It is true that we could still apply our semi-parametric approach at a higher level to model the dynamics of the stochastic clock, extending the possible dynamics well beyond the CEV and VG cases described above. Nevertheless, it seems worthwhile to check whether the semi-parametric approach to modeling the price process itself instead of its quadratic variation could be maintained.

To introduce the problem, let us stick to the most famous cases of *self-similar* copulas that we studied above: the Brownian motion and the FGM copulas. What is the difference between these two copulas? From what we know of Brownian motion, we can conjecture that the Brownian copula represents the temporal dependence of a process with independent increments. It is actually this feature that makes it possible to impose the martingale condition by simply modeling the marginal distribution. But what could be said about the FGM copula? Is it also leading to the description of a process with independent increments? We will see that the answer is no. This suggests the idea of providing different representations of processes with independent and dependent increments. This is the original contribution of our research group and for this reason we call it the Bologna approach. The key idea is to apply copula functions to the dependence structure of a process and its increments. Having specified this dependence structure we can recover both the marginal distribution of the process at each date and the copula function linking the levels of the process at two dates. We can then apply the latter in the DNO approach that we discussed above to recover a Markov process with a prescribed dependence structure of the increments. In what follows, we show how to accomplish these tasks, and how this enables us to impose the martingale condition on the process.

3.4.1 *C*-Convolution

The problem we want to address is the following. Given two dates $s < t$, we have the process X and its increment

$$X_t = X_s + Y,$$

where Y is the increment recorded at time t. We want to assign a dependence structure on X_s and Y and recover: (i) the distribution of X_t; (ii) the copula linking X_t and X_s. More generally, we face the problem of the distribution of a variable which is the sum of two variables $(X+Y)$ of which we know the joint distribution (we have dropped the subscripts to simplify the notation). It is well known that if X and Y are independent, the distribution of the sum is given by an operation called the *convolution*, of the two variables. Since we are interested in the general case in which X and Y are dependent, we will generalize this definition to that of *C-convolution*, where C stands for the copula function linking X and Y. Of course, the standard convolution should be recovered for $C = \Pi$. The key point to work out the result is again the link between copula functions and the conditional probability.

Proposition 3.4.1. *Let X and Y be two real-valued random variables on the same probability space* $(\Omega, \mathcal{F}, \mathbb{P})$ *with corresponding copula* $C_{X,Y}$ *and continuous marginals* F_X *and* F_Y*. Then,*

$$C_{X,X+Y}(u, v) = \int_0^u D_1 C_{X,Y}\left(w, F_Y(F_{X+Y}^{-1}(v) - F_X^{-1}(w))\right) dw \qquad (3.56)$$

and

$$F_{X+Y}(t) = \int_0^1 D_1 C_{X,Y}\left(w, F_Y(t - F_X^{-1}(w))\right) dw. \qquad (3.57)$$

Proof.

$$F_{X,X+Y}(s,t) = \mathbb{P}(X \le s, X+Y \le t)$$

$$= \int_{-\infty}^{s} \mathbb{P}(X+Y \le t | X = x)\, dF_X(x)$$

$$= \int_{-\infty}^{s} \mathbb{P}(Y \le t - x | X = x)\, dF_X(x)$$

$$= \int_{-\infty}^{s} D_1 C_{X,Y}\left(F_X(x), F_Y(t-x)\right)\, dF_X(x)$$

$$= \int_{0}^{F_X(s)} D_1 C_{X,Y}\left(w, F_Y(t - F_X^{-1}(w)))\right)\, dw,$$

where we made the substitution $w = F_X(x) \in (0,1)$.

Then, the copula function linking X and $X+Y$ is

$$C_{X,X+Y}(u,v) = F_{X,X+Y}\left(F_X^{-1}(u), F_{X+Y}^{-1}(v)\right)$$

$$= \int_{0}^{u} D_1 C_{X,Y}\left(w, F_Y(F_{X+Y}^{-1}(v) - F_X^{-1}(w)))\right)\, dw.$$

Moreover

$$F_{X+Y}(t) = \lim_{s \to +\infty} F_{X,X+Y}(s,t) = \int_{0}^{1} D_1 C_{X,Y}\left(w, F_Y(t - F_X^{-1}(w)))\right)\, dw.$$

□

Pure mathematicians are left with a question mark: since $D_1 C(u,v)$ exists almost everywhere in $[0,1]^2$ (see Section 2.15), it could happen that there exist some t's such that $D_1 C(u,v)$ is not defined on the curve $v = F_Y(t - F_X^{-1}(u))$ for $u \in (a,b)$ (this, being a curve, defines a null set). Hence (3.57) doesn't hold for such t's. We prove below that this is never the case apart from a null set of t's: as a consequence, $F_{X+Y}(t)$ can be figured out for all t's by using right continuity. Anyway, if we assume continuous and strictly increasing marginal c.d.f.'s this property is obviously satisfied and so people not interested in different and more singular situations may skip this technical lemma.

Lemma 3.4.2. *Let $C(u,v)$ be a copula function and F_X and F_Y be two continuous c.d.f.s, then $\forall t$ a.e. $D_1 C\left(\omega, F_Y(t - F_X^{-1}(\omega))\right)$ is well defined $\forall \omega$ a.s.*

Proof. Let us assume that the thesis is false and define

$$N = \{(\omega,v) : D_1 C(\omega,v) \text{ is not defined}\}.$$

It follows that an interval (t_0, t_1) with $t_0 < t_1$ exists, such that $\forall t \in (t_0, t_1)$ a.e. there exists $A_t = (a_t, b_t) \subset [0,1]$, $a_t < b_t$, for which, $\forall \omega \in A_t$ a.e. $(\omega, F_Y(t - F_X^{-1}(\omega))) \in N$. Then $(a,b) \subset [0,1]$ exists for which an interval $(\hat{t}, \hat{t} + h) \subset (t_0, t_1)$ is defined, such that $(a,b) \subset A_t, \forall t \in (\hat{t}, \hat{t} + h)$ a.e.

If $\hat{\omega} \in (a,b)$ exists such that $\exists [s, s + \epsilon] \subset (\hat{t}, \hat{t} + h)$ where $F_Y(t - F_X^{-1}(\hat{\omega}))$ is strictly increasing with respect to t, this would contradict the existence a.e. of the function $v \to D_1 C(u,v)$ (see Nelsen (2006), Theorem 2.2.7).

On the other hand, if $\forall \omega \in (a, b)$, $F_Y(t - F_X^{-1}(\omega))$ is constant with respect to $t \in (\hat{t}, \hat{t} + h)$, putting $v_t(\omega) = F_Y(t - F_X^{-1}(\omega))$, we get

$$F_Y(\hat{t} - F_X^{-1}(\omega)) = F_Y(\hat{t} + h - F_X^{-1}(\omega)) \quad \forall \omega \in (a, b)$$
$$\Rightarrow \mathbb{P}(\hat{t} - F_X^{-1}(\omega) < Y \leq \hat{t} + h - F_X^{-1}(\omega)) = 0 \quad \forall \omega \in (a, b)$$
$$\Rightarrow \mathbb{P}(\hat{t} - F_X^{-1}(b) < Y \leq \hat{t} + h - F_X^{-1}(a)) = 0$$
$$\Rightarrow \bar{v} = F_Y(\hat{t} - F_X^{-1}(b)) = F_Y(\hat{t} + h - F_X^{-1}(a))$$
$$= F_Y(t - F_X^{-1}(\omega)) \quad \forall t \in (\hat{t}, \hat{t} + h), \forall \omega \in (a, b)$$
$$\Rightarrow \forall t \in (\hat{t}, \hat{t} + h), v_t(\omega) = \bar{v}.$$

But in this case $D_1 C(u, v)$ will not exist at all points $(u, \bar{v})$ with $u \in (a, b)$, and this will contradict the existence for all u a.e. of $D_1 C(u, v) \, \forall v$ (see Nelsen (2006), Theorem 2.2.7). $\square$

Going back to the result above, we see that we have implicitly defined the concept of *C-convolution* proposed above, and this concept holds in full generality. We report the formal definition below.

Definition 3.4.1. Let F, H be two continuous c.d.f.s and C a copula function. We define the **C-convolution** of H and F as the c.d.f.

$$H \overset{C}{*} F(t) = \int_0^1 D_1 C\left(w, F(t - H^{-1}(w))\right) dw.$$

We may also add an interesting result that may be useful to aggregate or decompose *C-convolutions*.

Remark 3.4.1. The C-convolution operator is closed with respect to mixtures of copula functions. In fact, it is trivial to show that for all bivariate copula functions A and B, if $C(u, v) = \lambda A(u, v) + (1 - \lambda)B(u, v)$ for $\lambda \in [0, 1]$, then, for all c.d.f.s H and F,

$$H \overset{C}{*} F = H \overset{\lambda A + (1-\lambda)B}{*} F = \lambda H \overset{A}{*} F + (1 - \lambda)H \overset{B}{*} F. \tag{3.58}$$

So, the first part of our problem is solved. Given marginals of X and Y and their copula function, we recover the distribution of $X+Y$. The proposition above also describes the dependence structure between X and $X+Y$. For completeness, we may also prove that this actually defines a copula. So, the proposition could actually be used to construct new copula functions.

Proposition 3.4.3. *Let F, G, H be three continuous c.d.f.s, $C(w, \lambda)$ a copula function and*

$$\tilde{C}(u, v) = \int_0^u D_1 C\left(w, F(G^{-1}(v) - H^{-1}(w))\right) dw.$$

$\tilde{C}(u, v)$ is a copula function iff

$$G = H \overset{C}{*} F. \tag{3.59}$$

Proof. Let us assume (3.59) to hold. Since there exists a probability space and two random variables X and Y with joint distribution function $F(x, y) = C(F_X(x), F_Y(y))$, thanks to Proposition 3.4.1, $\tilde{C}$ is a copula function.

Vice versa, let $\tilde{C}$ be a copula function. Necessarily $\tilde{C}(1, v) = v$ holds. But

$$\tilde{C}(1, v) = \int_0^1 D_1 C \left(w, F \left(G^{-1}(v) - H^{-1}(w) \right) \right) dw$$

$$= H \overset{C}{*} F \left(G^{-1}(v) \right)$$

and

$$H \overset{C}{*} F \left(G^{-1}(v) \right) = v$$

for all $v \in (0, 1)$ if and only if $G = H \overset{C}{*} F$. $\square$

Remark 3.4.2. Notice that, unlike *C-convolution*, the corresponding copula is not closed with reference to a mixture of copulas. Nevertheless, we have

$$\tilde{C}(u, v) = \lambda \int_0^u D_1 A \left(w, F((H \overset{C}{*} F)^{-1}(v) - H^{-1}(w)) \right) dw$$

$$+ (1 - \lambda) \int_0^u D_1 B \left(w, F((H \overset{C}{*} F)^{-1}(v) - H^{-1}(w)) \right) dw, \qquad (3.60)$$

with $H \overset{C}{*} F$ given by (3.58).

We are now ready to apply the approach above to the representation of stochastic processes. Define X_s with marginal distribution F_s and Y with distribution F_Y. Then, we may recover the distribution of $X_t = X_s + Y$ as

$$F_t(x_t) = \int_0^1 D_1 C \left(w, F_Y(x_t - F_s^{-1}(w)) \right) dw$$

and the copula $C_{st}(u_s, u_t)$ as

$$C_{st}(u_s, u_t) = \int_0^{u_s} D_1 C \left(w, F_Y(F_t^{-1}(u_t) - F_s^{-1}(w)) \right) dw.$$

Then, the copula $C_{st}(u_s, u_t)$ can be applied in the DNO approach to construct a Markov process. Notice that unlike in the standard DNO approach we are not free to specify the distribution of the process at each time, except for the first element. Nevertheless, this reduced degree of freedom is compensated by the possibility of selecting the distributions of the increments of the process.

Example 3.4.1. *Examples of C-convolutions. We show a Monte Carlo simulation of random observations from our C-convolution F_{X+Y}. The simulation procedure is based on the fact that $F_{X+Y}(X + Y) \overset{d}{=} U(0, 1)$. So, we generate a sequence of pseudo-random numbers $r_1, \ldots, r_n$ from $U(0, 1)$ and then we compute $z_i = F_{X+Y}^{-1}(r_i)$, $i = 1, \ldots, n$. The sequence $z_1, \ldots, z_n$ is the desired random sample from F_{X+Y}. The realizations depend on the copula function $C_{X,Y}$ and on the marginal distributions F_X and F_Y. Figures 3.3 and 3.4 display the histograms of the simulated values when the copula $C_{X,Y}$ and its parameter change: in both figures we report the independence case (that is, $C_{X,Y}(u, v) = \Pi(u, v)$) to underline how Frank and Clayton copulas affect the distribution of the sum. We report the copula parameter in terms of the corresponding Kendall's τ. The marginal distributions F_X and F_Y are standard Gaussian.*

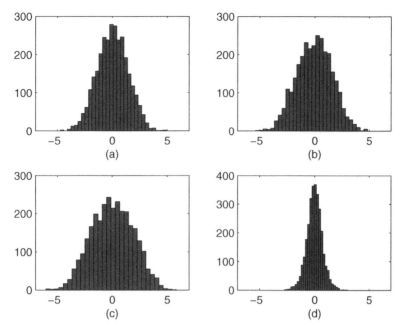

Figure 3.3 Histograms of 3000 simulated values from the C-convolution F_{X+Y} when the marginal distributions F_X and F_Y are standard Gaussian and the copula $C_{X,Y}$ is a Frank copula: (a) independence $(\tau = 0)$; (b) $\tau = 0.2$; (c) $\tau = 0.6$; (d) $\tau = -0.6$.

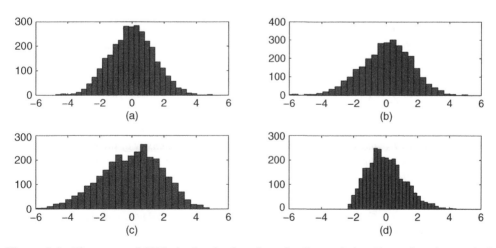

Figure 3.4 Histograms of 3000 simulated values from the C-convolution F_{X+Y} when the marginal distributions F_X and F_Y are standard Gaussian and the copula $C_{X,Y}$ is a Clayton copula: (a) independence $(\tau = 0)$; (b) $\tau = 0.2$; (c) $\tau = 0.6$; (d) $\tau = -0.2$.

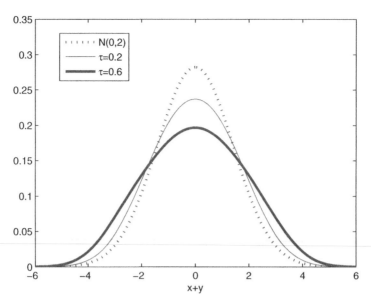

Figure 3.5 Numerical approximation of the density of the C-convolution, f_{X+Y}, when the marginal distributions F_X and F_Y are standard Gaussian and the copula $C_{X,Y}$ is a Frank copula.

Example 3.4.2. _The density of the C-convolution._ *The density function of the C-convolution, f_{X+Y}, may be obtained by differentiating F_{X+Y} with respect to t. We get*

$$f_{X+Y}(t) = \int_0^1 c_{X,Y}(w, F_Y(t - F_X^{-1}(w))) f_Y(t - F_X^{-1}(w))\, dw,$$

where $c_{X,Y}$ denotes the copula density, that is, $c_{X,Y} = D_{1,2}C_{X,Y}$. Generally, this integral function does not have a closed form. So, a discrete approximation by numerical integration is necessary. Among several methods of numerical integration, the simplest way is the following: given a partition $\{w_0, \ldots, w_n\}$ of the interval $[0, 1]$ such that $0 \le w_0 \le \cdots \le w_n \le 1$, a simple approximation of f_{X+Y} at the point t is given by

$$f_{X+Y}(t) \simeq \sum_{i=1}^n c_{X,Y}(w_{i-1}, F_Y(t - F_X^{-1}(w_{i-1}))) f_Y(t - F_X^{-1}(w_{i-1}))(w_i - w_{i-1}).$$

The approximation improves as n *increases. For the sake of illustration, we consider standard Gaussian margins F_X and F_Y. Figures 3.5 and 3.6 display the density of the C-convolution in the case where $C_{X,Y}$ is Frank or Clayton and compare the shape of the C-convolution density with the independence case.*

Example 3.4.3. _Simulation from C_{X+Y}._ *The conditional sampling method may be used to simulate data from the C-convolution copula C_{X+Y} given the copula $C_{X,Y}$ and the margins F_X and F_Y. The algorithm is the following:*

1. Simulate two independent r.vs (z, w) from $U(0, 1)$.
2. Set $u = z$.
3. Set $w = D_1 C_{X,X+Y}(z, v)$ and solve for v.
4. The desired pair is (u, v).

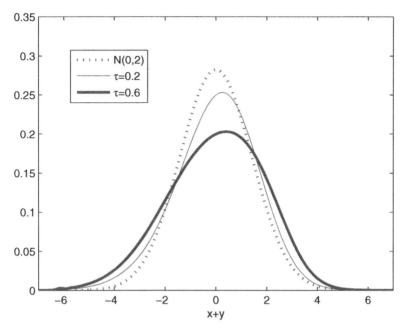

Figure 3.6 Numerical approximation of the density of the C-convolution, f_{X+Y}, when the marginal distributions F_X and F_Y are standard Gaussian and the copula $C_{X,Y}$ is a Clayton copula.

Notice that in our case $D_1 C_{X,X+Y}(z, v) = D_1 C_{X,Y}(z, F_Y(F_{X+Y}^{-1}(v) - F_X^{-1}(z)))$. Figures 3.7 and 3.8 compare simulations from $C_{X,X+Y}$ and from $C_{X,Y}$ for different values of the dependence parameter.

Example 3.4.4. *The copula $C_{X,X+Y}$. The copula $C_{X,X+Y}$ represents the dependence structure between X and $X+Y$ induced by the dependence structure between X and Y and different copulas $C_{X,Y}$ affect such a dependence in different ways. We compute by numerical integration, the level $C_{X,X+Y}(u, v)$ on a $N \times N$ grid of $[0, 1] \times [0, 1]$: $\{(u_k, v_h), k = 1, \ldots, N, h = 1, \ldots, N\}$ such that $0 \leq u_0 \leq \cdots \leq u_N \leq 1$ and $0 \leq v_0 \leq \cdots \leq v_N \leq 1$. The discrete approximation of $C_{X,X+Y}$ at a point (u_k, v_h) is given by (as described in Example 3.4.2)*

$$C_{X+Y}(u_k, v_h) \simeq \sum_{i=1}^{n_k} D_1 C_{X,Y}(w_{i-1}^{u_k}, F_Y(F_{X+Y}^{-1}(v_h) - F_X^{-1}(w_{i-1}^{u_k})))(w_i^{u_k} - w_{i-1}^{u_k}),$$

where $w_0^{u_k}, \ldots, w_{n_k}^{u_k}$ is a partition of the interval $[0, u_k]$. Figure 3.9 displays the distance of $C_{X,X+Y}(u, v)$ from the independence copula $\Pi(u, v)$ when the copula function $C_{X,Y}$ is Clayton with different values of the dependence parameter.

Example 3.4.5. *Propagation of the C-convolution. In this example we consider the construction of the density function of a discrete time stochastic process $X = (X_t)_{t\geq 1}$ by using the C-convolution iteratively. In fact, the distribution of X_t, say $F_t(\cdot)$, can be recovered by an iterative application of the C-convolution to X_{t-1} and ΔX_t for $t = 2, \ldots,$ given an initial*

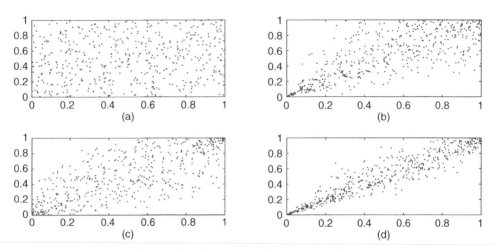

Figure 3.7 Comparison of simulations from $C_{X,Y}$ and $C_{X,X+Y}$ when $C_{X,Y}$ is a Clayton copula and the marginal distributions F_X and F_Y are Gaussian. (a) $(u, v) \sim C_{X,Y}$ with $\tau = 0.1$; (b) $(u, v) \sim C_{X,Y}$ with $\tau = 0.6$; (c) $(u, v) \sim C_{X,X+Y}$ when $C_{X,Y}$ as in (a); (d) $(u, v) \sim C_{X,X+Y}$ when $C_{X,Y}$ as in (b).

distribution $F_1(\cdot)$ of X_1; if the increment has a stationary distribution F and C is the invariant copula linking X_{t-1} and ΔX_t, we have

$$F_t(x_t) = (F_{t-1} \overset{C}{*} F)(x_t) = \int_0^1 D_1 C(w, F(x_t - F_{t-1}^{-1}(w))) \, dw, \quad t = 2, \ldots$$

The density function is given by

$$f_t(x_t) = \int_0^1 c(w, F(x_t - F_{t-1}^{-1}(w))) f(x_t - (F_{t-1}^{-1}(w))) \, dw,$$

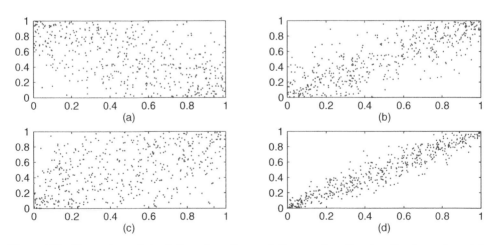

Figure 3.8 Comparison of simulations from $C_{X,Y}$ and $C_{X,X+Y}$ when $C_{X,Y}$ is a Frank copula and the marginal distributions F_X and F_Y are Gaussian. (a) $(u, v) \sim C_{X,Y}$ with $\tau = -0.4$; (b) $(u, v) \sim C_{X,Y}$ with $\tau = 0.6$; (c) $(u, v) \sim C_{X,X+Y}$ when $C_{X,Y}$ as in (a); (d) $(u, v) \sim C_{X,X+Y}$ when $C_{X,Y}$ as in (b).

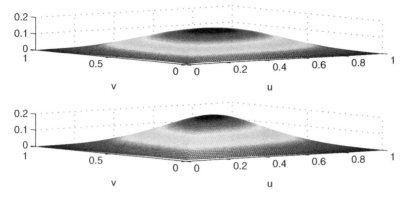

Figure 3.9   Graph of $C_{X,X+Y}(u, v) - \Pi(u, v)$ when the marginal distributions are Gaussian and $C_{X,Y}$ is a Clayton copula with $\tau = 0.1$ (top) and $\tau = 0.6$ (bottom).

where f is the density of F. Notice that in this case a numerical approximation of the integral is necessary for each time t. Figures 3.10 and 3.11 show the effect of the Clayton and Frank copulas (with different levels of dependence) on f_t when the distribution of the increment is invariant and N(2, 1).

3.4.2   Markov Processes with Independent Increments

As said above, Π-convolution coincides with the standard convolution. However, our approach directly produces the general shape of the copula functions that must be used to generate a process with independent increments.

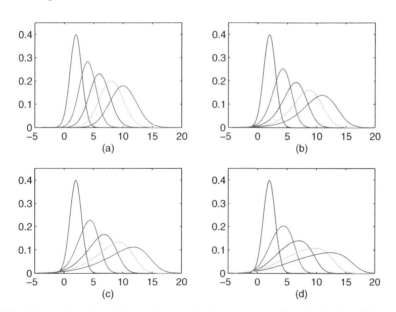

Figure 3.10   Density function f_t for $t = 1, \ldots, 5$ in the case where the distribution of the increment is stationary and $N(2, 1)$. (a) C is the independence copula; (b) C is Clayton with $\tau = 0.2$; (c) C is Clayton with $\tau = 0.4$; (d) C is Clayton with $\tau = 0.6$.

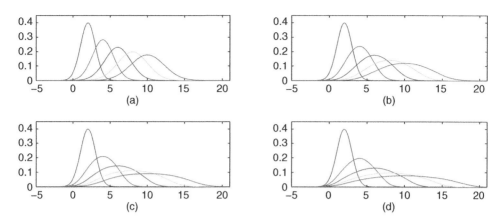

Figure 3.11   Density function f_t for $t = 1, \ldots, 5$ in the case where the distribution of the increment is stationary and $N(2, 1)$. (a) C is the independence copula; (b) C is Frank with $\tau = 0.2$; (c) C is Frank with $\tau = 0.4$; (d) C is Frank with $\tau = 0.6$.

Definition 3.4.2. A Markov process is said to be with independent increments if and only if its temporal dependence structure is given by a sequence of copulas

$$C_{st}(u_s, u_t) = \int_0^{u_s} F_{s,t}(F_t^{-1}(u_t) - F_s^{-1}(w))\, dw,$$

with $F_t = F_{s,t} \overset{\Pi}{*} F_s$.

Notice that again we are free to choose the distribution of the first element of the Markov chain and the distribution of increments.

Brownian Motion and Stable Processes

For practical applications it is very useful to work with distributions which are closed under convolution, meaning that F_t and F_s are part of the same distributions family. The first example that comes to mind is obviously the Brownian motion, in which case the distribution of increments is Gaussian. Here we want to give another example, extending the analysis to the more general family of *stable distributions*.

Example 3.4.6. The α-stable process. *Let Z be a random variable, Z has an α-stable distribution with parameters $\alpha, \mu, \beta, \gamma$ (with $0 < \alpha \leq 2$, $\mu \in \mathbb{R}$, $\beta \in [-1, 1]$, $\gamma \geq 0$) if its characteristic function is of the following type:*

$$\phi_Z(\lambda) = e^{i\mu\lambda - \gamma^\alpha |\lambda|^\alpha (1 - i\beta sign(\lambda) W(\alpha, \lambda))}$$

with

$$W(\alpha, \lambda) = \begin{cases} tg\left(\frac{\pi\alpha}{2}\right), & \text{if } \alpha \neq 1 \\ -\frac{2}{\pi}log|\lambda|, & \text{if } \alpha = 1. \end{cases}$$

$\alpha = 2$ corresponds to the normal distribution. The α-stable distribution is symmetric around the origin if $\mu = 0$, $\beta = 0$.

It follows that

$$\phi_X(\lambda) = e^{-\gamma^\alpha |\lambda|^\alpha} = e^{-(\gamma|\lambda|)^\alpha} = \phi_{\gamma Z}(\lambda),$$

where Z is α-stable distributed with $\mu = 0$, $\beta = 0$, $\gamma = 1$. This way, $X \stackrel{d}{=} \gamma Z$ and $F_X(x) = \Phi_Z\left(\frac{x}{\gamma}\right)$, where Φ_Z is the c.d.f. of Z. If (X, Y) is an α-stable vector with independent components, then by simply applying the standard convolution formula it is easy to find that

$$C_{X,X+Y}(u, v) = \int_0^u \Phi_Z\left(\frac{\Phi_Z^{-1}(v) - \rho\Phi_Z^{-1}(w)}{(1 - \rho^\alpha)^{\frac{1}{\alpha}}}\right) dw,$$

where $\rho = \frac{\gamma_X}{\gamma_{X+Y}}$.

   A symmetric α-stable process X_t is a Lévy process with increments following a symmetric α-stable distribution with a given parameter $\gamma > 0$. It is characterized by a characteristic function of type

$$\phi_{X_t}(\lambda) = e^{-\gamma^\alpha t|\lambda|^\alpha},$$

that is, X_t has an α-stable distribution with parameter $\gamma_t = \gamma t^{1/\alpha}$.

   By the above arguments, the 2-copula associated with (X_s, X_t) is

$$C_{st}(u, v) = \int_0^u \Phi_Z\left(\frac{\Phi_Z^{-1}(v) - \rho_{st}\Phi_Z^{-1}(w)}{(1 - \rho_{st}^\alpha)^{\frac{1}{\alpha}}}\right) dw,$$

where $\rho = \frac{\gamma s^{1/\alpha}}{\gamma t^{1/\alpha}} = \left(\frac{s}{t}\right)^{1/\alpha}$.

   Similarly, as for the Brownian copula, since the stable process is a Markov process, the corresponding 2-copulas must satisfy

$$C_{st} = C_{s\tau} * C_{\tau t},$$

for all $s < \tau < t$. If we consider the family of α-stable copula functions

$$C^\rho(u, v) = \int_0^u \Phi_Z\left(\frac{\Phi_Z^{-1}(v) - \rho\Phi_Z^{-1}(w)}{(1 - \rho^\alpha)^{\frac{1}{\alpha}}}\right) dw.$$

with $\rho \in (0, 1)$, like in the Brownian copula case, we can find that

$$C^{\rho_1} * C^{\rho_2} = C^{\rho_1\rho_2}$$

for all $\rho_1, \rho_2 \in (0, 1)$.

   Again, like in the Gaussian copula case (see Section 3.2.3), we could construct stochastic process Y_t whose 2-level copulas are of the α-stable type. Let ρ_{st} be the parameter associated with the copula C_{st}. The required closure property implies that, for $s < \tau < t$,

$$\rho_{st} = \rho_{s\tau}\rho_{\tau t}.$$

The same conclusions as in the remark of Section 3.2.3 hold.

3.4.3 Markov Processes with Dependent Increments

Most of the interest in our approach is for models with dependent increments. A limitation of the technique is that in almost all cases the construction of the process would require simulation. Just for the sake of illustration, we report the formulas for the Clayton copula.

Example 3.4.7. *The Clayton copula*

$$C(u, v) = \max\left([u^{-\theta} + v^{-\theta} - 1]^{-1/\theta}, 0\right) \quad \theta \in [-1, \infty) \setminus \{0\}.$$

Since $D_1 C(u, v) = \max\left([u^{-\theta} + v^{-\theta} - 1]^{-1/\theta-1} u^{-1-\theta}, 0\right)$

$$H \overset{C}{*} F(t) = \int_0^1 \max\left([w^{-\theta} + F(t - H^{-1}(w))^{-\theta} - 1]^{-1/\theta-1} w^{-1-\theta}, 0\right) dw$$

and

$$\tilde{C}(u, v) = \int_0^u \max\left([w^{-\theta} + F((H \overset{C}{*} F)^{-1}(v) - H^{-1}(w))^{-\theta} - 1]^{-1/\theta-1} w^{-1-\theta}, 0\right) dw.$$

It is clear that all one can do with this is to resort to numerical integration, or to use the simulation procedure that we will report below. The reason is that the Clayton copula is not closed with respect to the C-convolution operation. The same holds for the Gumbel and the Frank copulas.

Example 3.4.8. *The Gumbel copula*

$$C(u, v) = uv \exp(-\theta \ln u \ln v), \quad \theta \in (0, 1].$$

Since $D_1 C(u, v) = (1 - \theta) v \exp(-\theta \ln u \ln v)$

$$H \overset{C}{*} F(t) = (1 - \theta) \int_0^1 F(t - H^{-1}(w)) \exp(-\theta \ln w \ln F(t - H^{-1}(w))) dw$$

and

$$\tilde{C}(u, v) = (1 - \theta) \int_0^u F((H \overset{C}{*} F)^{-1}(v) - H^{-1}(w))$$

$$\times \exp(-\theta \ln w \ln F((H \overset{C}{*} F)^{-1}(v) - H^{-1}(w))) dw.$$

Example 3.4.9. *The Frank copula*

$$C(u, v) = -\frac{1}{\theta} \ln\left(1 + \frac{(e^{-\theta u} - 1)(e^{-\theta v} - 1)}{e^{-\theta} - 1}\right), \quad \theta \in \mathbb{R} \setminus \{0\}.$$

Since $D_1 C(u, v) = \frac{e^{-\theta u}(e^{-\theta v} - 1)}{e^{-\theta} - 1 + (e^{-\theta u} - 1)(e^{-\theta v} - 1)}$

$$H \overset{C}{*} F(t) = \int_0^1 \frac{e^{-\theta w}(e^{-\theta F(t - H^{-1}(w))} - 1)}{e^{-\theta} - 1 + (e^{-\theta w} - 1)(e^{-\theta F(t - H^{-1}(w))} - 1)} dw$$

and

$$\tilde{C}(u, v) = \int_0^u \frac{e^{-\theta w}(e^{-\theta F((H \overset{C}{*} F)^{-1}(v) - H^{-1}(w))} - 1)}{e^{-\theta} - 1 + (e^{-\theta w} - 1)(e^{-\theta F((H \overset{C}{*} F)^{-1}(v) - H^{-1}(w))} - 1)} dw.$$

Here we report an example in which the closure property with respect to the C-convolution is verified instead.

The Gaussian Process

We now show that if the relationship between increments is Gaussian, then the relationship between levels is also Gaussian:

$$C(u, v) = \int_0^u \Phi\left(\frac{\Phi^{-1}(v) - \rho\Phi^{-1}(w)}{\sqrt{1 - \rho^2}}\right), \quad \rho \in (-1, 1), \tag{3.61}$$

where Φ is the standard normal cumulative distribution function.

Since $D_1 C(u, v) = \Phi\left(\frac{\Phi^{-1}(v) - \rho\Phi^{-1}(u)}{\sqrt{1-\rho^2}}\right)$

$$H \overset{C}{*} F(t) = \int_0^1 \Phi\left(\frac{\Phi^{-1}(F(t - H^{-1}(w))) - \rho\Phi^{-1}(w)}{\sqrt{1 - \rho^2}}\right) dw$$

and

$$\tilde{C}(u, v) = \int_0^u \Phi\left(\frac{\Phi^{-1}(F((H \overset{C}{*} F)^{-1}(v) - H^{-1}(w))) - \rho\Phi^{-1}(w)}{\sqrt{1 - \rho^2}}\right) dw.$$

If we consider a bivariate random vector (X, Y) normally distributed such that $F_X \overset{d}{=} N(\mu, \sigma^2)$ and $F_Y \overset{d}{=} N(m, s^2)$ and ρ is the covariance coefficient, then (3.61) is the copula associated with (X, Y) and $F_{X+Y} = F_X \overset{C}{*} F_Y \overset{d}{=} N(\mu + m, \sigma^2_{X+Y} = \sigma^2 + s^2 + 2\sigma s\rho)$. Hence

$$\tilde{C}(u, v) = C_{X,X+Y}(u, v) = \int_0^u \Phi\left(\frac{\Phi^{-1}(F_Y((F_X \overset{C}{*} F_Y)^{-1}(v) - F_X^{-1}(w))) - \rho\Phi^{-1}(w)}{\sqrt{1 - \rho^2}}\right) dw$$

$$= \int_0^u \Phi\left(\frac{\Phi^{-1}\left(\Phi\left(\frac{F_{X+Y}^{-1}(v) - F_X^{-1}(w) - m}{s}\right)\right) - \rho\Phi^{-1}(w)}{\sqrt{1 - \rho^2}}\right) dw$$

$$= \int_0^u \Phi\left(\frac{F_{X+Y}^{-1}(v) - F_X^{-1}(w) - m - s\rho\Phi^{-1}(w)}{s\sqrt{1 - \rho^2}}\right) dw$$

$$= \int_0^u \Phi\left(\frac{\sigma_{X+Y}\Phi^{-1}(v) - \Phi^{-1}(w)(\sigma + s\rho)}{s\sqrt{1 - \rho^2}}\right) dw$$

$$= \int_0^u \Phi\left(\frac{\Phi^{-1}(v) - \hat{\rho}\Phi^{-1}(w)}{\sqrt{1 - \hat{\rho}^2}}\right) dw,$$

where $\hat{\rho} = \frac{\sigma + s\rho}{\sigma_{X+Y}} = \frac{Cov(X, X+Y)}{\sigma(\sigma_{X+Y})}$.

It may be proved that the same property can be extended to *elliptical* copulas. On the contrary, the extension to the family of stable distributions does not work.

Elliptical Processes

Let us now consider the case in which the level and the increment are linked by a two-dimensional elliptical copula. The main definitions and properties of n-dimensional elliptical copulas were introduced in Section 2.11. By (2.33), the elliptical copula in the two-dimensional case is

$$C(u, v) = \int_{-\infty}^{F_X^{-1}(u)} \int_{-\infty}^{F_Y^{-1}(v)} \sqrt{ac - b^2} g(as^2 + 2bst + ct^2) \, ds \, dt, \qquad (3.62)$$

where we assumed, without any loss of generality, that both marginal distributions have zero mean. Recall that the parameters a, b, c are so that the symmetric matrix $\Sigma^{-1} = \begin{pmatrix} a & b \\ b & c \end{pmatrix}$ is positive definite. Since

$$D_1 C(u, v) = \sqrt{ac - b^2} \frac{1}{f_X(F_X^{-1}(u))} \int_{-\infty}^{F_Y^{-1}(v)} g(aF_X^{-1}(u)^2 + 2bF_X^{-1}(u)t + ct^2) \, dt,$$

$$H \overset{C}{*} F(t) = \sqrt{ac - b^2} \int_0^1 \frac{1}{f_X(F_X^{-1}(w))} \int_{-\infty}^{F_Y^{-1}(F(t - H^{-1}(w)))} g(aF_X^{-1}(w)^2$$
$$+ 2bF_X^{-1}(w)t + ct^2) \, dt \, dw$$

and

$$\tilde{C}(u, v) = \sqrt{ac - b^2} \int_0^u \frac{1}{f_X(F_X^{-1}(w))} \int_{-\infty}^{F_Y^{-1}(F(H\overset{C}{*}F^{-1}(v) - H^{-1}(w)))} g(aF_X^{-1}(w)^2$$
$$+ 2bF_X^{-1}(w)t + ct^2) \, dt \, dw.$$

If $H = F_X$ and $F = F_Y$, then $H \overset{C}{*} F = F_{X+Y}$ and we have

$$\tilde{C}(u, v) = \sqrt{ac - b^2} \int_0^u \frac{1}{f_X(F_X^{-1}(w))} \int_{-\infty}^{F_{X+Y}^{-1}(v) - F_X^{-1}(w)} g(aF_X^{-1}(w)^2 + 2bF_X^{-1}(w)t + ct^2) \, dt \, dw$$

$$= \sqrt{ac - b^2} \int_{-\infty}^{F_X^{-1}(u)} \int_{-\infty}^{F_{X+Y}^{-1}(v) - s} g(as^2 + 2bst + ct^2) \, dt \, ds$$

$$= \sqrt{ac - b^2} \int_{-\infty}^{F_X^{-1}(u)} \int_{-\infty}^{F_{X+Y}^{-1}(v)} g(as^2 + 2bs(\hat{t} - s) + c(\hat{t} - s)^2) \, d\hat{t} \, ds$$

$$= \sqrt{ac - b^2} \int_{-\infty}^{F_X^{-1}(u)} \int_{-\infty}^{F_{X+Y}^{-1}(v)} g((a + c - 2b)s^2 + 2\hat{t}s(b - c) + c\hat{t}^2) \, d\hat{t} \, ds,$$

and this is of the same type as (3.62) with associated matrix $\begin{pmatrix} a + c - 2b & b - c \\ b - c & c \end{pmatrix}$. The elliptical copulas set contains Gaussian copulas and Student's t copulas as particular cases. It is trivial to check that Gaussian copulas can be recovered by considering

$$g(z) = \frac{1}{2\pi} e^{-\frac{z^2}{2}}.$$

Student's t with m degrees of freedom copulas are those copulas of type (2.33) with

$$g(z) = \frac{\Gamma(\frac{m+2}{2})}{\pi m \Gamma(\frac{m}{2})} \sqrt{ac - b^2} \left(1 + \frac{ax^2 + 2bxy + cy^2}{m}\right)^{-\frac{m+2}{2}}.$$

Simulating Processes with Dependent Increments

An approach similar to that used to simulate Markov processes can be applied to the simulation of dependent increment processes. Again, the key technique is conditional sampling, as described in the quasi-algorithm reported below. The input is given by a sequence of distributions of increments that for the sake of simplicity we assume stationary, $F_{Y_i} = F_Y$, and a temporal dependence structure that we consider stationary as well, $C_{X_i, Y_{i+1}}(u, v) = C(u, v)$. We also assume $X_0 = 0$. We describe a procedure to generate the iteration of an n-step trajectory:

1. $i = 1$.
2. Generate u from the uniform distribution.
3. Compute $X_i = F_Y^{-1}(u)$.
4. Use conditional sampling to generate v from $D_1 C(u, v)$.
5. Compute $Y_{i+1} = F_Y^{-1}(v)$.
6. $X_{i+1} = X_i + Y_{i+1}$.
7. Compute the distribution $F_{X_{i+1}}(t)$ by C-convolution.
8. Compute $u = F_{X_{i+1}}(X_{i+1})$.
9. $i = 1 + 1$.
10. If $i < n + 1$ go to step 4, else End.

Figure 3.12 shows a number of simulated trajectories generated by using this quasi-algorithm. The dependence structure is given by a Clayton copula.

3.4.4   Extracting Dependent Increments in Markov Processes

We saw that if one decides to focus on the dependence structure of increments one loses the degree of freedom of choosing the marginal distribution of the process at selected times and a sequence of copula functions, as done in the DNO approach. If one wants to stick to the latter

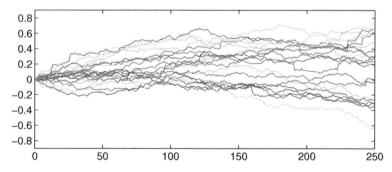

Figure 3.12   Example of trajectories of dependent increment Markov processes for Clayton copula whose parameter is consistent with a Kendall's τ equal to 0.1. The marginal distributions are Gaussian with (annualized) volatility 0.2.

approach, it may be possible to recover the dependence structure of increments consistent with the specification chosen.

In order to show how the procedure works, we introduce what we could call the *canonical representation* of the Markov process. We remember that when we use the canonical representation of a copula function we actually restrict the analysis to cases in which the copula is *absolutely continuous*.

Remark 3.4.3. It is straightforward to check that

$$c_{X,X+Y}(u,v) = \frac{c_{X,Y}\left(u, F_Y\left(F_{X+Y}^{-1}(v) - F_X^{-1}(u)\right)\right) f_Y\left(F_{X+Y}^{-1}(v) - F_X^{-1}(u)\right)}{f_{X+Y}\left(F_{X+Y}^{-1}(v)\right)}.$$

Moreover, recall that

$$f_{X+Y}(t) = \int_0^1 c_{X,Y}\left(w, F_Y(t - F_X^{-1}(w))\right) f_Y(t - F_X^{-1}(w))\, dw,$$

where lowercase letters denote the densities of copula functions and c.d.f.s.

So, the remark above enables us to link the density of the copula representing the dependence of increments to the density of the copula describing the Markov process. This can be used to check if a density copula of increments generates a given copula selected to represent the Markov process. A word of caution is needed to recognize that, differently from what happens with standard copula densities, for which the marginals are uniformly distributed, here marginal densities must enter into the definition of the copula of increments. Here we report an example of one of the *self-similar* copulas presented in the DNO approach, the FGM.

Example 3.4.10. *Let F, H, and G be three cumulative distribution functions on* $\mathbb{R}$*, with corresponding densities f, h, and g. Let us consider a copula C whose density is*

$$c(u,v) = \left[1 + \theta(1 - 2u)(1 - 2G(F^{-1}(v) + H^{-1}(u)))\right] \frac{g(F^{-1}(v) + H^{-1}(u))}{f(F^{-1}(v))}. \tag{3.63}$$

Then

$$\tilde{C}(u,v) = \int_0^u D_1 C(w, F(G^{-1}(v) - H^{-1}(w)))\, dw = uv(1 + \theta(1 - u)(1 - v))$$

and

$$G = H \overset{C}{*} F.$$

The point is easily verified using the remark above:

$$\tilde{c}(u,v) = c(u, F(G^{-1}(v) - H^{-1}(u))) \frac{f(G^{-1}(v) - H^{-1}(u))}{g(G^{-1}(v))},$$

and by substituting (3.63) we trivially get

$$\tilde{c}(u,v) = 1 + \theta(1 - 2u)(1 - 2v),$$

which is the density of the FGM copula.

Moreover, using the same remark and equation (3.63):

$$(H \overset{c}{*} F)'(t) = \int_0^1 c(w, F(t - H^{-1}(w))) f(t - H^{-1}(w)) \, dw$$

$$= \int_0^1 [1 + \theta(1 - 2w)(1 - 2G(t))] g(t) \, dw$$

$$= g(t) \left[1 + \theta(1 - 2G(t)) \int_0^1 (1 - 2w) \, dw \right]$$

$$= g(t).$$

In general, it is possible to recover the copula of increments from the copula linking the levels, assuming the marginal c.d.f.s of levels and the increment are given. More precisely, from (3.56) we get

$$D_1 C_{X,X+Y}(u, v) = D_1 C_{X,Y} \left(u, F_Y(F_{X+Y}^{-1}(v) - F_X^{-1}(u)) \right).$$

Setting $v' = F_Y(F_{X+Y}^{-1}(v) - F_X^{-1}(u))$ we get $v = F_{X+Y}(F_Y^{-1}(v') + F_X^{-1}(u))$ and

$$D_1 C_{X,X+Y}(u, F_{X+Y}(F_Y^{-1}(v') + F_X^{-1}(u))) = D_1 C_{X,Y} \left(u, v' \right).$$

Integrating both sides we get

$$C_{XY}(u, v') = \int_0^u D_1 C_{X,X+Y}(w, F_{X+Y}(F_Y^{-1}(v') + F_X^{-1}(w))) \, dw,$$

which is a copula function if and only if

$$F_Y(t) = \int_0^1 D_1 C_{X,X+Y}(w, F_{X+Y}(t + F_X^{-1}(w))) \, dw.$$

As noticed above, given the copula between levels $C_{X,X+Y}$, the copula linking level and increments also depend on F_X, F_Y, and F_{X+Y}.

3.4.5   Martingale Processes

We are now going to add another restriction to our representation of the Markov process, that is the martingale condition. Formally, we want to choose the stochastic process $\{X_t\}_{t \geq 0}$ such that

$$\mathbb{E}[f(X_s)(X_t - X_s)] = 0, \tag{3.64}$$

for all s and $t > s$ and all Borel measurable functions f.

It is now clear that modeling increments makes it easier to address the martingale problem, since it sets the focus on the conditional distribution of the increment $Y = X_t - X_s$ for all times t and s:

$$\mathbb{P}(X_t - X_s \leq y | X_s). \tag{3.65}$$

Of course the solution to the problem is very easy if we limit ourselves to processes with independent increments. In this case it is sufficient to ensure that all the increments of the Markov process have zero mean.

Proposition 3.4.4. *Any process whose increments* $Y \equiv X_t - X_s$, *are independent of the corresponding initial values* X_s ($C_{X_s,Y}(u, v) \equiv uv$) *and the distribution of* F_Y *have zero mean for all t is a martingale.*

Proof. In this case

$$\mathbb{E}[X_t - X_s|X_s] = \mathbb{E}[X_t - X_s] = 0$$

and the process is a martingale.   □

The problem becomes more involved if we allow for dependent increments. In this case, the martingale condition implies restrictions both on the shape of the distribution of increments and on the copula function. The restriction is formalized in the following theorem.

Theorem 3.4.5. *Let* $X = (X_t)_{t \geq 0}$ *be a Markov process and set* $Y_{(st)} = X_t - X_s$. X *is a martingale if and only if for all t, s, s < t:*

1. $F_{Y_{(st)}}$ *has finite mean.*
2. $\int_0^1 F_{Y_{(st)}}^{-1}(v) D_1 C_{X_s,Y_{(st)}}(u, dv) = 0, \quad \forall u \in [0, 1]$ *a.e.*

Proof. X is a Markov process, to which we impose the condition $\mathbb{E}[X_t - X_s|X_s] = 0$ for every s, t. But

$$\mathbb{E}[X_t - X_s|X_s] = \int_{-\infty}^{+\infty} z d\mathbb{P}(X_t - X_s \leq z|X_s)$$

$$= \int_{-\infty}^{+\infty} z D_1 C_{X_s,Y_{(st)}}\left(F_{X_s}(X_s), F_{Y_{(st)}}(dz)\right)$$

$$= \int_0^1 F_{Y_{(st)}}^{-1}(dv) D_1 C_{X_s,Y_{(st)}}\left(F_{X_s}(X_s), dv\right).$$

The thesis follows setting $F_{X_s}(X_s) = u$.   □

If we restrict the selection of distributions of increments to the set of symmetric distributions, we can formally work out the restriction that has to be imposed on the dependence of increments.

Proposition 3.4.6. *The martingale condition is satisfied for every symmetric distribution of increments* $F_{Y_{(st)}}$ *if and only if the copula between the increments and the levels has the symmetry property*

$$\check{C}(u, v) \equiv u - C(u, 1 - v) = C(u, v). \tag{3.66}$$

Proof. For simplicity we set $C_{X_s,Y_{(st)}} = C$ and $F_{Y_{(st)}} = F$, F being a symmetric distribution,

$$\int_0^1 F^{-1}(v) D_1 C(u, dv) = \int_0^{\frac{1}{2}} F^{-1}(v) D_1 C(u, dv) + \int_{\frac{1}{2}}^1 F^{-1}(v) D_1 C(u, dv)$$

$$= \int_0^{\frac{1}{2}} F^{-1}(v) D_1 C(u, dv) + \int_{\frac{1}{2}}^0 F^{-1}(1 - \rho) D_1 C(u, d(1 - \rho))$$

$$= \int_0^{\frac{1}{2}} F^{-1}(v) D_1 C(u, dv) + \int_0^{\frac{1}{2}} F^{-1}(\rho) D_1 C(u, d(1 - \rho))$$

$$= \int_0^{\frac{1}{2}} F^{-1}(v)(D_1 C(u, dv) + D_1 C(u, d(1 - v))) = 0, \quad \forall u \in (0, 1).$$

The last condition is satisfied for every symmetric distribution F iff (notice that, in the last integral, $F^{-1}(v) < 0$ in a given non-empty interval)

$$D_1(C(u, dv) + C(u, d(1 - v))) = 0 \quad \forall u, v \in (0, 1).$$

It may be easily verified that this condition is satisfied if and only if

$$C(u, v) + C(u, 1 - v) = u,$$

which is the symmetry condition required for the copula.   □

Notice that there are almost infinite possibilities to build copulas with the symmetry described above. A simple technique would be to select any copula $A(u,v)$ and define its symmetric counterpart $\check{A}(u, v) \equiv u - A(u, 1 - v)$. Then, construct a mixture copula $C(u.v) \equiv 0.5A + 0.5\check{A}$. It is an easy exercise to prove that $C(u,v)$ is endowed with the property required in the proposition above (see Klement *et al.* (2002) for an application of this technique to general concepts of symmetry). So, the choice is extremely vast, and almost everything can be done. Not everything, however, as the counterexample below demonstrates.

Remark 3.4.4. Assume you want to construct a martingale process with symmetric and dependent increments, with dependence modeled by an FGM copula. It is easy to show that this is unfeasible. Select $A_\theta(u, v) = uv + \theta uv(1 - u)(1 - v)$ as required. Now compute

$$\check{A}_\theta(u, v) = u - u(1 - v) - \theta uv(1 - u)(1 - v) = uv - \theta uv(1 - u)(1 - v) = A_{-\theta}(u, v).$$

But now, by the property of the FGM copula discussed above (or by direct computation), we have

$$0.5A_\theta(u, v) + 0.5\tilde{A}_\theta(u, v) = 0.5A_\theta(u, v) + 0.5A_{-\theta}(u, v) = A_0(u, v) = uv,$$

and the increments turn out to be independent.

Simulating Martingale Processes with Dependent Increments

Simulating a martingale process with symmetric marginal and dependent increments as described in Proposition 3.4.6 is very similar to the procedure used for dependent increment processes in general. The only difficulty stems from the simulation of the mixture copula. This is very easy to do if we consider that if a couple (X, Y) is extracted from a given copula $C(u,v)$, then the couple $(X, -Y)$ is extracted from the symmetric counterpart $u - C(u, 1 - v)$. Here we report the algorithm modified accordingly:

1. $i = 1$.
2. Generate u from the uniform distribution.
3. Compute $X_i = F_Y^{-1}(u)$.
4. Generate ξ from the uniform distribution.
5. Use conditional sampling to generate v from $D_1 C(u, v)$.
6. Compute $Y_{i+1} = F_Y^{-1}(v)$.
7. If $\xi \leq 0.5$, $Y_{i+1} = -Y_{i+1}$.
8. $X_{i+1} = X_i + Y_{i+1}$.
9. Compute the distribution $F_{X_{i+1}}(t)$ by C-convolution.
10. Compute $u = F_{X_{i+1}}(X_{i+1})$.
11. $i = 1 + 1$.
12. If $i < n + 1$ go to step 4, else End.

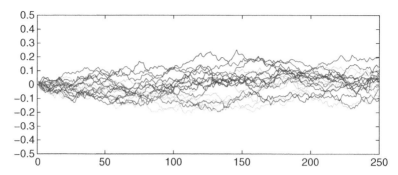

Figure 3.13 Example of trajectories of martingale Markov processes with dependent increments for Clayton copula whose parameter is consistent with a Kendall's τ equal to 0.1. The marginal distributions are standard Gaussian with (annualized) volatility 0.2.

Figure 3.13 shows a number of simulated trajectories of a martingale with dependent increments generated by using the quasi-algorithm above. The dependence structure is given by a Clayton copula.

3.5   MULTIVARIATE PROCESSES

We are now going to extend the DNO copula-based representation to multivariate Markov processes. This extension was actually done by Ibragimov (2005, 2009), with application to higher-order Markov processes. That is, univariate processes with the property

$$\mathbb{P}(X_t \leq x_t | \mathcal{F}_s) = \mathbb{P}(X_t \leq x_t | X_s, X_{s-1}, \ldots, X_{s-k}). \tag{3.67}$$

But actually, the problem is conceptually equivalent to a multivariate Markov process of dimension k. Again, extension of the DNO approach exploits the principle of conditional independence. So, to introduce the analysis we refresh the concept of conditional probability with respect to k variables:

$$\mathbb{P}(U_1 \leq u_1, \ldots, U_{m-k} \leq u_{m-k} | U_{m-k+1} = u_{m-k+1}, \ldots, U_m = u_m) = \frac{\frac{\partial^2 C(u_1, \ldots, u_m)}{\partial u_{m-k+1} \ldots \partial u_m}}{\frac{\partial^2 C(1, \ldots, u_{m-k+1} \ldots u_m)}{\partial u_{m-k+1} \ldots \partial u_m}}.$$

3.5.1   Multivariate Markov Processes

For the sake of simplicity, we show how to extend the DNO approach to a bivariate system. Let $m, n \geq 2$ and A and B be, respectively, m- and n-dimensional copulas. Set

$$A_{1,\ldots,m|m-1,m}(u_1, \ldots, u_{m-2}, \xi, \eta) = \frac{\frac{\partial^2 A(u_1, \ldots, u_{m-2}, \xi, \eta)}{\partial \xi \partial \eta}}{\frac{\partial^2 A(1, \ldots, 1, \xi, \eta)}{\partial \xi \partial \eta}}$$

and

$$B_{1,\ldots,n|1,2}(\xi, \eta, u_3, \ldots, u_n) = \frac{\frac{\partial^2 B(\xi, \eta, u_3, \ldots, u_n)}{\partial \xi \partial \eta}}{\frac{\partial^2 B(\xi, \eta, 1, \ldots, 1)}{\partial \xi \partial \eta}}.$$

If

$$A(1, \ldots, 1, \xi, \eta) = B(\xi, \eta, 1, \ldots, 1) = C(\xi, \eta),$$

where C is a bivariate copula, we can define the $\star^2$ product of the copulas A and B as the copula $D = A \star^2 B : [0, 1]^{m+n-2} \to [0, 1]$ given by

$$D(u_1, \ldots, u_{m+n-2}) = \int_0^{u_{m-1}} \int_0^{u_m} A_{1,\ldots,m|m-1,m}(u_1, \ldots, u_{m-2}, \xi, \eta)$$
$$\cdot B_{1,\ldots,n|1,2}(\xi, \eta, u_3, \ldots, u_n) dC(\xi, \eta). \tag{3.68}$$

The $\star^2$ operator is a particular case of the $\star^k$ operator defined in Ibragimov (2009).

Now we provide a vector representation of a bivariate Markov process. Let $(X, Z) = \{(X_t, Z_t)\}_{t \geq 0}$ be an $\mathbb{R}^2$-valued stochastic process defined on the probability space $(\Omega, \mathcal{F}, \mathbb{P})$. Let $(\mathcal{F}_t^{X,Z})_{t \geq 0}$ be its natural filtration.

By definition, (X, Z) is a Markov process if, for all i, such that $t_1 < \ldots < t_i < \ldots < t_{n+1}$ and all $(x_{n+1}, z_{n+1}) \in \mathbb{R}^2$

$$\mathbb{P}(X_{n+1} \leq x_{n+1}, Z_{n+1} \leq z_{n+1} | X_n, Z_n, X_{n-1}, Z_{n-1}, \ldots, X_1, Z_1)$$
$$= \mathbb{P}(X_{n+1} \leq x_{n+1}, Z_{n+1} \leq z_{n+1} | X_n, Z_n). \tag{3.69}$$

Let $C_{t_1, t_2, \ldots, t_n}(u_1, v_1, \ldots, u_n, v_n)$ denote the $2n$-dimensional copulas corresponding to the joint distribution of the random vector $(X_1, Z_1, X_2, Z_2, \ldots, X_n, Z_n)$.

We set

$$C_{t_1, \ldots, t_n | t_n}(u_1, v_1, \ldots, u_n, v_n) = \frac{\frac{\partial^2 C(u_1, v_1, \ldots, u_{n-1}, v_{n-1}, \xi, \eta)}{\partial \xi \, \partial \eta}}{\frac{\partial^2 C(1, \ldots, 1, \xi, \eta)}{\partial \xi \, \partial \eta}}$$

and

$$C_{t_1, \ldots, t_n | t_1}(\xi, \eta, \ldots, u_n, v_n) = \frac{\frac{\partial^2 C(\xi, \eta, \ldots, u_n, v_n)}{\partial \xi \, \partial \eta}}{\frac{\partial^2 C(1, \ldots, 1, \xi, \eta)}{\partial \xi \, \partial \eta}}.$$

We are now ready to provide the extension of the DNO approach.

Theorem 3.5.1. *An $\mathbb{R}^2$-valued stochastic process (X, Z) is a Markov process if and only if for all i, such that $t_1 < \ldots < t_i < \ldots < t_n$*

$$C_{t_1, t_2, \ldots, t_n} = C_{t_1, t_2} \star^2 C_{t_2, t_3} \star^2 \cdots \star^2 C_{t_{n-1}, t_n}. \tag{3.70}$$

Proof. It is trivial to show that property (3.69) holds if and only if

$$\mathbb{P}(X_1 \leq x_1, Z_1 \leq z_1, \ldots, X_{n-1} \leq x_{n-1}, Z_{n-1} \leq z_{n-1}, X_{n+1} \leq x_{n+1}, Z_{n+1} \leq z_{n+1} | X_n, Z_n)$$
$$= \mathbb{P}(X_1 \leq x_1, Z_1 \leq z_1, \ldots, X_{n-1} \leq x_{n-1}, Z_{n-1} \leq z_{n-1} | X_n, Z_n)$$
$$\cdot \mathbb{P}(X_{n+1} \leq x_{n+1}, Z_{n+1} \leq z_{n+1} | X_n, Z_n). \tag{3.71}$$

That is,

$$\mathbb{P}(X_1 \leq x_1, Z_1 \leq z_1, \ldots, X_{n-1} \leq x_{n-1}, Z_{n-1} \leq z_{n-1}, X_{n+1} \leq x_{n+1}, Z_{n+1} \leq z_{n+1} | X_n, Z_n)$$
$$= C_{1,\ldots,n|t_n}\left(F_{X_1}(x_1), F_{Z_1}(z_1), \ldots, F_{X_{n-1}}(x_{n-1}), F_{Z_{n-1}}(z_{n-1}), F_{X_n}(X_n), F_{Z_n}(Z_n)\right)$$
$$\cdot C_{n,n+1|n}\left(F_{X_n}(X_n), F_{Z_n}(Z_n), F_{X_{n+1}}(x_{n+1}), F_{Z_{n+1}}(z_{n+1})\right). \tag{3.72}$$

Integrating (3.72) over $X_n^{-1}((-\infty, x_n)) \times Z_n^{-1}((-\infty, z_n))$, we get

$$
\begin{aligned}
&C_{1,\ldots,n,n+1}(F_{X_1}(x_1), F_{Z_1}(z_1), \ldots, F_{X_{n+1}}(x_{n+1}), F_{Z_{n+1}}(z_{n+1})) \\
&= \int_{-\infty}^{x_n} \int_{-\infty}^{z_n} C_{1,\ldots,n|n}\left(F_{X_1}(x_1), F_{Z_1}(z_1), \ldots, F_{X_{n-1}}(x_{n-1}), F_{Z_{n-1}}(z_{n-1}), F_{X_n}(x), F_{Z_n}(z)\right) \\
&\qquad \cdot C_{n,n+1|n}\left(F_{X_n}(x), F_{Z_n}(z), F_{X_{n+1}}(x_{n+1}), F_{Z_{n+1}}(z_{n+1})\right) dF_{(X_n,Z_n)}(x,z) \\
&= \int_{-\infty}^{F_{X_n}(x_n)} \int_{-\infty}^{F_{Z_n}(z_n)} C_{1,\ldots,n|n}\left(F_{X_1}(x_1), F_{Z_1}(z_1), \ldots, F_{X_{n-1}}(x_{n-1}), F_{Z_{n-1}}(z_{n-1}), \xi, \eta\right) \\
&\qquad \cdot C_{n,n+1|n}\left(\xi, \eta, F_{X_{n+1}}(x_{n+1}), F_{Z_{n+1}}(z_{n+1})\right) dC_n(\xi,\eta) \\
&= C_{1,\ldots,n} \star^2 C_{n,n+1}\left(F_{X_1}(x_1), F_{Z_1}(z_1), \ldots, F_{X_{n+1}}(x_{n+1}), F_{Z_{n+1}}(z_{n+1}).\right).
\end{aligned}
\tag{3.73}
$$

By induction, we obtain (3.70).

Conversely, suppose that (3.70) holds. We have

$$
\begin{aligned}
&\mathbb{P}(X_1 \le x_1, Z_1 \le z_1, \ldots, X_{n+1} \le x_{n+1}, Z_{n+1} \le z_{n+1}) \\
&= C_{1,\ldots n,n+1}\left(F_{X_1}(x_1), F_{Z_1}(z_1), \ldots, F_{X_{n+1}}(x_{n+1}), F_{Z_{n+1}}(z_{n+1})\right) \\
&= \int_{-\infty}^{F_{X_n}(x_n)} \int_{-\infty}^{F_{Z_n}(z_n)} C_{1,\ldots,n|n}\left(F_{X_1}(x_1), F_{Z_1}(z_1), \ldots, F_{X_{n-1}}(x_{n-1}), F_{Z_{n-1}}(z_{n-1}), \xi, \eta\right) \\
&\qquad \cdot C_{n,n+1|n}\left(\xi, \eta, F_{X_{n+1}}(x_{n+1}), F_{Z_{n+1}}(z_{n+1})\right) dC_n(\xi,\eta) \\
&= \mathbb{E}\left[\mathbb{P}(X_1 \le x_1, Z_1 \le z_1, \ldots, X_{n-1} \le x_{n-1}, Z_{n-1} \le z_{n-1}| X_n, Z_n) \right. \\
&\qquad \left. \cdot \mathbb{P}(X_{n+1} \le x_{n+1}, Z_{n+1} \le z_{n+1}| X_n, Z_n) \mathbf{1}_{\{X_n \le x_n, Z_n \le z_n\}}\right],
\end{aligned}
$$

from which (3.71) follows.   □

3.5.2 Granger Causality and the Martingale Condition

Let us now come to the martingale condition in a multivariate setting. The martingale restriction requirement is paramount in pricing techniques based on the no-arbitrage assumption and is also common in all statistical analysis of financial markets based on the efficient market hypothesis. In full generality, in a bivariate system we would like to impose the condition

$$
\begin{aligned}
\mathbb{E}[X_t - X_s | X_s, X_t] &= 0, \\
\mathbb{E}[Z_t - Z_s | Z_s, Z_t] &= 0.
\end{aligned}
\tag{3.74}
$$

Addressing the problem in full generality looks quite involved. Throughout the applications in this book, we would rely on simplifying assumptions, imposing some restrictions on the four-variable copula linking $X_t - X_s, Z_t - Z_s, X_s$, and Z_s. As in the univariate model, assuming independent increments would help. Anyway, in a multivariate setting independent increments may be given different interpretations. The most general definition could be called *vector independent increments*, meaning

$$
\mathbb{P}(X_t - X_s \le x, Z_t - Z_s \le z | X_s, Z_s) = \mathbb{P}(X_t - X_s \le x, Z_t - Z_s \le z).
\tag{3.75}
$$

In this case, just like in the univariate setting, it would suffice to assume that the increments of both variables have zero mean. But independent increments may be understood with a different meaning, assuming that all information sufficient to describe the dynamics of one variable is represented by the history of that variable. Formally, we define

$$\mathbb{P}(X_t - X_s \leq x | X_s, Z_s) = \mathbb{P}(X_t - X_s \leq x | X_s),$$
$$\mathbb{P}(Z_t - Z_s \leq z | X_s, Z_s) = \mathbb{P}(Z_t - Z_s \leq z | Z_s).$$

This provides an intermediate case between that of *vector independence* and that of dependent increments. Each increment can depend on the past history of the corresponding variable, and enlarging the information set to include other variables would not increase the explanatory power for that increment. This concept coincides with a concept which is very familiar to econometricians, namely *Granger causality*. Non-Granger causality in finance was popular in the mid-to-late 1990s. Although there were some applied papers published (Hiemstra and Jones, 1994; Silvapulle and Moosa, 1999), this area did not take off, since many academics were not happy with the test statistics developed based on the probability integral. Econometricians are now beginning to develop copula-based methods for testing non-linear causality, and the copula approach has a natural extension in this direction. We could rephrase our assumption stating that each and every variable is not *Granger-caused* by any of the others. For this reason, we define this independence assumption as Granger independence. Clearly, it is intuitive that if each of the variables is a martingale with respect to its own history, it remains so even if the information set is enlarged to include the history of other variables. Actually, what sounds intuitive needs to be proved, and it is a very involved concept in financial mathematics. It is called the *H-condition* and means that if a variable is a martingale with respect to its natural filtration (that is, the filtration generated by its own history), it remains so even with respect to the enlargement of the filtration. The proof would proceed in two steps. In the first, one proves that under no Granger causality if a process is Markov with respect to its own filtration, then it remains Markov with respect to the enlarged filtration as well. In the second, one uses a result due to Brèmaud and Yor (1978) which states that if a Markov process is a martingale with respect to its own filtration and it remains Markov with respect to the enlargement of the filtration, then it remains a martingale with respect to the larger filtration. For this reason, most of the multivariate equity applications that we will see in the following chapters will be based on this concept of *Granger-independent* increments.

4

Copula-based Econometrics
of Dynamic Processes

Parallel with applications in finance, copula functions have also been applied to econometrics. Differently from finance, most of the applications have been devoted to explore temporal dependence, rather than spatial dependence, and even in cases in which the tool was applied to a multivariate setting, the interest was mainly on the dynamics. The application of copulas has made it possible to extend the standard dynamic models proposed in a linear framework to a more general setting. The setting is not so general as to be fully non-parametric, and for this reason is called semi-parametric. In this chapter we present a general review of these applications up to the new model proposed in this book.

4.1 DYNAMIC COPULA QUANTILE REGRESSIONS

The characterization of the dependence between random variables at a given quantile is crucial in financial econometrics, where the distributions of log-returns have fat tails and are non-elliptic and where the tail dependency has to be considered. Bouyè and Salmon (2009) consider a general approach to non-linear quantile regression modeling where the form of the quantile regression is implied by the copula linking the assets involved. They refer to this relationship as a *c-quantile* regression to distinguish it from a quantile regression, which may have been assumed to be linear or estimated non-parametrically as proposed and developed by Koenker and Basset (1978). In this way, the non-normality of the joint distribution of the variables involved is considered.

First, the copula pth quantile (simply c-quantile) curve of Y conditionally on X is defined. Let $C(., .; \theta)$ be a parametric copula and let Θ be the parameter space.

Definition 4.1.1. The pth c-quantile curve of y conditional on x is defined by the implicit equation

$$p = D_1 C(F_X(x), F_Y(y); \theta), \quad \theta \in \Theta. \tag{4.1}$$

If $D_1 C$ is partially invertible in its second argument and if we denote it by $D_1 C_v^{-1}$, we can solve equation (4.1) by y in order to capture the relationship between X and Y:

$$y = F_Y^{-1}(D_1 C_v^{-1}(F_X(x), p; \theta)) \doteq \mathbf{q}(x, p; \theta), \tag{4.2}$$

where F_Y^{-1} is the pseudo-inverse of F_Y. Three important properties of $\mathbf{q}(x, p; \theta)$ are the following:

- if $0 < p_1 \leq p_2 < 1$ then $\mathbf{q}(x, p_1; \theta) \leq \mathbf{q}(x, p_2; \theta)$;
- let $x_1 \leq x_2$, if $C(u,v)$ is concave in u then $\mathbf{q}(x_1, p; \theta) \leq \mathbf{q}(x_2, p; \theta)$;
- let $x_1 \leq x_2$, if $C(u,v)$ is convex in u then $\mathbf{q}(x_1, p; \theta) \geq \mathbf{q}(x_2, p; \theta)$.

Moreover, in the particular case of radially symmetric random variables the next theorem shows the existence of a remarkable relationship between the pth c-quantile curve and the $(1 - p)$th c-quantile curve.

Theorem 4.1.1 *(Bouyè and Salmon, 2009).   If two random variables X and Y are radially symmetric about (a, b), then*

$$\mathbf{q}(a - x, p; \theta) + \mathbf{q}(a + x, 1 - p; \theta) = 2b. \tag{4.3}$$

Example 4.1.1. *Consider the bivariate Gaussian copula with correlation coefficient θ:*

$$C(u, v; \theta) = \int_{-\infty}^{\Phi^{-1}(u)} \int_{-\infty}^{\Phi^{-1}(v)} \frac{1}{2\pi\sqrt{1 - \theta^2}} e^{\frac{2\theta xy - x^2 - y^2}{2(1 - \theta^2)}} \, dx \, dy.$$

Its first partial derivative is

$$D_1 C(u, v; \theta) = \Phi\left(\frac{\Phi^{-1}(v) - \theta\Phi^{-1}(u)}{\sqrt{1 - \theta^2}}\right).$$

So that, solving $p = \Phi\left(\frac{\Phi^{-1}(v) - \theta\Phi^{-1}(u)}{\sqrt{1 - \theta^2}}\right)$ for v and setting $u = F_X(x)$ and $v = F_Y(y)$ we find the pth c-quantile curve

$$y = F_Y^{-1}\left[\Phi\left(\theta\Phi^{-1}(F_X(x)) + \sqrt{1 - \theta^2}\Phi^{-1}(p)\right)\right].$$

Example 4.1.2. *Consider the Clayton copula with strictly positive parameter θ, $C(u, v; \theta) = (u^{-\theta} + v^{-\theta} - 1)^{-\frac{1}{\theta}}$, for which*

$$D_1 C(u, v; \theta) = (1 + u^\theta(v^{-\theta} - 1))^{-\frac{1+\theta}{\theta}}.$$

Solving $p = (1 + u^\theta(v^{-\theta} - 1))^{-\frac{1+\theta}{\theta}}$ for v gives

$$v = ((p^{-\frac{\theta}{1+\theta}} - 1)u^{-\theta} + 1)^{-\frac{1}{\theta}},$$

which provides us with the c-quantile relating v and u for different values of p. Setting $u = F_X(x)$ and $v = F_Y(y)$ we find the conditional c-quantiles for Y conditional on X:

$$y = F_Y^{-1}\left(((p^{-\frac{\theta}{1+\theta}} - 1)F_X(x)^{-\theta} + 1)^{-\frac{1}{\theta}}\right).$$

Example 4.1.3. *If we consider, in general, Archimedean copulas $C(u, v) = \varphi^{-1}(\varphi(u) + \varphi(v))$, since $D_1 C(u, v) = \frac{\varphi'(u)}{\varphi'(\varphi^{-1}(\varphi(u) + \varphi(v)))}$, the quantile regression curve is given by*

$$y = F_Y^{-1}\left(\varphi^{-1}[\varphi\left(\varphi'^{-1}\left(\frac{1}{p}\varphi'(F_X(x))\right)\right) - \varphi(F_X(x))]\right).$$

Now, let $\mathbf{y}_T = (y_1, \ldots, y_T)$ be a random sample on Y and $\mathbf{x}_T = ((x_t^1)_{t=1,\ldots,T}, \ldots, (x_t^k)_{t=1,\ldots,T})$ a random k-vector sample on X.

Definition 4.1.2. The pth copula quantile regression $\mathbf{q}(\mathbf{x}_t, p; \theta)$ is a solution to the following problem:

$$\min_\theta \left(\sum_{t \in \mathcal{T}_p} p|y_t - \mathbf{q}(\mathbf{x}_t, p; \theta)| + \sum_{t \in \mathcal{T}_{1-p}} (1 - p)|y_t - \mathbf{q}(\mathbf{x}_t, p; \theta)|\right), \tag{4.4}$$

where $\mathcal{T}_p = \{t : y_t \geq \mathbf{q}(\mathbf{x}_t, p; \theta)\}$ and $\mathcal{T}_{1-p}$ is its complement.

By postulating given margins for variables X and Y and their copula, Bouyè and Salmon implicitly assume a specific parametric functional for $\mathbf{q}(\mathbf{x}_t, p; \theta)$ and, as a consequence, the probability level p appears as an argument of the function $\mathbf{q}$ itself. Finally, after estimating the margins via empirical distribution functions, they estimate the copula quantile parameter θ by solving problem (4.4), which is an $\mathbb{L}^1$ norm estimator. As we will see in Section 4.3, Chen and Fan (2006) provide conditions for the consistency of this semi-parametric approach to the estimation of copula-based time series models. This c-quantile approach enables us to examine the dependency between assets at any given quantile, including extreme quantiles. Often, one may not be interested in dependency in the far extremes where highly infrequent but potentially disastrous joint losses may occur and one may be more interested in the dependency of large but not extreme losses that can arise, and it is in this latter case that the c-quantile approach should provide a better measure of association between the assets.

4.2 COPULA-BASED MARKOV PROCESSES: NON-LINEAR QUANTILE AUTOREGRESSION

Chen *et al.* (2009a) employ parametric copula functions to generate quantile autoregression (QAR) models which are non-linear in parameters. These models have several advantages over the linear QAR models (see Koenker and Xiao, 2006) since, by construction, the copula-based non-linear QAR models are globally plausible with monotone conditional quantile functions over the entire support of the conditioning variables. Moreover, copula-based Markov models provide a rich source of potential non-linear dynamics describing temporal dependence (and tail dependence). In particular, they distinguish the temporal dependence from the specification of the marginal (stationary) distribution of the response. Stationarity of the processes considered implies that only one marginal distribution is required for the specification in addition to the choice of a copula.

The data generating process (DGP) is specified in the following definition.

Definition 4.2.1. (Stationary copula-based Markov process). Let $\{X_t : t = 1, \ldots, n\}$ be a sample of a stationary first-order Markov process generated from $\{G^*, C(\cdot, \cdot; \theta^*)\}$, where $G^*(\cdot)$ is the true invariant distribution which is absolutely continuous w.r.t. Lebesgue measure on $\mathbb{R}$ and $C(\cdot, \cdot; \theta^*)$ is the parametric copula of (X_{t-1}, X_t) up to unknown value θ^* which is absolutely continuous w.r.t. Lebesgue measure on $[0, 1]^2$ and is neither the Frèchet–Hoeffding upper nor lower bound.

Example 4.2.1. *Simulation of a stationary copula-based Markov process.* We simulate a strictly stationary copula-based Markov process $X = (X)_{t=1,\ldots,T}$ as described in Definition 4.2.1 above. The simulation procedure is the dynamic version of the conditional sampling method. First, we generate n i.i.d. $U(0, 1)$ samples $(\zeta_1, \ldots, \zeta_n)$ and then for each i we solve $u_i = h^{-1}(\zeta_i | u_{i-1})$, where $h(u_i | u_{i-1}) \equiv D_1 C(u_{i-1}, u_i)$. The desired sample is $(u_1, \ldots, u_n)$. To simulate the copula-based Markov time series $(x_1, \ldots, x_n)$ it suffices to apply a generalized inverse of the stationary distribution G, that is $x_i = G^{-1}(u_i)$. Figures 4.1–4.4 present time series plots in the case of a Clayton copula with $\tau = 0.8$ and $\tau = 0.2$ and with marginal distributions given by the standard Gaussian distribution (Figures 4.1, 4.2) and the Student's t distribution with degree of freedom equal to 5 (Figures 4.3, 4.4). We notice the asymmetric dependence structure of the time series, which becomes stronger as θ increases.

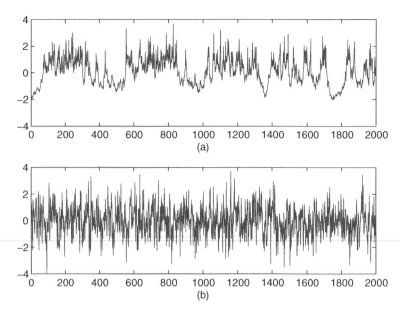

Figure 4.1 Simulated trajectories from a copula-based Markov process: (a) C is a Clayton copula with $\tau = 0.8$ ($\theta = 8$); (b) C is a Clayton copula with $\tau = 0.2$ ($\theta = 0.5$). The stationary marginal distribution is $N(0, 1)$.

Example 4.2.2. *In this example we present a simulation of a bivariate copula-based Markov process with cross-sectional and temporal dependence structure. More precisely, a process $(X^1, X^2) = (X_t^1, X_t^2)_{t=1,\dots,T}$ such that:*

- *Each marginal process is a univariate strictly stationary copula-based Markov process, that is X_t^1 is generated by $(G_1, C_1(\cdot, \cdot, \theta))$ and X_t^2 is generated by $(G_2, C_2(\cdot, \cdot, \theta))$.*
- *The marginal components X_t^1 and X_t^2 are linked by a (cross-sectional) invariant copula function $C_{1,2}(\cdot, \cdot, \theta_{1,2})$, where $\theta_{1,2}$ denotes the cross-sectional parameter.*

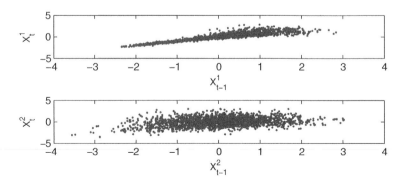

Figure 4.2 Scatter plots of realizations of the time series generated by the copula-based Markov process: (a) C is a Clayton copula with $\tau = 0.8$ ($\theta = 8$); (b) C is a Clayton copula with $\tau = 0.2$ ($\theta = 0.5$). The stationary marginal distribution is $N(0, 1)$.

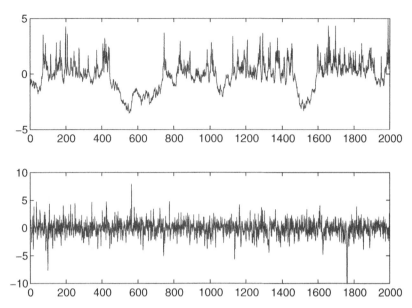

Figure 4.3 Simulated trajectories from a copula-based Markov process: (a) C is a Clayton copula with $\tau = 0.8$ ($\theta = 8$); (b) C is a Clayton copula with $\tau = 0.2$ ($\theta = 0.5$). The stationary marginal distribution is $t_{(5)}$.

The simulation procedure is again the dynamic conditional sampling which is used to generate $(u_1^1, \ldots, u_n^1)$ and $(u_1^2, \ldots, u_n^2)$, but in this case for each i the couple (u_i^1, u_i^2) is linked by the cross-sectional copula $C_{1,2}$. The stationary marginal distributions G_1 and G_2 are $t_{(5)}$. So, the desired sample is given by $(x_i^1, x_i^2) = (G_1^{-1}(u_i^1), G_2^{-1}(u_i^2))$, $i = 1, \ldots, n$. Figures 4.5–4.7 depict the histograms of the marginal simulated values $(x_1^1, \ldots, x_n^1)$ and $(x_1^2, \ldots, x_n^2)$ and the scatter plots of (x_i^1, x_i^2) for different selected copulas C_1, C_2, and $C_{1,2}$. For the sake of illustration, Figures 4.8–4.10 report the non-parametric estimation of the joint densities of the simulated bivariate trajectories.

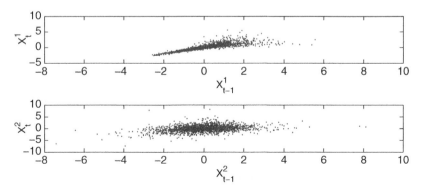

Figure 4.4 Scatter plots of realizations of the time series generated by the copula-based Markov process: (a) C is a Clayton copula with $\tau = 0.8$ ($\theta = 8$); (b) C is a Clayton copula with $\tau = 0.2$ ($\theta = 0.5$). The stationary marginal distribution is $t_{(5)}$.

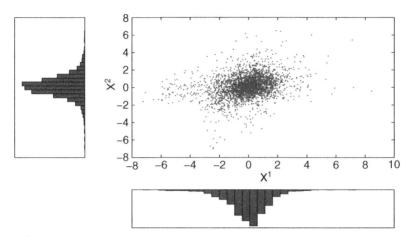

Figure 4.5 Dependence and marginal behavior of a simulated bivariate copula-based Markov process (X^1, X^2) when C_1 and C_2 are Clayton copulas with parameter $\tau = 0.6$ ($\theta = 3$) and $C_{1,2}$ is a Student's t copula with parameter $\theta_{1,2} = (\rho_{1,2}, \nu_{1,2}) = (0.3, 10)$.

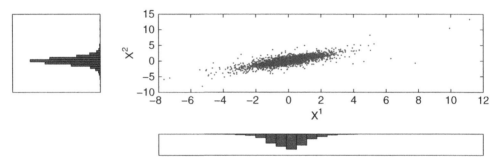

Figure 4.6 Dependence and marginal behavior of a simulated bivariate copula-based Markov process (X^1, X^2) when C_1 and C_2 are Clayton copulas with parameter $\tau = 0.2$ ($\theta = 0.5$) and $C_{1,2}$ is a Student's t copula with parameter $\theta_{1,2} = (\rho_{1,2}, \nu_{1,2}) = (0.8, 3)$.

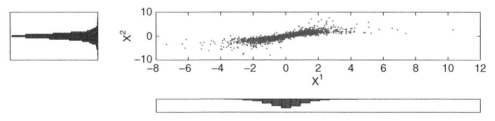

Figure 4.7 Dependence and marginal behavior of a simulated bivariate copula-based Markov process (X^1, X^2) when C_1 and C_2 are Clayton copulas with parameter $\tau = 0.2$ ($\theta = 0.5$) and $C_{1,2}$ is a Frank copula with parameter $\tau = 0.8$ ($\theta_{1,2} = 18.1915$).

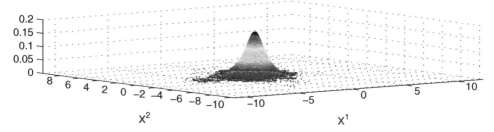

Figure 4.8 Empirical joint density of a simulated bivariate copula-based Markov process (X^1, X^2) when C_1 and C_2 are Clayton copulas with parameter $\tau = 0.6$ ($\theta = 3$) and $C_{1,2}$ is a Student's t copula with parameter $\theta_{1,2} = (\rho_{1,2}, \nu_{1,2}) = (0.3, 10)$.

We know that the (true) joint distribution of (X_{t-1}, X_t) is $F^*(x_{t-1}, x_t) = C^*(G^*(x_{t-1}), G^*(x_t); \theta^*)$, and that the copula function $C^*(\cdot, \cdot; \theta^*)$ is a bivariate probability distribution function with uniform marginals. As we know, the conditional distribution of X_t given $X_{t-1} = y$ is obtained by differentiation of $C^*(\cdot, \cdot; \theta^*)$ with respect to its first argument:

$$\mathbb{P}(X_t \le x | X_{t-1} = y) = \left. \frac{\partial C^*(u, v; \theta^*)}{\partial u} \right|_{u=G^*(y), v=G^*(x)} \doteq D_1 C^*(G^*(y), G^*(x)).$$

For any $\tau \in (0, 1)$, solving $\tau = D_1 C^*(G^*(y), G^*(x))$ for x (in terms of τ), we obtain the τth conditional quantile function of X_t given $X_{t-1} = y$:

$$Q_{X_t}(\tau | y) = G^{*,-1}(D_1 C_v^{*,-1}(\tau, G^*(y))) = G^{*,-1}(h^*(y)) \doteq H^*(y),$$

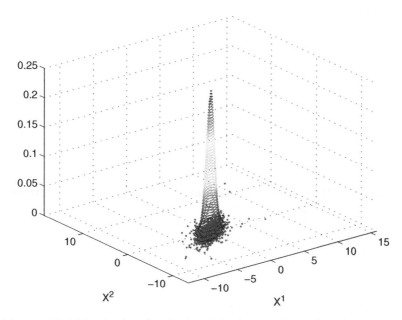

Figure 4.9 Empirical joint density of a simulated bivariate copula-based Markov process (X^1, X^2) when C_1 and C_2 are Clayton copulas with parameter $\tau = 0.2$ ($\theta = 0.5$) and $C_{1,2}$ is a Student's t copula with parameter $\theta_{1,2} = (\rho_{1,2}, \nu_{1,2}) = (0.8, 3)$.

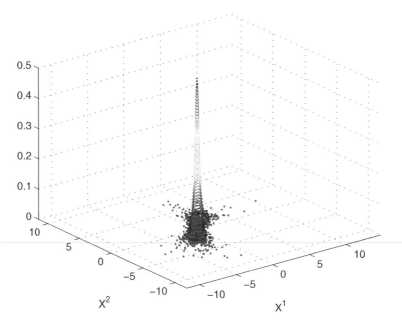

Figure 4.10 Empirical joint density of a simulated bivariate copula-based Markov process (X^1, X^2) when C_1 and C_2 are Clayton copulas with parameter $\tau = 0.2$ ($\theta = 0.5$) and $C_{1,2}$ is a Frank copula with parameter $\tau = 0.8$ ($\theta_{1,2} = 18.1915$).

where $h^*(y) = D_1 C_v^{*,-1}(\tau, G^*(y))$ and $D_1 C_v^{*,-1}(\cdot; u)$ denote the partial inverse of $D_1 C^*(u, v)$ with respect to $v = G^*(x_t)$. Chen *et al.* (2009) model parametrically both $C(\cdot, \cdot; \theta)$ and $G(\cdot; \alpha)$. The τth conditional quantile function of X_t, $Q_{X_t}(\tau|y)$, becomes a function of the unknown parameters θ and α, i.e.

$$Q_{X_t}(\tau|y) = G^{-1}(D_1 C_v^{-1}(\tau; G(y; \alpha), \theta), \alpha) = G^{-1}(h(y; \theta, \alpha), \alpha).$$

Denoting $\gamma = (\theta, \alpha)$, we write

$$Q_{X_t}(\tau|y) = H(y; \gamma).$$

Notice that the proposed QAR model may allow for different parameters over different τ values, so the vector of unknown parameters can be viewed as a function of τ, $\gamma(\tau) = (\alpha(\tau), \theta(\tau))$, and the following non-linear QAR model can be considered:

$$Q_{X_t}(\tau|X_{t-1}) = G^{-1}(h(X_{t-1}, \theta(\tau), \alpha(\tau)), \theta(\tau)) = H(X_{t-1}; \gamma(\tau)). \quad (4.5)$$

The estimation of the vector of parameters $\gamma(\tau)$ of the copula-based QAR model (4.5) is based on the following non-linear quantile autoregression:

$$\min_{\gamma \in \Gamma} \sum_t \rho_\tau(X_t - H(X_{t-1}; \gamma)),$$

where $\rho_\tau = u(\tau - \mathbf{1}(u < 0))$ is the usual check function (Koenker and Bassett, 1978). The solution is the estimate of $\gamma(\tau)$:

$$\hat{\gamma}(\tau) = \arg\min \sum_t \rho_\tau(X_t - H(X_{t-1}; \gamma)),$$

and the τth conditional quantile of X_t given X_{t-1} can be estimated by $\hat{Q}_{X_t}(\tau | X_{t-1} = y) = H(y; \hat{\gamma}(\tau)) = G^{-1}(D_1 C_v^{-1}(\tau, G(y; \hat{\alpha}(\tau)); \hat{\theta}(\tau)); \hat{\alpha}(\tau))$. Under some regularity conditions (see Chen et al., 2009), $\hat{\gamma}(\tau)$ is consistent for $\gamma(\tau)$ and its asymptotic distribution is Gaussian with convergence rate $\sqrt{n}$ as established in the following proposition. Notice that differently from the case of semi-parametric estimation of a copula-based Markov model (i.e., Chen and Fan, 2006), in this framework any β-mixing decay rate condition on $\{X_t : t = 1, \ldots, n\}$ is needed.

Proposition 4.2.1 (Chen et al., 2009). *In the framework described in Definition 4.2.1, if the copula-based Markov process is ergodic and assumptions A1–A6 in Chen et al. (2009) are satisfied, for any fixed $\tau \in]0, 1[$ we have*

$$\hat{\gamma}(\tau) - \gamma(\tau) = o_p(1)$$

and

$$\sqrt{n}(\hat{\gamma}(\tau) - \gamma(\tau)) \xrightarrow{d} N(0, \tau(1 - \tau)V^{-1}(\tau)\Sigma(\tau)V^{-1}(\tau)),$$

where

$$\mathbb{E}\left[g^*(Q_{X_t}(\tau | X_{t-1})) \left(\frac{\partial H(X_{t-1}; \gamma(\tau))}{\partial \gamma} \right) \left(\frac{\partial H(X_{t-1}; \gamma(\tau))}{\partial \gamma} \right)' \right]$$

and

$$\Sigma(\tau) = \mathbb{E}\left[\left(\frac{\partial H(X_{t-1}; \gamma(\tau))}{\partial \gamma} \right) \left(\frac{\partial H(X_{t-1}; \gamma(\tau))}{\partial \gamma} \right)' \right],$$

where g^* is the true invariant density function of the process.

4.3   COPULA-BASED MARKOV PROCESSES: SEMI-PARAMETRIC ESTIMATION

This section introduces semi-parametric estimators of a copula-based Markov model and establishes their asymptotic properties under easily verifiable conditions. The main contribution is due to Chen and Fan (2006). In that paper, the authors study a class of univariate copula-based semi-parametric stationary Markov models in which copulas are parameterized but the invariant marginal distributions are left unspecified. In these models there are only two parameters: the copula dependence parameter θ and the invariant marginal distribution function $G(\cdot)$, which can be estimated by a non-parametric method (i.e., the empirical distribution function or the kernel smoothed estimator).

Let $X = (X_t)_{t \in \mathbb{Z}}$ be a stationary Markov process of order 1 as in Definition 4.2.1 and let $H(x,y)$ be the joint distribution function of X_{t-1} and X_t. We know that H completely determines the probabilistic structure of X. If C is the copula between X_{t-1} and X_t, we model a stationary Markov process by specifying the marginal distribution of X_t and the copula C instead of the joint distribution H, which will be $H(x_1, x_2) = C(G(x_1), G(x_2); \theta)$.

If the copula of X_{t-1} and X_t is either the upper bound $C(u, v) = \min(u, v)$ or the lower bound $C(u, v) = \max(u + v - 1, 0)$, then X_t is almost surely a monotonic function of X_{t-1}: then resulting time series is deterministic and we would have $X_t = X_{t-1}$ for the upper bound and $X_t = G^{*-1}(1 - G^*(X_{t-1}))$ for the lower bound.

The conditional density of X_t given X_{t-1} is given by

$$h^*(X_t | X_{t-1}) = g^*(X_t) c(G^*(X_{t-1}), G^*(X_t); \theta^*), \qquad (4.6)$$

where $c(\cdot, \cdot; \theta^*)$ is the copula density and $g^*(\cdot)$ is the density of the true invariant distribution $G^*(\cdot)$.

A semi-parametric copula-based time series model is completely determined by (G^*, θ^*). The unknown invariant marginal distribution G^* can be estimated by $G_n(\cdot)$, the rescaled empirical distribution function defined as

$$G_n(x) = \frac{1}{n+1} \sum_{t=1}^{n} \mathbf{1}_{\{X_t \leq x\}}. \qquad (4.7)$$

Since the log-likelihood function is given by

$$\ell(\theta) = \frac{1}{n} \sum_{t=1}^{n} \log g^*(X_t) + \frac{1}{n} \sum_{t=1}^{n} \log c(G^*(X_{t-1}), G^*(X_t); \theta),$$

ignoring the first term and replacing G^* with G_n we obtain the semi-parametric estimator $\tilde{\theta}$ of θ^*:

$$\tilde{\theta} = \arg\max_{\theta} \tilde{L}(\theta), \qquad \tilde{L}(\theta) = \frac{1}{n} \sum_{t=1}^{n} \log c(G_n(X_{t-1}), G_n(X_t); \theta).$$

The main difficulty in establishing the asymptotic properties of the semi-parametric estimator $\tilde{\theta}$ is that the score function and its derivatives could blow up to infinity near the boundaries. To overcome this difficulty, Chen and Fan (2006) first established convergence of $G_n(\cdot)$ in a weighted metric and then used it to establish the consistency and asymptotic normality of $\tilde{\theta}$. Let $\tilde{U}_n(z) = G_n(G^{*-1}(z))$, $z \in]0, 1[$ and let $w(\cdot)$ be a continuous function on $[0, 1]$ which is strictly positive on $]0, 1[$, symmetric at $z = 1/2$, and increasing on $]0, 1/2]$ and suppose that X_t is β-mixing (see Section 4.5), then (Chen and Fan (2006), Lemma 4.1):

$$\sup_{z \in [0,1]} \left| \frac{\tilde{U}_n(z) - z}{w(z)} \right| = o_{a.s}(1), \qquad \sup_{y} \left| \frac{G_n(y) - G^*(y)}{w(G^*(y))} \right| = o_{a.s}(1),$$

provided $\beta_t = O(t^{-b})$ for some $b > 0$ and $\int_0^1 \frac{1}{w(z)} \log(1 + \frac{1}{w(z)}) dz < \infty$. Weight functions of the form $w(z) = [z(1 - z)]^{1-\xi}$ for all $z \in]0, 1[$ and some $\xi \in]0, 1[$ approach zero when z approaches 0 or 1. Hence, such results allow us to handle unbounded score functions.

The maximum likelihood estimator $\tilde{\theta}$ is consistent and asymptotically Gaussian with rate of convergence $\sqrt{n}$ (Chen and Fan (2006), Propositions 4.2 and 4.3). In particular, let $l(u, v; \theta) = \log c(u, v; \theta)$ and denote $l_\theta(u, v; \theta) = \frac{\partial l(u, v; \theta)}{\partial \theta}$, $l_{\theta,\theta}(u, v; \theta) = \frac{\partial^2 l(u, v; \theta)}{\partial \theta \partial \theta}$, $l_{\theta,u}(u, v; \theta) = \frac{\partial^2 l(u, v; \theta)}{\partial \theta \partial u}$, and $l_{\theta,v}(u, v; \theta) = \frac{\partial^2 l(u, v; \theta)}{\partial \theta \partial v}$. Then, two conclusions hold:

1. Under some regularity conditions (C1–C2 and C4–C5 in Chen and Fan (2006), pp. 317–318) and if X_t is β-mixing with mixing decay rate $\beta_t = O(t^{-b})$ for some $b > 0$, we have $\|\tilde{\theta} - \theta^*\| = o_p(1)$.

2. Under some regularity conditions (A1–A3 and A5–A6 in Chen and Fan (2006), pp. 318–319) and if X_t is β-mixing with mixing decay rate $\beta_t = O(t^{-b})$ for some $b > \gamma/(\gamma - 1)$ with $\gamma > 1$, we have $\sqrt{n}(\tilde{\theta} - \theta^*) \to N(0, B^{-1}\Sigma B)$, where $B = -\mathbb{E}[l_{\theta,\theta}(U_{t-1}, U_t; \theta^*)]$ and $\Sigma = \lim_{n \to \infty} Var(\sqrt{n}A_n^*)$, where

$$A_n^* = \frac{1}{n-1}\sum_{t=2}^{n}[l_\theta(U_{t-1}, U_t; \theta^*) + W_1(U_{t-1}) + W_2(U_t)],$$

$$W_1(U_{t-1}) = \int_0^1 \int_0^1 [\mathbf{1}\{U_{t-1} \le u - u\}]l_{\theta,u}(u, v; \theta^*)c(u, v; \theta^*)du dv,$$

$$W_2(U_t) = \int_0^1 \int_0^1 [\mathbf{1}\{U_t \le v - v\}]l_{\theta,v}(u, v; \theta^*)c(u, v; \theta^*)du dv,$$

where $U_t = G^*(X_t)$.

The assumption of a β-mixing condition can be replaced with a so-called strong mixing condition but in this case the existence of finite higher-order moments of the score function and its partial derivatives will be stronger than those for β-mixing processes. As many copula models have score functions blowing up at a fast rate, it is essential to maintain minimal requirements for the existence of finite higher-order moments and this is a sufficient motivation to assume β-mixing instead of strong mixing.

An alternative estimation procedure for the copula parameter θ^* and the invariant distribution G^* is proposed by Chen et al. (2009b). They call it sieve maximum likelihood estimation and it consists of approximating the unknown marginal density by a flexible parametric family of densities with increasing complexity (sieves) and then maximizing the joint likelihood w.r.t. the unknown copula parameter and the sieve parameters of the approximating marginal density. The sieve MLEs of any smooth functionals of (θ^*, G^*) are root-n consistent, asymptotically normal, and efficient.

We write the true conditional density of X_t given X_{t-1} as in (4.6). Let

$$h(X_t|X_{t-1}; g, \theta) = g(X_t)c(G(X_{t-1}), G(X_t); \theta)$$

denote any candidate conditional density of X_t given X_{t-1}. Here, the parameters are g and θ. The log-likelihood associated with $h(X_t|X_{t-1}; \theta, g)$ is given by

$$l(\theta, g; X_t, X_{t-1}) = \log g(X_t) + \log c(G(X_{t-1}), G(X_t); \theta)$$
$$\equiv \log g(X_t) + \log c\left(\int \mathbf{1}_{\{y \le X_{t-1}\}}g(y)dy, \int \mathbf{1}_{\{y \le X_t\}}g(y)dy; \theta\right),$$

and the joint log-likelihood function of the data $\{X_t, t = 1, \dots, n\}$ is

$$L_n(\theta, g) = \frac{1}{n}\sum_{t=2}^{n} l(\theta, g; X_t, X_{t-1}) + \frac{1}{n}\log g(X_1).$$

Let Θ be the parameter space for the copula parameter θ and let $\mathcal{G}$ be the parameter space for g. Then, the approximate sieve MLE $\hat{\gamma}_n = (\hat{\theta}_n, \hat{g}_n)$ is defined as

$$L_n(\hat{\theta}_n, \hat{g}_n) \ge \max_{\theta \in \Theta, g \in \mathcal{G}_n} L_n(\theta, g) - O_p(\delta_n^2),$$

where δ_n is a positive sequence such that $\delta_n = o(1)$ and $\mathcal{G}_n$ denotes the sieve space (i.e., a sequence of finite-dimensional parameter spaces that becomes dense as $n \to \infty$ in $\mathcal{G}$). The sieves used by Chen et al. (2009b) for approximating the invariant density function are linear.

Two examples are given by

$$\mathcal{G}_n = \left\{ g_{K_n} \in \mathcal{G} : g_{K_n} = \left(\sum_{k=1}^{K_n} a_k A_k(y) \right)^2, \int g_{K_n}(y) dy = 1 \right\}, \quad K_n \to \infty, \frac{K_n}{n} \to 0,$$

to approximate a square root density, and

$$\mathcal{G}_n = \left\{ g_{K_n} \in \mathcal{G} : g_{K_n} = \exp \left(\sum_{k=1}^{K_n} a_k A_k(y) \right), \int g_{K_n}(y) dy = 1 \right\}, \quad K_n \to \infty, \frac{K_n}{n} \to 0,$$

to approximate a log-density, where $\{A_k : k \geq 1\}$ consists of known basis functions and $\{a_k : k \geq 1\}$ is the collection of unknown sieve coefficients.

We conclude this section with the two main theorems concerning consistency and asymptotic normality.

Theorem 4.3.1 *(Chen et al., 2009).*   *Under some regularity conditions (Assumptions 3.1–3.2 in Chen et al. (2009)), if $K_n \to \infty$ and $\frac{K_n}{n} \to 0$, we have $\|\hat{\gamma}_n - \gamma\| = o_p(1)$.*

Let $\rho : \Theta \times \mathcal{G} \to \mathbb{R}$ be a smooth functional and $\rho(\hat{\gamma}_n)$ be the plug-in sieve MLE of $\rho(\gamma)$.

Theorem 4.3.2 *(Chen et al., 2009).*   *Under some regularity conditions (Assumptions 4.1–4.7 in Chen et al. (2009)), if X_t is strictly stationary β-mixing, we have $\sqrt{n}(\rho(\hat{\gamma}_n) - \rho(\gamma)) \to_d N(0, \|\frac{\partial \rho(\gamma^*)}{\partial \gamma}\|^2)$.*

Example 4.3.1. *In this example we show the performance of the semi-parametric estimator $\tilde{\theta}$ in finite samples by a Monte Carlo simulation of the copula-based Markov process. We expect the performance to depend on the choice of the copula function and on the choice of the marginal distribution. We simulate 3000 paths of several sizes of the model $\{G^*, C(\cdot, \cdot; \theta^*)\}$ for three families of copula functions C (Clayton, Frank, Gaussian) with three different values of the parameter θ^* (consistent with three levels of dependence: $\tau = 0.2$, $\tau = 0.4$, and $\tau = 0.8$) and two marginal distributions G^* (standard Gaussian and Student's t). For each simulated path we estimate $G_n(\cdot)$ via 4.7 and we use that to find the maximum likelihood estimate $\tilde{\theta}$. Tables 4.1–4.3 report the average values and the relative mean squared error of the MLEs. We notice that the marginal distribution does not have a significant impact on estimates, while the choice of the parameter of the copula is important. In fact, if the dependence structure increases (that is, θ is higher), estimates are less precise. The accuracy improves as the sample size increases (Figure 4.11).*

Example 4.3.2. Application to empirical data.   *We provide an empirical application to three market index time series: NIKKEI225, FTSE100, and NAS100. Figures 4.12 and 4.13 show the log-prices and the returns of these indices. The cross-sectional dependence structure is displayed in Figures 4.14–4.19. For each pair of market indices we report the scatter plots and the empirical bivariate densities. We can notice a very weak dependence between NIKKEI225 and FTSE100 and between FTSE100 and NAS100 and only a slightly higher dependence between NIKKEI225 and NAS100.*

The time series of each market index can be considered as a realization of a copula-based Markov model. Then one can apply the semi-parametric estimation technique to estimate the parameter of the generating copula C. Table 4.4 shows estimates of the copula-based Markov model for four different families of copulas. We can observe that the dependence structure for the three indices proposed is rather weak, but in all cases it is significantly different from zero (Figure 4.20).

Table 4.1   Monte Carlo simulation: average values and mean squared errors of the maximum likelihood estimates for the Clayton copula

Parameter	Sample size	$G^* \sim N(0, 1)$		$G^* \sim t_{(3)}$	
		Mean	MSE	Mean	MSE
$\theta^* = 0.5$	$N = 500$	0.5021	0.0106	0.5027	0.0100
$(\tau = 0.2)$	$N = 1000$	0.5023	0.0054	0.5003	0.0054
	$N = 1500$	0.5015	0.0036	0.5032	0.0038
$\theta^* = 1.3333$	$N = 500$	1.2812	0.0709	1.2889	0.0722
$(\tau = 0.4)$	$N = 1000$	1.3059	0.0390	1.3048	0.0420
	$N = 1500$	1.3182	0.0257	0.3039	0.0265
$\theta^* = 8$	$N = 500$	5.2672	10.1608	5.3068	10.0550
$(\tau = 0.8)$	$N = 1000$	6.2283	6.6428	6.2440	6.7655
	$N = 1500$	6.6327	5.3144	6.2683	6.6734

Table 4.2   Monte Carlo simulation: average values and mean squared errors of the maximum likelihood estimates for the Frank copula

Parameter	Sample size	$G^* \sim N(0, 1)$		$G^* \sim t_{(3)}$	
		Mean	MSE	Mean	MSE
$\theta^* = 1.8609$	$N = 500$	1.8419	0.0835	1.8604	0.0830
$(\tau = 0.2)$	$N = 1000$	1.8551	0.0443	1.8459	0.0410
	$N = 1500$	1.8461	0.0266	1.8510	0.0278
$\theta^* = 4.1615$	$N = 500$	4.1075	0.1267	4.1039	0.1288
$(\tau = 0.4)$	$N = 1000$	4.1337	0.0650	4.1337	0.0614
	$N = 1500$	4.1491	0.0409	4.1421	0.0410
$\theta^* = 18.1915$	$N = 500$	16.1085	7.1618	16.1359	7.0865
$(\tau = 0.8)$	$N = 1000$	17.0401	2.6100	17.0690	2.4853
	$N = 1500$	17.4264	1.3431	17.4324	1.3539

Table 4.3   Monte Carlo simulation: average values and mean squared errors of the maximum likelihood estimates for the Gaussian copula

Parameter	Sample size	$G^* \sim N(0, 1)$		$G^* \sim t_{(3)}$	
		Mean	MSE	Mean	MSE
$\theta^* = 0.3090$	$N = 500$	0.2975	0.0020	0.2954	0.0019
$(\tau = 0.2)$	$N = 1000$	0.3039	9.2307×10^{-4}	0.3035	8.9190×10^{-4}
	$N = 1500$	0.3045	5.9237×10^{-4}	0.3048	6.4994×10^{-4}
$\theta^* = 0.5878$	$N = 500$	0.5660	0.0018	0.5669	0.0017
$(\tau = 0.4)$	$N = 1000$	0.5778	7.2604×10^{-4}	0.5759	7.2604×10^{-4}
	$N = 1500$	0.5809	4.8889×10^{-4}	0.5808	4.8504×10^{-4}
$\theta^* = 0.9511$	$N = 500$	0.9168	0.0015	0.9170	0.0015
$(\tau = 0.8)$	$N = 1000$	0.9348	3.9252×10^{-4}	0.9344	4.0475×10^{-4}
	$N = 1500$	0.9401	2.0137×10^{-4}	0.9404	1.9538×10^{-4}

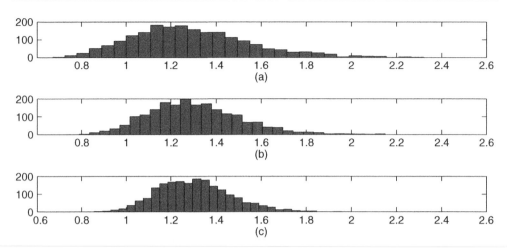

Figure 4.11 Monte Carlo simulation: histogram of ML estimates in the case where C is a Clayton copula with $\theta^* = 1.333$ and G^* is $t_{(3)}$: (a) $N = 500$; (b) $N = 1000$; (c) $N = 1500$.

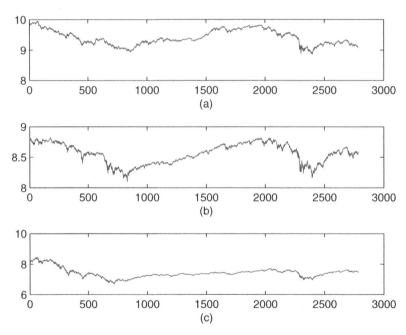

Figure 4.12 Empirical data. Log-prices: (a) NIKKEI225; (b) FTSE100; (c) NAS100.

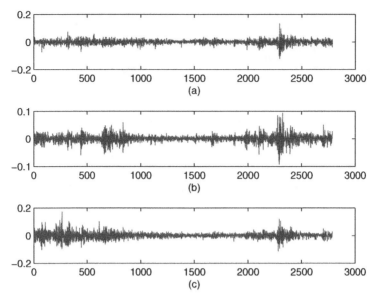

Figure 4.13   Empirical data. Log-returns: (a) NIKKEI225; (b) FTSE100; (c) NAS100.

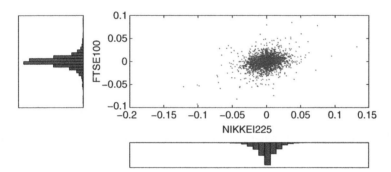

Figure 4.14   Empirical data. Log-returns of NIKKEI225 and FTSE100: dependence and marginal behavior.

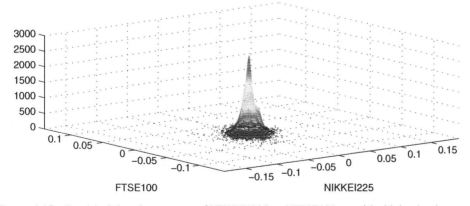

Figure 4.15   Empirical data. Log-returns of NIKKEI225 and FTSE100: empirical joint density.

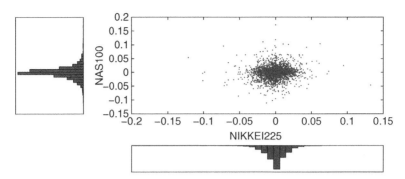

Figure 4.16 Empirical data. Log-returns of NIKKEI225 and NAS100: dependence and marginal behavior.

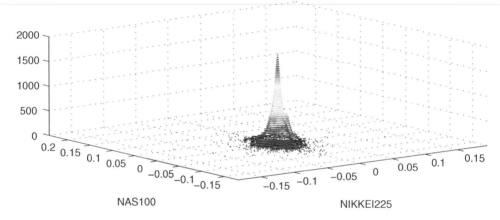

Figure 4.17 Empirical data. Log-returns of NIKKEI225 and NAS100: empirical joint density.

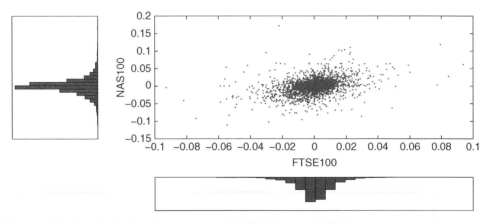

Figure 4.18 Empirical data. Log-returns of FTSE100 and NAS100: dependence and marginal behavior.

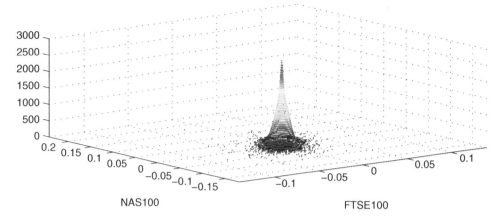

Figure 4.19   Empirical data. Log-returns of FTSE100 and NAS100: empirical joint density.

Table 4.4   Semi-parametric estimates of the (temporal) dependence structure for three market indices: NIKKEI225, FTSE100, and NAS100

	NIKKEI225	FTSE100	NAS100
Clayton	$0.0087 \ (5.991 \times 10^{-4})$	$0.0209 \ (5.4670 \times 10^{-4})$	$0.0184 \ (9.8662 \times 10^{-4})$
Frank	$-0.1385 \ (0.0023)$	$-0.2896 \ (0.0026)$	$-0.3262 \ (0.0026)$
Gaussian	$-0.0228 \ (2.2479 \times 10^{-4})$	$-0.0541 \ (9.0892 \times 10^{-5})$	$-0.0645 \ (2.6384 \times 10^{-4})$
Gumbel	$1.0086 \ (1.291 \times 10^{-5})$	$1.0059 \ (1.8031 \times 10^{-4})$	$1.0020 \ (1.2114 \times 10^{-4})$

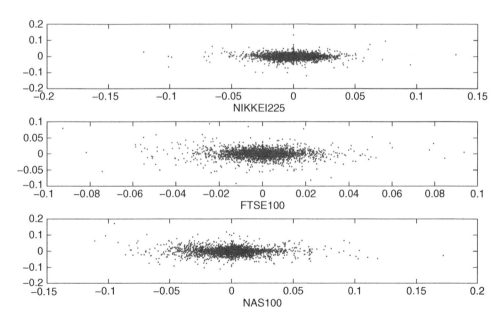

Figure 4.20   Empirical data. Temporal dependence structure.

4.4 COPULA-BASED MARKOV PROCESSES: NON-PARAMETRIC ESTIMATION

An important issue in risk management is the non-linear dependence between risks. It relates to the whole joint distribution of the variables and the main problem concerns the tail of the joint distribution. To overcome serious drawbacks of the different approaches which have been proposed in the econometric literature to describe non-linear dependence, Gagliardini and Gourièroux (2007) explore an approach in which the joint density is constrained and depends on a small number of one-dimensional functional parameters; that is, each parameter is a function of one variable. The joint density is parameterized by means of a function A defined on a subset of $\mathbb{R}$. Functional parameters instead of scalar parameters provide a higher flexibility and a better data fit and, moreover, allow for a much richer description of non-linear dependence compared with the standard finite-dimensional case.

Gagliardini and Gourièroux (2007) consider a family of distributions which are parameterized in terms of the one-dimensional functional parameter A. The distributions are continuous w.r.t. the Lebesgue measure with density function $f(x, y; A)$. The family of densities $f(x, y; A)$ can be specified in various ways. For example:

$$f(x, y; A) = f_{X|Y}(x|y; A) f_Y(y|A),$$

where $f_{X|Y}(x|y; A)$ is a family of conditional densities of X given Y and $f_Y(y|A)$ is a family of marginal densities of Y, both parameterized by A. Another example is provided by specifying the copula density $c(u, v; A)$ and the marginal densities $f_X(x|A)$, $f_Y(y|A)$. In fact, in this case

$$f(x, y; A) = c(F_X(x; A), F_Y(y; A); A) f_X(x|A) f_Y(y|A),$$

where $F_X(x; A)$ and $F_Y(y; A)$ are the marginal distributions.

In a time series framework, the bivariate joint density of X_{t-1}, X_t can be used to study the risk dynamics. If X_t is a stationary homogeneous Markov process, the dynamics is fully characterized by $F_{X_{t-1}, X_t}(x_{t-1}, x_t)$ and $F_{X_{t-1}}$ and F_{X_t} are equal because of stationarity. In this case the density $f_{X_{t-1}, X_t}(x_{t-1}, x_t; A)$ is parameterized by the two-dimensional functional parameter $A = (f, a)$, where f is the stationary marginal density and a the functional parameter which characterizes the copula of X_{t-1}, X_t; that is, the non-linear serial dependence.

Example 4.4.1. *Gourièroux and Robert (2006) introduced the stochastic unit root model to study the links between long memory and heavy tails. This process is a Markov process with two random regimes, defined as*

$$X_t = \begin{cases} X_{t-1} + \varepsilon_t, & \text{with} \quad \text{probability} \quad \eta(X_{t-1}); \\ \varepsilon_t, & \text{with} \quad \text{probability} \quad 1 - \eta(X_{t-1}), \end{cases}$$

where ε_t are i.i.d. and independent from X_{t-1}, with density g and η is a function with values in $]0, 1]$ which characterizes the non-linear serial dependence of X_t. Gourièroux and Robert show that the conditional density is given by

$$f(y|x; A) = \eta(x) g(y - x) + [1 - \eta(x)] g(y)$$

and the model parameter is $A = (\eta, g)$.

The non-parametric estimation of the function A follows the same lines as in the standard finite-dimensional parametric framework, but it has to be generalized to take into account the functional nature of the parameter. Let $f(x, y; A)$ be a family of bivariate densities, where the functional parameter A belongs to an open set $\mathcal{A}$ of all functions from $\mathbb{R}$ to $\mathbb{R}$ (or $\mathbb{R}^d$) endowed with the $\mathbb{L}^2(\lambda)$-norm ($\|\cdot\|_{\mathbb{L}^2(\lambda)}$) corresponding to the Lebesgue measure λ. The function $A_0 \in \mathcal{A}$ denotes the true value of the functional parameter and $f(\cdot, \cdot; A_0)$ the corresponding true p.d.f. The derivative of the log-density with respect to A, $D_A \log f(x, y; A)$, is a Hadamard derivative (van der Vaart and Wellner, 1996). It is assumed to exist and be defined by

$$\log f(x, y; A + h) - \log f(x, y; A) = \langle D_A \log f(x, y; A), h \rangle + R(x, y; A, h),$$

for $A, A + h \in \mathcal{A}$, where $D_A \log f(x, y; A)$ is a linear mapping from $\mathbb{L}^2(\lambda)$ to $\mathbb{R}$ which associates to $h \in \mathbb{L}^2(\lambda)$ the quantity $\langle D_A \log f(x, y; A), h \rangle \in \mathbb{R}$. Analogous to the information matrix is the information operator I, which is the covariance operator of the functional score at A_0, $D_A \log f(x, y; A_0)$, and is a bounded, non-negative, and self-adjoint operator from $\mathbb{L}^2(\lambda)$ into itself with a continuous inverse I^{-1} (by assumption I is invertible). The authors introduce a class of constrained non-parametric estimators in the form of minimum chi-squared estimators and study their consistency and asymptotic distribution in two possible frameworks: when the observations $(X_t, Y_t)_{t=1,\dots,T}$ are i.i.d. and when $(X_t, Y_t)_{t=1,\dots,T}$ is a time series with serial dependence. In the i.i.d. case, the minimum chi-squared estimator of parameter A is defined by

$$\hat{A}_T = \arg \min_{A \in \Theta \subset \mathcal{A}} Q_T(A) = \int_0^1 \int_0^1 \frac{[\hat{f}_T(x, y) - f(x, y; A)]^2}{\hat{f}_T(x, y)} w_T(x, y) dx dy,$$

where w_T is a smooth function converging pointwise to one when $T \to \infty$ and $\hat{f}_T(x, y)$ is a kernel estimator of the unconstrained bivariate density function (Rosenblatt, 1956)

$$\hat{f}_T(x, y) = \frac{1}{Th_T^2} \sum_{t=1}^T K\left(\frac{x - X_t}{h_T}\right) K\left(\frac{y - Y_t}{h_T}\right),$$

where K is a kernel and h_T is a bandwidth, which, under some regularity conditions (see the assumptions in Appendix B.1 in Gagliardini and Gourièroux (2007)), is consistent for its mean and asymptotically normal

$$\sqrt{Th_T^2}(\hat{f}_T(x, y) - \mathbb{E}[\hat{f}_T(x, y)]) \xrightarrow{d} N(0, \sigma^2(x, y; A_0)),$$

with $\sigma^2(x, y; A_0) = f(x, y; A_0) \left(\int K(s) ds\right)^2$. The constrained estimator of the bivariate density is then $f_T(x, y; \hat{A}_T)$. Consistency and asymptotic normality of $\hat{A}_T$ are contained in the following propositions.

Proposition 4.4.1 *(Gagliardini and Gourièroux, 2007).* *Under some regularity conditions (see Appendix B in Gagliardini and Gourièroux (2007)), if the observations $(X_t, Y_t)_{t=1,\dots,T}$ are i.i.d. we have:*

- *If the bandwidth h_T is such that $h_T = c_T T^{-\alpha}$ with $\lim_{T\to\infty} c_T = c > 0$ and $0 \le \alpha \le 1/2$, the chi-squared estimator $\hat{A}_T$ is consistent in $\mathbb{L}^2(\lambda)$: that is, $\|\hat{A}_T - A_0\|_{\mathbb{L}^2(\lambda)} \xrightarrow{\mathbb{P}} 0$.*
- *If the bandwidth h_T is such that $h_T = c_T T^{-\alpha}$ with $\lim_{T\to\infty} c_T = c > 0$ and $\frac{1}{4m}\left(1 + \frac{2m-1}{4m^2+2m+1}\right) \le \alpha \le 1/4\left(1 - \frac{1}{2}\frac{2m-1}{4m^2+2m+1}\right)$, where m is the degree of differentiability of the density f, the chi-squared estimator $\hat{A}_T$ is λ-a.s. pointwise asymptotically normal*

$$\sqrt{Th_T}(\hat{A}_T(s) - A_0(s) - I^{-1}\mathbb{E}[\psi_T(s; A)]) \longrightarrow$$

$$N\left(0, \left(\int K^2(s) ds\right) \alpha_0(s; A)^{-1}\right), \quad \lambda - a.s., \quad s \in [0, 1],$$

where

$$\psi_T(s; A) = \int (\hat{f}_T(s, y) - f(s, y; A))w_T(s, y)\gamma_0(s, y; A)dy$$

$$+ \int (\hat{f}_T(x, s) - f(x, s; A))w_T(x, s)\gamma_1(x, s; A)dx$$

$$+ \int \int (\hat{f}_T(x, y) - f(x, y; A))w_T(x, y)\gamma_2(x, y, s; A)dxdy,$$

with $\gamma_0, \gamma_1, \gamma_2$ $\mathbb{R}$-valued functions and

$$\alpha_0(s; A) = \int \gamma_0^2(s, y; A)f(s, y; A)dy + \int \gamma_1^2(x, s; A)f(x, s; A)dx.$$

The previous results are easily extended to the time series framework. The key assumption in this case is that the process $(X_t)_{t=1,...,T}$ is a strictly stationary, homogeneous Markov process, with transition density $f(x_t, x_{t-1}; A)$, and β-mixing coefficients satisfying $\beta_k = O(\rho^k), \rho < 1$. Similarly to the i.i.d. case, the minimum chi-squared estimator is defined by minimizing a chi-squared divergence between the conditional distribution in the family and its unconstrained kernel estimator

$$\hat{A}_T = \arg \min_{A \in \Theta \subset \mathcal{A}} Q_T(A) = \int_0^1 \int_0^1 \frac{[\hat{f}_T(x|y) - f(x|y; A)]^2}{\hat{f}_T(x|y)} w_T(x, y)\hat{f}_{Y,T}(y)dxdy.$$

The regularity conditions are similar to the i.i.d. case, considering the conditional distribution $f(x|y; A)$ instead of the joint one $f(x, y; A)$, the conditional differential operator $D_A \log f(X|Y; A)$, and the conditional information operator $I_{X|Y}$.

4.5 COPULA-BASED MARKOV PROCESSES: MIXING PROPERTIES

An important issue that arises is: how does the form of the copula C relate to the strength of temporal dependence in the Markov process? This is crucial because statistical inference requires the application of the central limit theorem for dependent data and some form of weak dependence has to be satisfied. Let $\{X_t, t \in \mathbb{Z}\}$ be a stationary sequence of real random variables defined on a probability space $(\Omega, \mathcal{F}, \mathbb{P})$ and for $s, t \in \mathbb{Z} \cup \{-\infty, \infty\}$, let $\mathcal{F}_s^t \subseteq \mathcal{F}$ denote the σ-field generated by the random variables $\{X_r : s \leq r \leq t\}$. The following definition is taken from Beare (2009).

Definition 4.5.1. The β-mixing coefficients $\{\beta_k : k \in \mathbb{N}\}$ corresponding to the sequence of random variables $\{X_t\}$ are given by

$$\beta_k = \frac{1}{2} \sup_{m \in \mathbb{Z}} \sup_{\{A_i\}, \{B_j\}} \sum_{i=1}^I \sum_{j=1}^J |\mathbb{P}(A_i \cap B_j) - \mathbb{P}(A_i)\mathbb{P}(B_j)|,$$

where the second supremum is taken over all finite partitions $\{A_1, \ldots, A_I\}$ and $\{B_1, \ldots, B_J\}$ of Ω such that $A_i \in \mathcal{F}_{-\infty}^m$ for each i and $B_j \in \mathcal{F}_{m+k}^\infty$ for each j.

We say that the process $\{X_t, t \in \mathbb{Z}\}$ is β-mixing if $\beta_k \to 0$ as $k \to \infty$. Beare (2009) gives sufficient conditions on the copula function C to ensure that a copula-based Markov process is β-mixing. Before introducing them, we recall the definition of the maximal correlation of the copula C.

Definition 4.5.2. The maximal correlation ρ_C of the copula C is defined as

$$\rho_C = \sup_{f,g} \left| \int_0^1 \int_0^1 f(u)g(v)C(du, dv) \right|,$$

where the supremum is taken over all $f, g \in \mathbb{L}^2([0, 1])$ such that $\int_0^1 f(u)du = \int_0^1 g(u)du = 0$ and $\int_0^1 f^2(u)du = \int_0^1 g^2(u)du = 1$.

The following theorem identifies sufficient conditions on C to ensure the β-mixing property.

Theorem 4.5.1 (Beare, 2009). *If the copula C is symmetric and absolutely continuous with square integrable density c and moreover if $\rho_C < 1$, then there exists $A < \infty$ and $\gamma > 0$ such that $\beta_k \leq Ae^{-\gamma k}$ for all k.*

The proof of the theorem uses the following preliminary bound:

$$\beta_k \leq \|c_k - 1\|_2,$$

where c_k is the density of the copula C_k between X_0 and X_k which is obtained, as known, by the product of Darsow *et al.* (1992):

$$C_k = C * C_{k-1}$$

for all k. Next, a spectral decomposition of c_k is used to show

$$\|c_k - 1\|_2 \leq \rho_C^{k-1} \|c - 1\|_2.$$

Let's remark that the geometric decay of $\|c_k - 1\|_2$ is even stronger than the β-mixing.

The condition $\rho_C < 1$ is satisfied for all absolutely continuous copulas with square integrable density such that c is positive a.e. on $[0, 1]$ as stated in Theorem 3.2 of Beare (2009). Commonly used parametric copula functions satisfying this condition include the FGM, Frank, and Gaussian copulas. Copulas exhibiting upper or lower tail dependence will not admit square integrable density and Theorem 4.5.1 cannot be applied directly. However, Chen *et al.* (2009) show that Markov processes generated via tail dependence copulas such as Clayton, Gumbel, and Student's t are geometric β-mixing. To prove this result they used the following definition of β-mixing.

Definition 4.5.3. (1) (Davydov, 1973) For a strictly stationary Markov process $\{X_t : t \in \mathbb{N}\}$, the β-mixing coefficients are given by

$$\beta_t = \int \sup_{0 \leq \phi \leq 1} |\mathbb{E}[\phi(X_{t+1})|X_1 = x_1] - \mathbb{E}[\phi(X_{t+1})]| dG_0(y),$$

where G_0 is the invariant marginal distribution of the process. $\{X_t\}$ is β-mixing if $\beta_t \to 0$ as $t \to \infty$ or is β-mixing with exponential decay rate if $\beta_t \leq \gamma e^{-\delta t}$ for some $\gamma > 0, \delta > 0$.

(2) (Chan and Tong, 2001) A strictly stationary Markov process $\{X_t : t \in \mathbb{N}\}$ is Harris ergodic if

$$\lim_{t \to \infty} \sup_{0 \leq \phi \leq 1} \mathbb{E}[\phi(X_{t+1})|X_1 = x_1] - \mathbb{E}[\phi(X_{t+1})]| = 0,$$

for almost all x_1; it is geometrically ergodic if there exists a measurable function W with $\int W(s)G_0(ds) < \infty$ and a constant $\kappa \in [0, 1[$ such that for all $t \geq 1$

$$\sup_{0 \leq \phi \leq 1} \mathbb{E}[\phi(X_{t+1})|X_1 = x_1] - \mathbb{E}[\phi(X_{t+1})] \leq \kappa^t W(x_1).$$

Definition 4.5.4. Let $\{X_t : t \in \mathbb{N}\}$ be an irreducible Markov process with transition measure $\mathbb{P}^n(x; A) = \mathbb{P}(X_{t+n} \in A|X_t = x)$, $n \geq 1$. A non-null set S is called small if there exists a positive integer n, a constant $b \geq 0$, and, a probability measure ν such that $\mathbb{P}^n(x; A) \geq b\nu(A)$ for all $x \in S$ and for all measurable sets A.

Theorem 4.5.2 *(Chan and Tong, 2001).* *Let $\{X_t : t \in \mathbb{N}\}$ be an irreducible and aperiodic Markov process. Suppose there exist a small set S and a non-negative measurable function L which is bounded away from 0 to ∞ on S and constants $r > 1$, $\gamma > 0$, $K > 0$ such that*

$$r\mathbb{E}[L(X_{t+1})|X_t = x] \leq L(x) - \gamma, \quad \forall x \in S^c$$

and

$$\int_{S^c} L(w)\mathbb{P}(x, dw) < K, \quad \forall x \in S.$$

Then $\{X_t\}$ is geometrically ergodic.

Another way to characterize the serial persistence of $\{X_t\}$ is given by ρ-mixing coefficients.

Definition 4.5.5. The ρ-mixing coefficients $\{\rho_k : k \in \mathbb{N}\}$ corresponding to the sequence $\{X_t : t \in \mathbb{N}\}$ are given by

$$\rho_k = \sup_{m \in \mathbb{Z}} \sup_{f,g} |Corr(f, g)|,$$

where the second supremum is taken over all square integrable random variables f and g measurable w.r.t. $\mathcal{F}_{-\infty}^m$ and $\mathcal{F}_{m+k}^\infty$ respectively, with positive and finite variance and where $Corr(f,g)$ denotes the correlation between f and g.

This definition was originally stated by Kolmogorov and Rozanov (1960), while the following definition of α-mixing has been heavily employed in econometrics.

Definition 4.5.6. The α-mixing coefficients $\{\alpha_k : k \in \mathbb{N}\}$ corresponding to the sequence $\{X_t : t \in \mathbb{N}\}$ are given by

$$\alpha_k = \sup_{m \in \mathbb{Z}} \sup_{A \in \mathcal{F}_{-\infty}^m, B \in \mathcal{F}_{m+k}^\infty} |\mathbb{P}(A \cap B) - \mathbb{P}(A)\mathbb{P}(B)|.$$

It was shown (see e.g. Proposition 3.11 in Bradley, 2007) that $2\alpha_k \leq \beta_k$ and $4\alpha_k \leq \rho_k$: this ensures that α-mixing is a weaker dependence condition than both β- and ρ-mixing:

$$\beta_k \to 0 \quad or \quad \rho_k \to 0 \quad \Rightarrow \quad \alpha_k \to 0, \quad k \to \infty.$$

Neither of the β- and ρ-mixing conditions is implied by the other: that is

$$\beta_k \to 0 \quad \nRightarrow \quad \rho_k \to 0$$

and

$$\rho_k \to 0 \quad \nRightarrow \quad \beta_k \to 0.$$

Central limit theorems are available for ρ-mixing processes that involve weaker assumptions on moments and mixing rates than do those available for β-mixing processes.

A simple condition on C is stated by the following theorem to ensure that the coefficients ρ_k decay at a geometric rate.

Theorem 4.5.3   *(Beare, 2009).*   *If $\rho_C < 1$, then there exist $A < \infty$ and $\gamma > 0$ such that $\rho_k \leq Ae^{-\gamma k}$ for all k.*

Beare (2009) (Theorem 4.2) provides a simple condition to verify that $\rho_C < 1$ for a specific copula function: the density of the absolutely part of C must be bounded away from zero on a set of measure one. Two examples of copulas satisfying this condition are given by the t-copula and Marshall–Olkin copula. Consequently, stationary Markov processes constructed using t-copulas or Marshall–Olkin copulas will satisfy ρ- and α-mixing conditions at geometric rate.

Unfortunately, the density of several copulas widely used in applied work, such as the Clayton and Gumbel copulas, is not bounded away from zero. For such copulas and for those for which ρ_C is unknown, one may seek to verify condition $\rho_C < 1$ numerically by using the following result.

Theorem 4.5.4   *(Beare, 2009).*   *Given a copula function C, for $n \in \mathbb{N}$, let K_n be the $n \times n$ matrix with $(i.j)$th element given by*

$$K_n(i, j) = C\left(\frac{i}{n}, \frac{j}{n}\right) - C\left(\frac{i}{n}, \frac{j-1}{n}\right) - C\left(\frac{i-1}{n}, \frac{j}{n}\right) + C\left(\frac{i-1}{n}, \frac{j-1}{n}\right) - \frac{1}{n^2}.$$

Let ϱ_n be the maximum eigenvalue of K_n. Then $n\varrho_n \to \rho_C$ as $n \to \infty$.

4.6   PERSISTENCE AND LONG MEMORY

Stationary Markov processes are regarded as canonical examples of short-memory stochastic processes because the value of such a process at a given time depends only on the value at the previous time. It can be interesting to study the long memory and the persistence in a time series, that is, the dependence between X_t and X_{t+h} with $h > 1$. As many authors have observed (see, among others, Embrechts *et al.* (2002), McNeil *et al.* (2005), Granger (2003)), the autocorrelation function is problematic in many settings – including the departure from Gaussianity and elliptic distributions that is common in financial market data – and not appropriate in describing persistence in unconditional distributions. A characterization of stochastic processes based on copulas is an approach to overcome these problems: for example, Ibragimov (2009) establishes necessary and sufficient copula-based conditions for Markov processes to exhibit m-dependence, r-dependence and conditional symmetry.

The definition of long memory processes $(X_t)_t$ is given by Granger (2003). He uses the (copula-linked) Hellinger measure of dependence $H(t, h) = H_{X_t, X_{t+h}}$ between r.v.s X_t and X_{t+h}:

$$H(t, h) = 1 - \int_{-\infty}^{+\infty} \int_{-\infty}^{+\infty} f_{X_t, X_{t+h}}^{1/2}(x, y) f_{X_t}^{1/2}(x) f_{X_{t+h}}^{1/2}(y) dx dy.$$

Definition 4.6.1. A process $(X_t)_t$ is a long memory process if for some constant $A > 0$

$$H(t, h) \sim Ah^{-p}, \quad h \to \infty,$$

where $p > 0$; it is a short memory process if

$$H(t, h) = O(e^{-Ah}), \quad h \to \infty.$$

Ibragimov and Lentzas (2009) study the long memory properties of copula-based stationary Markov processes using a number of dependence measures between X_t and X_{t+h} expressed in terms of their copula $C_{t,t+h}(u, v)$, which is obtained by iterating the product copula

$$C_{t,t+h} = C_{t,t+1} * C_{t,t+h-1} = C * C_{t,t+h-1}, \quad h = 2, \ldots,$$

where C is the invariant copula between two consecutive r.v.s of the process X_t and X_{t+1} for any t. Since the copula-based Markov process is stationary, every measure of temporal dependence is a function of h only. The most commonly used dependence measures and their expressions in terms of $C_{t,t+h}$ are the following (here $c_{t,t+h}$ is the density of $C_{t,t+h}$):

- $\phi^2(h) = \phi^2_{X_t, X_{t+h}} = \int_0^1 \int_0^1 c^2_{t,t+h}(u, v) du dv - 1$ (the Pearson's ϕ^2 coefficient);
- $\delta(h) = \delta_{X_t, X_{t+h}} = \int_0^1 \int_0^1 c_{t,t+h}(u, v) \log c_{t,t+h}(u, v) du dv$ (the relative entropy);
- $H(h) = H_{X_t, X_{t+h}} = \frac{1}{2} \int_0^1 \int_0^1 [c^{1/2}_{t,t+h} - 1]^2(u, v) du dv - 1$ (the Hellinger measure of dependence);
- $\kappa(h) = \kappa_{X_t, X_{t+h}} = \int_0^1 \int_0^1 [C_{t,t+h}(u, v) - uv] du dv$;
- $\lambda^2(h) = \lambda_{X_t, X_{t+h}} = \int_0^1 \int_0^1 [C_{t,t+h}(u, v) - uv]^2 du dv$;
- $v(h) = v_{X_t, X_{t+h}} = \sup_{u,v \in [0,1]} |C_{t,t+h}(u, v) - uv|$.

Example 4.6.1. *The particular case where the starting copula $C_{t,t+1}(u, v; \theta)$ is the Gaussian copula with the correlation coefficient θ is rather interesting. In fact, in this case, several authors explicitly computed some of the previous dependence measures. For example:*

- $\phi^2(1) = \theta^2/(1 - \theta^2)$ *and* $\delta(1) = -0.5 \log(1 - \theta^2)$ *(Joe, 1989);*
- $H(1) = 1 - \frac{(1-\theta^2)^{5/4}}{\left(1-\frac{\theta^2}{2}\right)^{3/2}}$ *(Granger et al., 2004);*
- $\kappa(1) = \frac{1}{2\pi} \arcsin(|\theta|/2)$ *and* $v(1) = \frac{1}{2\pi} \arcsin(|\theta|)$ *(Schweizer and Wolff, 1981).*

On the other hand, if $C_{t,t+1}(u, v; \theta)$ is an FGM copula we have $\phi^2(1) = \lambda^2(1) = \frac{\theta^2}{9}$, $\kappa(1) = \frac{|\theta|}{4}$, and $v(1) = |\theta|$.

We say that the process (X_t) exhibits ϕ^2-short memory if, for some constant $A > 0$,

$$\phi^2(h) = O(e^{-Ah}),$$

or exhibits ϕ^2-long memory if

$$\phi^2(h) \sim Ah^{-p},$$

with $p > 0$. The same definitions of short and long-memory are available for the other dependence measures. Proposition 1 in Ibragimov and Lentzas (2009) shows that ϕ^2-short memory is stronger than the other, whereas the following theorem shows when a copula-based Markov process exhibits short memory.

Theorem 4.6.1 (*Ibragimov and Lentzas, 2009*). *Let $(X_t)_t$ be a stationary Markov process based on a Gaussian copula $C^G(u, v; \theta)$ on an EFGM copula $C^{EFGM}(u, v; \theta)$. Then the process exhibits short memory in the sense that its measures of dependence satisfy, for some constant $A > 0$,*

$$\phi^2(h), \delta(h), H(h), \kappa(h), \lambda^2(h), v(h) = O(e^{-Ah}), \quad h \to \infty.$$

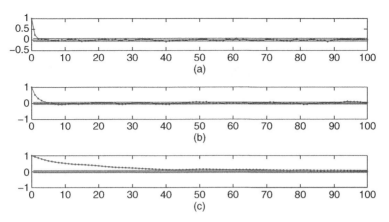

Figure 4.21 Autocorrelation function of a simulated copula-based Markov process generated by a Clayton copula with three different levels of dependence: (a) $\tau = 0.2$, (b) $\tau = 0.4$, (c) $\tau = 0.8$.

As for the Clayton copula, the authors empirically show that if the Markov process is based on the Clayton copula it exhibits long memory.

Example 4.6.2. *Consider again the simulation of a stationary copula-based Markov process. Let $(x_1, \ldots, x_n)$ be the generated sample. For the sake of illustration, we compute the autocorrelation function*

$$\rho_{x_{t-h}, x_t} = \frac{Cov(x_{t-h}, x_t)}{\sqrt{Var(x_{t-h})Var(x_t)}}$$

for different choices of copula functions with different parameters to capture the effect on the persistence. Figure 4.21 reports the autocorrelation function for the case where the process is generated by the Clayton copula and Figure 4.22 compares three different copulas (Clayton,

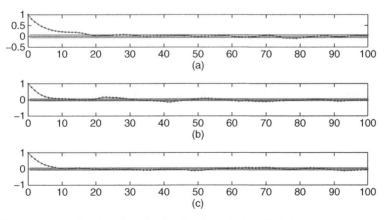

Figure 4.22 Autocorrelation function of a simulated copula-based Markov process generated by three copula functions with the same dependence structure given by Kendall's τ equal to 0.6: (a) Clayton, (b) Frank, (c) Gaussian.

Frank, and Gaussian). The Clayton copula seems to generate more persistence than Frank and Gaussian copulas.

4.7   *C*-CONVOLUTION-BASED MARKOV PROCESSES: THE LIKELIHOOD FUNCTION

In this section we construct the likelihood function of a Markov process generated by the *C*-convolution. A *C*-convolution-based Markov process $X = (X_t)_{t \geq 1}$ is *not stationary* because its distribution F_{X_t}, say F_t, is obtained by an iterative application of the *C*-convolution operator (Proposition 3.4.1) given the distribution of increments F_{Δ_t} and the copula between X_{t-1} and ΔX_t, say C_t:

$$F_t = F_{t-1} \overset{C_t}{*} F_{\Delta_t}, \quad t = 1, 2, \ldots$$

This is the key difference between a (strictly stationary) copula-based Markov process and the *C*-convolution-based Markov process. However, the study of the temporal dependence structure induced by this model is rather difficult, and is beyond the scope of this book.

As a particular case, we can consider the fact that the distribution of increments is stationary, $F_{\Delta_t} = F_\Delta$, for all t. Then, since $F_{X_1} = F_\Delta$ the DGP of this model is $(F_\Delta^*, C_t(\cdot, \cdot, ; \theta^*))$. We have in fact a *stationary increments* Markov process.

Now, consider the construction of the likelihood function of a *C*-convolution-based Markov process. Since X is Markov, the likelihood function needs the transition probability of X at the points t and $t-1$. Denote such a probability by $\mathbb{P}_t(x_t|x_{t-1}; \theta) = \mathbb{P}(X_t \leq x_t | X_{t-1} = x_{t-1}; \theta)$. We emphasize the dependence on the copula parameter θ, which plays a crucial role in estimation. We assume that $\mathbb{P}_t(x_t|x_{t-1}; \theta)$ has density $f_t(x_t|x_{t-1}; \theta)$ for each θ and for $t = 1, \ldots, T$. Denote by $(x_1, \ldots, x_T)$ a sample from X. Then, it is easy to construct the log-likelihood function of $(x_1, \ldots x_T)$; it is given by

$$\ell(\theta) = \log f(x_1; \theta) + \sum_{t=2}^{T} \log f_t(x_t|x_{t-1}; \theta),$$

where $f(x_1; \theta)$ is the marginal density of X_1. Since the first term is negligible (see Billingsley, 1986), the log-likelihood is simply given by

$$\ell(\theta) = \sum_{t=2}^{T} \log f_t(x_t|x_{t-1}; \theta).$$

But, the transition density $f_t(x_t|x_{t-1}; \theta)$ is the joint density between X_{t-1} and X_t divided by the marginal density of X_{t-1}:

$$f_t(x_t|x_{t-1}; \theta) = \frac{f_{X_{t-1}, X_t}(x_{t-1}, x_t; \theta)}{f_{X_{t-1}}(x_{t-1}; \theta)}, \quad t = 2, \ldots, T.$$

Moreover, the joint density $f_{X_{t-1}, X_t}(x_{t-1}, x_t; \theta)$ is

$$f_{X_{t-1}, X_t}(x_{t-1}, x_t; \theta) = c_{X_{t-1}, X_t}(F_{X_{t-1}}(x_{t-1}), F_{X_t}(x_t); \theta) f_{X_{t-1}}(x_{t-1}; \theta) f_{X_t}(x_t; \theta),$$

and hence

$$f_t(x_t|x_{t-1}; \theta) = c_{X_{t-1}, X_t}(F_{X_{t-1}}(x_{t-1}), F_{X_t}(x_t); \theta) f_{X_t}(x_t; \theta), \quad t = 2, \ldots, T.$$

Now, the three factors of $f_{X_t, X_{t-1}}(x_{t-1}, x_t; \theta)$ may be recovered by applying iteratively the C-convolution operator in Proposition 3.4.1. In fact, we have the distribution of X_t (for notational simplicity we omit the dependence on θ)

$$F_{X_t}(x_t) = \int_0^1 D_1 C_t \left(w, F_{\Delta X_t}(x_t - F_{X_{t-1}}^{-1}(w)) \right) dw, \quad t = 2, \ldots, T,$$

the density of X_t

$$f_{X_t}(x_t) = \int_0^1 c_t \left(w, F_{\Delta X_t}(x_t - F_{X_{t-1}}^{-1}(w)) \right) f_{\Delta X_t}(x_t - F_{X_{t-1}}^{-1}(w)) dw, \quad t = 2, \ldots, T,$$

the copula between the levels X_{t-1} and X_t

$$C_{X_{t-1}, X_t}(u, v) = \int_0^u D_1 C_t \left(w, F_{\Delta X_t}(F_{X_t}^{-1}(v) - F_{X_{t-1}}^{-1}(w)) \right) dw, \quad t = 2, \ldots, T,$$

and its density

$$c_{X_{t-1}, X_t}(u, v) = c_t \left(u, F_{\Delta X_t}(F_{X_t}^{-1}(v) - F_{X_{t-1}}^{-1}(u)) \right) \frac{f_{\Delta X_t}(F_{X_t}^{-1}(v) - F_{X_{t-1}}^{-1}(u))}{f_{X_t}(F_{X_t}^{-1}(v))}, \quad t = 2, \ldots, T.$$

Then, we have

$$c_{X_{t-1}, X_t}(F_{X_{t-1}}(x_{t-1}), F_{X_t}(x_t)) = c_t \left(F_{X_{t-1}}(x_{t-1}), F_{\Delta X_t}(\Delta x_t) \right) \frac{f_{\Delta X_t}(\Delta x_t)}{f_{X_t}(x_t)}, \quad t = 2, \ldots, T.$$

Finally, the transition densities are

$$f_t(x_t | x_{t-1}; \theta) = c_t \left(F_{X_{t-1}}(x_{t-1}), F_{\Delta X_t}(\Delta x_t); \theta \right) f_{\Delta X_t}(\Delta x_t), \quad t = 2, \ldots, T.$$

Notice that in our C-convolution-based Markov process the transition densities involve the distribution of X_{t-1} and the distribution of the next increment ΔX_t.

So, the log-likelihood of the process is the following:

$$\ell(\theta) = \sum_{t=2}^{T} \log f_t(x_t | x_{t-1}; \theta) = \sum_{t=1}^{T} \log \left[c_t \left(F_{X_{t-1}}(x_{t-1}; \theta), F_{\Delta X_t}(\Delta x_t); \theta \right) f_{\Delta X_t}(\Delta x_t) \right]$$

$$\propto \sum_{t=1}^{T} \log \left[c_t \left(F_{X_{t-1}}(x_{t-1}; \theta), F_{\Delta X_t}(\Delta x_t); \theta \right) \right].$$

We now give some examples.

Example 4.7.1. *Suppose that C_t is a Clayton copula for each t: $C_t(u, v; \theta) = (u^{-\theta} + v^{-\theta} - 1)^{-\frac{1}{\theta}}$, $\theta \in]0, +\infty[$. The distribution of X_t is given by*

$$F_{X_t}(x_t; \theta) = \int_0^1 D_1 C_t(w, F_{\Delta X_t}(x_t - F_{X_{t-1}}^{-1}(w; \theta)); \theta) dw$$

$$= \int_0^1 w^{-\theta-1}(w^{-\theta} + [F_{\Delta X_t}(x_t - F_{X_{t-1}}^{-1}(w; \theta))]^{-\theta} - 1)^{-\frac{1}{\theta}-1} dw, \quad t = 1, \ldots, T,$$

whereas the transition densities are

$$f_t(x_t|x_{t-1};\theta) = (1+\theta)[F_{X_{t-1}}(x_{t-1};\theta)]^{-\theta-1} \left([F_{X_{t-1}}(x_{t-1};\theta)]^{-\theta} + [F_{\Delta X_t}(\Delta x_t)]^{-\theta} - 1\right)^{-\frac{1}{\theta}-2}$$
$$\times [F_{\Delta X_t}(\Delta x_t))]^{-\theta-1} f_{\Delta X_t}(\Delta x_t), \quad t = 2,\dots,T.$$

Finally, the log-likelihood is

$$\ell(\theta) \propto \sum_{t=2}^{T} \Bigg\{ -(1+\theta)\log F_{\Delta X_t}(\Delta x_t)$$
$$-\left(\frac{1}{\theta}+1\right)\log\left([F_{X_{t-1}}(x_{t-1};\theta)]^{-\theta} + [F_{\Delta X_t}(\Delta x_t)]^{-\theta} - 1\right) - (1+\theta)\log F_{X_{t-1}}(x_{t-1};\theta) \Bigg\}.$$

Example 4.7.2. *Suppose that C_t is a Frank copula for each t:* $C_t(u,v;\theta) = -\frac{1}{\theta}\log\left(1 + \frac{(e^{-\theta u}-1)(e^{-\theta v}-1)}{e^{-\theta}-1}\right)$, $\theta \in]0,+\infty[$. *Then*

$$F_{X_t}(x_t;\theta) = \int_0^1 \frac{e^{-\theta w}(e^{-\theta F_{\Delta X_t}(x_t-F_{X_{t-1}}^{-1}(w;\theta))} - 1)}{(e^{-\theta}-1) + (e^{-\theta w}-1)(e^{-\theta F_{\Delta X_t}(x_t-F_{X_{t-1}}^{-1}(w;\theta))} - 1)} dw, \quad t = 2,\dots,T,$$

and

$$f_t(x_t|x_{t-1};\theta) = \frac{\theta e^{-\theta F_{X_{t-1}}(x_{t-1};\theta)} f_{\Delta X_t}(\Delta x_t)(e^{-\theta F_{\Delta X_t}(\Delta x_t)} - 1)(1 - e^{-\theta})}{[(e^{-\theta}-1) + (e^{-\theta F_{X_{t-1}}(x_{t-1};\theta)} - 1)(e^{-\theta F_{\Delta X_t}(\Delta x_t)} - 1)]^2}, \quad t = 2,\dots,T.$$

The log-likelihood function is then

$$\ell(\theta) \propto T\log\theta + T\log(1-e^{-\theta}) + \sum_{i=2}^{T} \Big\{ -\theta(F_{X_{t-1}}(x_{t-1};\theta) + F_{\Delta X_t}(\Delta x_t))$$
$$- \log[(e^{-\theta}-1) + (e^{-\theta F_{X_{t-1}}(x_{t-1};\theta)} - 1)(e^{-\theta F_{\Delta X_t}(\Delta x_t)} - 1)]^2 \Big\}.$$

Example 4.7.3. *Suppose that C_t is a Gaussian copula for each t:*

$$C_t(u,v;\theta) = \int_{-\infty}^{\Phi^{-1}(u)} \int_{-\infty}^{\Phi^{-1}(v)} \frac{1}{2\pi\sqrt{1-\theta^2}} e^{\frac{2\theta xy - x^2 - y^2}{2(1-\theta^2)}} dx dy.$$

We have

$$F_{X_t}(x_t;\theta) = \int_0^1 N_{\theta\Phi^{-1}(w),(1-\theta^2)}(\Phi^{-1}(F_{\Delta X_t}(x_t - F_{X_{t-1}}^{-1}(w;\theta))))dw, \quad t = 2,\dots,T,$$

where $N_{\theta\Phi^{-1}(w),(1-\theta^2)}(\cdot)$ denotes a Gaussian distribution with mean $\theta\Phi^{-1}(w)$ and variance $(1-\theta^2)$;

$$f_t(x_t|x_{t-1};\theta) = f_{\rho\Phi^{-1}(F_{X_{t-1}}(x_{t-1};\theta)),(1-\theta^2)}(\Phi^{-1}(F_{\Delta X_t}(\Delta X_t))) \frac{f_{\Delta X_t}(\Delta x_t)}{\phi(\Phi^{-1}(F_{\Delta X_t}(\Delta x_t)))},$$
$$t = 2,\dots,T,$$

where $f_{\theta\Phi^{-1}(w),(1-\theta^2)}(\cdot)$ denotes the density function of $N_{\theta\Phi^{-1}(w),(1-\theta^2)}(\cdot)$. Then,

$$\ell(\theta) \propto \sum_{t=2}^{T} \log f_{\rho\Phi^{-1}(F_{X_{t-1}}(x_{t-1};\theta)),(1-\theta^2)}(\Phi^{-1}(F_{\Delta X_t}(\Delta x_t))).$$

Example 4.7.4. *As an example, we provide estimates of the parameter of the copula function linking X_{t-1} (lagged log-prices) and ΔX_t (log-returns) in the spirit of Example 4.3.2 for the same market indices: NIKKEI225, FTSE100, and NAS100. Notice that this is not the estimation of a C-convolution-based process because here the stationarity of the distributions of X_{t-1} and of ΔX_t is supposed whereas we know that the C-convolution generates a non-stationary Markov process. Therefore, the example has to be considered as a preliminary study of temporal dependencies between the levels and the increments. Table 4.5 contains semi-parametric estimates and the relative standard errors for Clayton, Frank, and Gaussian copulas, whereas Figure 4.23 depicts the dependence between lagged log-prices and log-returns for the three data sets.*

Table 4.5 Semi-parametric estimates of the (temporal) dependence structure between log-prices and log-returns for three market indices: NIKKEI225, FTSE100, and NAS100

	NIKKEI225	FTSE100	NAS100
Clayton	$0.0201\ (3.5748 \times 10^{-4})$	$0.0322\ (3.8481 \times 10^{-4})$	$0.0000\ (0.0000)$
Frank	$-0.2622\ (0.0022)$	$-0.2724\ (0.0024)$	$-0.2633\ (0.0028)$
Gaussian	$-0.0558\ (5.3757 \times 10^{-5})$	$-0.0655\ (1.2821 \times 10^{-5})$	$-0.0502\ (8.5456 \times 10^{-5})$

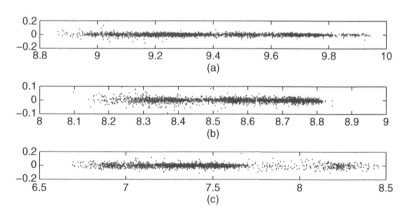

Figure 4.23 Dependence structure between log-prices and log-returns: (a) NIKKEI225; (b) FTSE100; (c) NAS100

5

Multivariate Equity Products

The main target of this chapter is to apply the copula-based representation of Markov processes discussed in Chapters 3 and 4 to the problem of pricing multivariate equity derivative contracts. The trading activity on equity markets has witnessed a trend that has progressively changed the focus from different elements of the probability distribution of returns. Using statistical concepts, it has been a progressive run to exploit higher and multidimensional moments of the distribution. The starting point of trading styles is based on exploitation of the first moment of the distribution, placing bets on expected excess returns. This is nowadays called *alpha* generating activity, or directional trading, which is bound to exploit information on excess returns of specific stocks or indices. While this remains the main activity in the asset management industry, investors active in the trading floors on higher frequencies have moved to exploit non-normality of the returns, which is particularly evident at the daily and intra-daily level. This new investment style has been implemented focusing on trading higher moments of the distribution. We have then seen a rapid development of volatility trading techniques and option trading strategies to exploit changes of even higher-order moments (the *trade-the-skew* strategy is the most well-known example). On another line of development, moments of higher-dimensional distribution have also become the focus of attention of traders and sophisticated investors. These developments have been accompanied by a progress of financial innovation at unprecedented pace. Volatility and correlation swaps on the one side and basket equity derivatives on the other are the most obvious examples of innovative products. On the side of quantitative techniques that have accompanied these processes, the switch to higher moments trading has spurred research into the application of more sophisticated continuous time processes to finance, such as Lévy processes or additive processes, while the focus on high-dimensional probabilities has motivated research on copula function applications. These developments evolved on clearly separated paths and never merged together, and did not even approach each other. The method focused on stochastic processes was applied to strategies bound to exploit changes in volatility, and more generally in the dynamics of univariate prices, while copula functions were entirely dedicated to the definition of strategies bound to exploit changes in correlation. The purpose of this chapter is to show that the latter tool, copula functions, can be applied successfully to both volatility and correlation trading. The reader interested in a more detailed analysis of standard applications to multivariate option pricing may be referred to Cherubini *et al.* (2004) for a review.

5.1 MULTIVARIATE EQUITY PRODUCTS

In standard jargon, we refer to multivariate or basket equity derivative products for option contracts (or portfolios of option contracts) which have a multivariate set of assets, namely stocks, as the underlying. The prices of these assets are linked by a payoff function that is defined in the contract, and that determines the amount to be paid or received by the parties by which it was underwritten. If we define a set of assets $\{S^1, \ldots, S^m\}$ and a set of times $\{t_1, \ldots, t_n\}$, the value of each of these contracts can be specified by defining the payoff function linking the prices of assets at each relevant date. In this chapter, we will adopt the notation S_j^i to denote the price of asset i at time t_j, and X_j^i to represent the logarithm of this price. When we talk about multivariate equity derivatives in a strict sense, we typically think of a contract which, at a given final date t_n, pays a sum which is computed on the basis of the prices of the underlying assets at that time, namely $f(S_n^1, \ldots, S_n^m)$. We call these products *European* multivariate equity derivatives and we will see how to apply copulas to this problem, and the application is standard. However, it is easy to see that the problem of pricing a product whose value depends on the price of a single asset at different times is conceptually identical, and the payoff can be represented by the function $f(S_1^i, \ldots, S_n^i)$. From our point of view, products like this one, which are actually called *path-dependent* contracts, could be considered as multivariate contracts, where the term "multivariate" is intended in a temporal dependence sense. The same copula function techniques could then fruitfully be applied to these products as well, once the proper copula function is chosen. Finally, the multivariate products that we find in real markets are somewhat mixed, with the payoff function involving the prices of baskets of assets at several dates. These products, that we could call *multivariate path-dependent* equity derivatives, are the most difficult to price, because they have larger dimensions than pure multivariate and path-dependent derivatives, and these dimensions include both a spatial and a temporal dependence perspective.

5.1.1   European Multivariate Equity Derivatives

European multivariate equity derivatives can be classified according to the specific shape of the payoff function $f(S_n^1, \ldots, S_n^m)$ defining the value of the contract at the expiration date t_n. A way to make this taxonomy easier is to consider options (call and put), and an aggregation function to collect the underlying assets into a single reference price for the computation of the payoff. Formally, we specify the payoff function as

$$f\left(S_n^1, \ldots, S_n^m\right) = \max\left(\omega\left(g\left(S_n^1, \ldots, S_n^m\right) - K\right), 0\right),$$

where K is the strike price of the option, ω is a binary variable defining whether the option is of the call type ($\omega = 1$) or put ($\omega = -1$). Function $g(.)$ is the aggregation function characterizing the multivariate contract. The simplest example that comes to mind is to specify the aggregation function as the sum, yielding *basket options*:

$$f(S_n^1, \ldots, S_n^m) = \max\left(\omega\left(\sum_{i=1}^m \frac{S_n^i}{m} - K\right), 0\right).$$

Example 5.1.1. *Spread options. As an example of a European multivariate equity derivative, we consider a simple spread option in the bivariate case. In particular, we simulate a bivariate stationary Markov process $(S^1, S^2) = (S_t^1, S_t^2)_{t=1,\ldots,T}$ in the spirit of Definition 4.2.1. That is:*

- *Each marginal process is a strictly stationary copula-based Markov process, so S_t^1 is generated by $(G_1, C_1(\cdot, \cdot; \theta))$ and S_t^2 is generated by $(G_2, C_2(\cdot, \cdot; \theta))$, where the margins G^1 and G^2 are Gaussian with mean and variance calibrated on daily log-prices.*
- *The marginal components S_t^1 and S_t^2 are linked by a (cross-sectional) invariant Student's t copula with parameters ρ and ν, $C_{1,2}(\cdot, \cdot; \rho, \nu)$.*

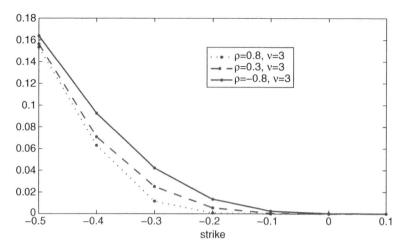

Figure 5.1   Monte Carlo price of a spread option at maturity $T = 25$ days (about 1 month) when C_1 and C_2 are Clayton copulas with $\theta = 8$ and $C_{1,2}$ is a Student's t copula.

We compute the price of a spread option on $(S_t^1, S_t^2)_{t=1,...,T}$ whose payoff at the expiration date is

$$SP_T = \max \left(S_T^1 - S_T^2 - K, 0 \right).$$

In Figures 5.1 and 5.2 we can observe the price dependence (as a function of the strike K) from the parameter of a cross-sectional copula function in the case of a spread on the levels for different maturity choices.

Example 5.1.2. *Basket equity note.* *Consider a five-year bond paying a single coupon at maturity yielding the arithmetic average return of a basket of three indices (S&P 500 for the*

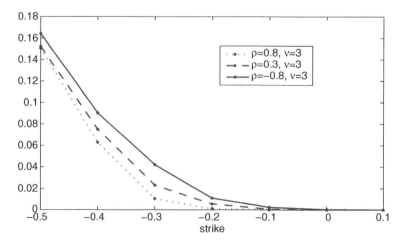

Figure 5.2   Monte Carlo price of a spread option at maturity $T = 125$ days (about 6 months) when C_1 and C_2 are Clayton copulas with $\theta = 8$ and $C_{1,2}$ is a Student's t copula.

USA, Nikkei 225 for Japan, and Eurostoxx 100 for Europe) with a guaranteed return equal to r_g:

$$\max \left(\sum_{i=1}^{m} \frac{s_n^i}{m}, 1 + r_g \right),$$

where s_n^i *denotes index i rescaled by its value at the origin of the contract (time* t_0):

$$s_n^i = \frac{S_n^i}{S_0^i}.$$

It is clear to see that this product can be decomposed into a bond paying interest r_g *at maturity and a long position in a basket call option:*

$$\max \left(\sum_{i=1}^{m} \frac{s_n^i}{m}, 1 + r_g \right) = 1 + r_g + \max \left(\sum_{i=1}^{m} \frac{s_n^i}{m} - (1 + r_g), 0 \right).$$

Other specifications of the aggregation function that are very common to find in the equity market are those assigning the minimum or maximum of the prices in the basket. These give rise to the so-called *rainbow options*, as they are called in market jargon. These options are also called call on min, put on max, and the like. To assess whether each of these products is long or short correlation, simply remember the AND/OR rule spelled out in Chapter 2. In structured finance, it is particularly frequent to find examples of call on min options, that today are also called Everest options after the name of the corresponding note.

Example 5.1.3. *Everest note. Consider a five-year bond paying a single coupon at maturity yielding the minimum return on a basket of three indices (S&P 500 for the USA, Nikkei 225 for Japan, and Eurostoxx 100 for Europe) with a guaranteed return equal to* r_g:

$$\max \left(\min_{1 \le i \le m} s_n^i, 1 + r_g \right),$$

where again $s_n^i = \frac{S_n^i}{S_0^i}$. *It is immediate to see that this product can be decomposed into a bond paying interest* r_g *at maturity and a long position in a call on min option:*

$$\max \left(\min_{1 \le i \le m} s_n^i, 1 + r_g \right) = 1 + r_g + \max \left(\min_{1 \le i \le m} s_n^i - (1 + r_g), 0 \right).$$

As a further generalization, one can extend the representation of the payoff function beyond the form $\max(x - y, 0)$ that denotes plain vanilla options. A choice that is widely used is that of digital options, like in the examples in Chapter 2. This leads to the class of products called *Altiplanos*.

Example 5.1.4. *Altiplano note. Consider a bond paying a sequence of coupons yielding a fixed return if the minimum of a basket of three indices (S&P 500 for the USA, Nikkei 225 for Japan, and Eurostoxx 100 for Europe) is higher than a given threshold B:*

$$r \cdot \mathbf{1}_{(B,+\infty)} \left(\min_{1 \le i \le m} s_n^i \right),$$

where $\mathbf{1}_{(B,+\infty)}(x)$ *is the indicator function returning value 1 if* $x > B$ *and 0 otherwise, and r is the coupon yield.*

Notice that differently from the previous cases, this structured product is endowed with a coupon stream instead of a single coupon paid at maturity. With respect to the other products, then, the Altiplano note foreshadows the issue of consistency of the prices of coupons, which is an issue of temporal consistency.

5.1.2 Path-dependent Equity Derivatives

For path-dependent products we could use the same taxonomy as that used for multivariate European products. We remember that the payoff function is now $f(S_1^i, \ldots, S_n^i)$. Breaking the function in the plain vanilla option payoff and an aggregation function $g(.)$, we have

$$f\left(S_1^i, \ldots, S_n^i\right) = \max\left(\omega\left(g\left(S_1^i, \ldots, S_n^i\right) - K\right), 0\right),$$

with K the strike price of the option and ω a binary variable defining option type as in the previous section. If the aggregating function is an average, we obtain *Asian options*. In the particular case of arithmetic average we have

$$\max\left(\omega\left(\sum_{j=1}^{n} \frac{S_j^i}{n} - K\right), 0\right).$$

Using the aggregating functions max(.) and min(.), we generate instead what are called *lookback* options:

$$\max\left(\omega\left(\max_{j}\left(S_j^i\right) - K\right), 0\right).$$

Furthermore, using the digital payoff function instead of the plain vanilla one we recover the pricing kernel of *barrier options*. So, for example, we have

$$\mathbf{1}_{(B,+\infty)}\left(\min_{j}\left(S_j^i\right)\right)$$

for barrier options of the *down* type, and

$$\mathbf{1}_{(-\infty,B]}\left(\max_{j}\left(S_j^i\right)\right)$$

for those of the *up* type.

Notice that path-dependent derivatives could actually be classified according to the kind of aggregation function used in the payoff definition. For these purposes two concepts are particularly relevant, namely those of *running maximum* and *running minimum*. Here we report the broadest definition, referred to our sets of assets and times:

$$M^i(t_r, t_j) \equiv \max\left\{S_k^i; t_r \leq t_k \leq t_j\right\},$$
$$m^i(t_r, t_j) \equiv \min\left\{S_k^i; t_r \leq t_k \leq t_j\right\},$$

where M^i and m^i denote the running maximum and minimum respectively of asset S^i. Notice that the definition refers to a period of time with respect to which the running minimum or maximum is measured. In the general case, when this period starts with the origin, that is t_1, we may omit the starting date. Anyway, it is not unusual to find cases in which reference to the first date cannot be avoided in the model of the pricing problem.

Notice that, as for the running maximum and minimum of the price, we could actually define a new variable representing application of the aggregation operator to the sequence of prices. So, we could for example define

$$\overline{S}^i(t_r, t_k) \equiv \sum_{j=r}^{k} \frac{S_j^i}{k - r + 1}, \tag{5.1}$$

where $\overline{S}^i(t_r, t_k)$ is the average from time t_r to time t_k. Again, we may simplify the notation by assuming that the first index is dropped if $t_r = t_1$. Actually, this is used to recover the relationship between newly issued Asian options and options issued r periods earlier.

Example 5.1.5. *Seasoned Asian options. Consider an Asian option issued at time t_0 with strike price equal to K and with r observations of the n already known. The problem is to evaluate the option. The payoff of the option issued at the starting date is a function of*

$$\overline{S}^i(t_1, t_n) - K = \sum_{j=1}^{n} \frac{S_j^i}{n} - K.$$

But, given an intermediate date t_r, this can be decomposed as

$$
\begin{aligned}
\overline{S}^i(t_1, t_n) - K &= \frac{r-1}{n} \overline{S}^i(t_1, t_{r-1}) + \frac{n-r+1}{n} \overline{S}^i(t_r, t_n) - K \\
&= \frac{n-r+1}{n} \left(\frac{r-1}{n-r+1} \overline{S}^i(t_1, t_{r-1}) + \overline{S}^i(t_r, t_n) - \frac{n}{n-r+1} K \right) \\
&= \frac{n-r+1}{n} \left(\overline{S}^i(t_r, t_n) - K_{r,k}^* \right),
\end{aligned}
$$

where

$$K_{r,n}^* = K + \frac{r-1}{n-r+1} \left(K - \overline{S}^i(t_1, t_{r-1}) \right).$$

Using the representation of running maxima and minima, it is also useful to represent products that are bound to exploit both spatial and temporal dependence.

Example 5.1.6. *Barrier Altiplano note. Consider a bond paying a sequence of coupons yielding a fixed return if the minimum of a basket of three indices (S&P 500 for the USA, Nikkei 225 for Japan, and Eurostoxx 100 for Europe) is higher than a given threshold B. The minimum across the assets is computed each year taking the minimum end-of-month values of the year. It is clear that this product is written on a basket including the running minima of the assets:*

$$r \cdot \mathbf{1}_{(B,+\infty)} \left(\min_i \left(\frac{m^i(t_{j-12}, t_j)}{S_0^i} \right) \right),$$

where r is the coupon yield.

5.2 RECURSIONS OF RUNNING MAXIMA AND MINIMA

We now show how to design a recursion of running maxima and minima of a univariate series. For the time being we assume we are endowed with information about:

- A sequence of prices of digital options with strike K and maturity t_j, $j = 1, 2, \ldots, n$, denoted $DC(t_j)$ for call options and $DP(t_j)$ for put options.
- A sequence of copula functions linking the prices of assets at two adjacent dates. We would identify these copulas from the subscripts, i.e. $C_{X_j, X_{j+1}}$.

From the first piece of information we recover the implied probability distribution of the underlying asset at each date. From the second we can specify the entire dynamics of its price. This is due to the fact that prices are assumed to be Markovian, consistent with the efficient market hypothesis. Actually, the Markovian requirement is not enough for our purposes, since it is well known that when we are facing a pricing problem, we must assume we are working under a risk measure such that asset prices are endowed with the martingale requirement. We will address this problem below. For now, the Markov assumption is sufficient to generate a very useful recursion between running minima and running maxima.

Let us first take the running maxima $M(t_j)$. Let us denote $F_j(K) = \mathbb{P}(S_j \leq K)$. These can be extracted from the prices of digital put options $DP(t_j)$ by dividing them by the risk-free discount factor.

We know that, because of the Markov property, the dynamics of asset prices can be represented as

$$
\begin{aligned}
\mathbb{P}(M(t_j) \leq K) &= \mathbb{P}(S_1 \leq K, S_2 \leq K, \ldots, S_j \leq K) \\
&= C_{S_1, S_1, \ldots, S_j}(F_1(K), F_2(K), \ldots, F_j(K)) \\
&= C_{S_1, S_2} \star C_{S_2, S_3} \cdots \star C_{S_{j-1}, S_j}(F_1(K), F_2(K), \ldots, F_j(K)).
\end{aligned}
$$

We also know that the $\star$ product operator is endowed with the associative property, that immediately yields a recursion formula:

$$
\begin{aligned}
\mathbb{P}(M(t_j) \leq K) &= C_{S_1, S_1, \ldots, S_{j-1}} \star C_{S_{j-1}, S_j}(F_1(K), F_2(K), \ldots, F_j(K)) \\
&= \int_0^{F_{j-1}(K)} D_{j-1} C_{S_1, S_1, \ldots, S_{j-1}}(F_1(K), \ldots, F_{j-2}(K), t) D_1 C_{S_{j-1}, S_j}(t, F_j(K)) dt \\
&= \int_0^{F_{j-1}(K)} \mathbb{P}(M(t_{j-2}) \leq K | S_{j-1} = t) D_1 C_{S_{j-1}, S_j}(t, F_j(K)) dt.
\end{aligned}
$$

The same result holds for the running minimum, of course. In this case we are interested in the survival copula, namely

$$
\mathbb{P}(m(t_j) > K) = \hat{C}_{S_1, \ldots, S_j}(1 - F_1(K), 1 - F_2(K), \ldots, 1 - F_j(K)).
$$

We may use an easy to prove result stating that the survival copula of a $\star$ product is equal to the $\star$ product of survival copulas, so that we have

$$
\mathbb{P}(m(t_j) > K) = \hat{C}_{S_1, S_2} \star \hat{C}_{S_2, S_3} \cdots \star \hat{C}_{S_{j-1}, S_j}(1 - F_1(K), 1 - F_2(K), \ldots, 1 - F_j(K))
$$

and the recursion is

$$
\begin{aligned}
\mathbb{P}(m(t_j) > K) &= \hat{C}_{S_1, S_1, \ldots, S_{j-1}} \star \hat{C}_{S_{j-1}, S_j}(1 - F_1(K), 1 - F_2(K), \ldots, 1 - F_j(K)) \\
&= \int_0^{1-F_{j-1}(K)} D_{j-1} \hat{C}_{S_1, S_1, \ldots, S_{j-1}}(1 - F_1(K), \ldots, 1 - F_{j-2}(K), t) D_1 \hat{C}_{S_{j-1}, S_j}(t, 1 - F_j(K)) dt \\
&= \int_0^{1-F_{j-1}(K)} \mathbb{P}(m(t_{j-2}) > K | S_{j-1} = t) D_1 \hat{C}_{S_{j-1}, S_j}(t, 1 - F_j(K)) dt.
\end{aligned}
$$

A particularly interesting case is that of symmetric Markov processes, which require

- symmetric marginal distributions,
- radially symmetric copulas.

These recursions can be used to recover the entire term structure of the prices of barrier options with an algorithm reminiscent of the bootstrapping procedure used to recover the term structure of interest rates from coupon bond prices. We describe it below.

Remark 5.2.1. *No-touch options bootstrapping* Assume we want to recover the entire term structure of *no-touch* options, that is options paying a fixed sum if the asset remains above a given threshold on a set of dates $\{t_1, t_2, \ldots, t_n\}$. We observe digital put options $DP(t_j)$ on each date and the term structure of risk-free discount factors $v(t_j)$. Furthermore, the temporal dependence structure is assumed to remain the same throughout the sample: $C_{S_{j-1}, S_j} = C$ for all j.

Algorithm
1. Initialize the system, setting $NT(t_1) = DP(t_1)$ and $C^1(u, v) = C(u, v)$.
2. Compute $NT(t_2)/v(t_2) = C^1(DP(t_1)/v(t_1), DP(t_2)/v(t_2))$.
3. For $j > 2$ to $j \leq n$ compute $C^j(u_1, \ldots, u_j) = C^{j-1} \star C(u_1, \ldots, u_j)$ and $NT(t_j)/v(t_j) = C^j(DP(t_1)/v(t_1), \ldots, DP(t_j)/v(t_j))$.

The Reflection Principle

In some special cases, in order to compute the distribution of the running maxima and minima, it is possible to avoid the above algorithm. In fact, when the underlying model satisfies the reflection principle, the distribution of the running maxima and running minima can simply be recovered from the distribution function of the level of the underlying process itself. We are going to show this result, analyzing the case of running maxima.

It is a general fact that the probability that the running maximum exceeds a level K is linked to the probability of the first passage time of the stochastic process Sj to the level K. In fact,

$$\mathbb{P}(M(t_j) \geq K) = \mathbb{P}(\exists k \leq j \text{ such that } S_k \geq K).$$

The random variable

$$T_K = \inf\{k : S_k \geq K\}$$

represents the first passage time to the level K, and

$$\mathbb{P}(M(t_j) \geq K) = \mathbb{P}(T_K \leq j).$$

If $S_k = e^{\sigma X_k}$, with $\sigma > 0$, $M^X(t_j) = \max_{0 \leq i \leq j} X_i$ and $\hat{K} = \frac{\ln K}{\sigma}$, we have that $\mathbb{P}(M(t_j) \geq K) = \mathbb{P}(M^X(t_j) \geq \hat{K})$ and $T_K = \inf\{k : X_k \geq \hat{K}\}$.

For stochastic processes satisfying the reflection principle, these probabilities turn out to be easily computable. The main reference for the validity of the reflection principle in discrete time is represented by the symmetric Bernoulli random walk. Let $(Y_i)_{i \geq 1}$ be a sequence of independent random variables with values on $\{-1, +1\}$ such that $\mathbb{P}(Y_i = -1) = \mathbb{P}(Y_i = 1) = \frac{1}{2}$. The stochastic process $(X_k)_{k \geq 0}$, defined as $X_i = \sum_{k=0}^{i} Y_k$, is called a symmetric Bernoulli random walk. If b is an integer in $\mathbb{Z}$, the first passage time to the level b is given by

$$T_b = \inf\{i \in \mathbb{N} : X_i = b\}.$$

If $b > 0$ (thanks to the symmetry of the process, similar arguments hold for $b < 0$), we have

$$\mathbb{P}(T_b \leq j) = \mathbb{P}(T_b \leq j, X_j \geq K) + \mathbb{P}(T_b \leq j, X_j < K) \qquad (5.2)$$

and

$$\mathbb{P}(T_b \leq j, X_j \geq b) = \mathbb{P}(X_j \geq b). \qquad (5.3)$$

Let us now consider a scenario ω for which $T_b(\omega) \leq j$ and $X_j(\omega) < b$. This means that the corresponding trajectory reaches the level b at the time $T_b(\omega)$ before time j, and at time j it is at a point less than b. But, for each trajectory of the Bernoulli random walk passing through b at time $T_b(\omega)$, there exists a trajectory passing through b at time $T_b(\omega)$ and then following a symmetric path. This argument allows us to conclude that

$$\mathbb{P}(T_b \leq j, X_j < b) = \mathbb{P}(T_b \leq j, X_j > b) = \mathbb{P}(X_j > b).$$

By (5.2) and (5.3), it follows that

$$\mathbb{P}(T_b \leq j) = 2\mathbb{P}(X_j > b) + \mathbb{P}(X_j = b).$$

Hence, if $M^X(t_j) = \max_{0 \leq i \leq j} X_i$

$$\mathbb{P}(M^X(t_j) \geq b) = \mathbb{P}(T_b \leq j) = 2\mathbb{P}(X_j > b) + \mathbb{P}(X_j = b) = 2\mathbb{P}(X_j \geq b) - \mathbb{P}(X_j = b),$$

from which

$$\mathbb{P}(M^X(t_j) > b) = \mathbb{P}(M(t_j) \geq b + 1) = 2\mathbb{P}(X_j \geq b + 1) - \mathbb{P}(X_j = b + 1)$$

and

$$\begin{aligned}
\mathbb{P}(M^X(t_j) \leq b) &= 1 - \mathbb{P}(M^X(t_j) \geq b + 1) \\
&= 1 - 2\mathbb{P}(X_j \geq b + 1) + \mathbb{P}(X_j = b + 1) \\
&= 1 - 2\mathbb{P}(X_j > b) + \mathbb{P}(X_j = b + 1) \\
&= 1 - 2(1 - \mathbb{P}(X_j \leq b)) + \mathbb{P}(X_j = b + 1) \\
&= -1 + 2\mathbb{P}(X_j \leq b) + \mathbb{P}(X_j = b + 1).
\end{aligned}$$

The above formula shows how, thanks to the reflection principle, it is possible to compute the distribution function of the running maximum through the distribution of the underlying process. The above result extends if we consider a suitable time change. Let τ_j be a stochastic process independent of a symmetric Bernoulli random walk X_j and such that $\tau_{j+1} - \tau_j \in \{0, 1\}$. Consider the time-changed stochastic process

$$Z_j = X_{\tau_j}.$$

Then the reflection principle still holds. In fact, let $M^Z(t_j) = \max\limits_{0 \le i \le j} Z_i$ and $M^X(t_j) = \max\limits_{0 \le i \le j} X_i$, then

$$
\begin{aligned}
\mathbb{P}(M^Z(t_j) \le b) &= \mathbb{P}(\max_{0 \le i \le j} X_{\tau_i} \le b) \\
&= \sum_{s=0}^{j} \mathbb{P}(\max_{0 \le i \le j} X_{\tau_i} \le b | \tau_j = s)\mathbb{P}(\tau_j = s) \\
&= \sum_{s=0}^{j} \mathbb{P}(\max_{0 \le r \le s} X_r \le b | \tau_j = s)\mathbb{P}(\tau_j = s) \\
&= \sum_{s=0}^{j} \mathbb{P}(\max_{0 \le r \le s} X_r \le b)\mathbb{P}(\tau_j = s) \\
&= \sum_{s=0}^{j} \mathbb{P}(M^X(t_s) \le b)\mathbb{P}(\tau_j = s) \\
&= \sum_{s=0}^{j} (2\mathbb{P}(X_s \le b) + \mathbb{P}(X_s = b+1) - 1)\mathbb{P}(\tau_j = s) \\
&= 2\mathbb{P}(X_{\tau_j} \le b) + \mathbb{P}(X_{\tau_j} = b+1) - 1.
\end{aligned}
$$

Remark 5.2.2. Let us consider the case of the running minima. Thanks to the symmetry of the Bernoulli random walk X_j,

$$\mathbb{P}(m^X(t_j) \le b) = \mathbb{P}(M^X(t_j) \ge -b) = 2\mathbb{P}(X_j \ge -b) - \mathbb{P}(X_j = -b) = 2\mathbb{P}(X_j \le b) - \mathbb{P}(X_j = b)$$

and, following the same technique used for the running maxima, assuming a time change τ_j we get

$$\mathbb{P}(m^Z(t_j) \le b) = 2\mathbb{P}(Z_j \le b) - \mathbb{P}(Z_j = b).$$

5.3 THE MEMORY FEATURE

We now show how to address, with the same recursion technique, a clause that is very often found in structured finance products with digital payoffs, and it is bound to modify the temporal structure of the derivative contracts embedded in them. It is called the *memory feature*, and it states that when the event triggering payment occurs, payments are made for all the previous periods in which the event had not taken place.

Example 5.3.1. *Altiplano note with memory. Consider a five-year bond paying a sequence of annual coupons yielding a fixed return if the S&P 500 index is higher than a given threshold B. The first coupon payment time at which the S&P index is above B, coupons are also paid for earlier coupons that had been worth zero.*

Assume, for the sake of simplicity, that all payments are made at maturity of the bond. It is easy to see that in this case the price can be recovered in a backward bootstrapping procedure starting from the last coupon payment, all the way back to the first.

More formally, we may start by defining $\mathcal{A}_j$ to be the set of all scenarios for which the event takes place at time t_j. The term $\overline{\mathcal{A}}_j$ denotes the complementary set. Assume the set of fixing dates is $\{t_1, t_2, \ldots, t_n\}$. We first observe that the product will pay coupons $c(n)$ if and only if the event takes place at time t_n. If we denote by $\mathbf{Q}(\mathcal{A}_n)$ the risk-neutral probability of the event taking place at time t_n, and by $v(t_n)$ the discount function, then the price of the coupon stream will be $v(t_n)c(n)\mathbf{Q}(\mathcal{A}_n)$.

Let us now move back to the case $\mathcal{A}_{n-1} \cap \overline{\mathcal{A}}_n$. In this scenario the event takes place at time t_{n-1} but does not take place at time t_n, so that the product pays a coupon stream of $c(n-1)$. Notice that the corresponding risk-neutral probability can be written as

$$\mathbf{Q}(\mathcal{A}_{n-1} \cap \overline{\mathcal{A}}_n) = \mathbf{Q}(\overline{\mathcal{A}}_n) - \mathbf{Q}(\overline{\mathcal{A}}_{n-1} \cap \overline{\mathcal{A}}_n).$$

Extending the argument by induction, we have that the risk-neutral probability of a scenario of a coupon stream equal to $c(n-j)$ is given by

$$\mathbf{Q}(\mathcal{A}_{n-j} \cap \overline{\mathcal{A}}_{n-j+1} \ldots \cap \overline{\mathcal{A}}_n) = \mathbf{Q}(\overline{\mathcal{A}}_{n-j+1} \cap \ldots \cap \overline{\mathcal{A}}_n) - \mathbf{Q}(\overline{\mathcal{A}}_{n-j} \cap \overline{\mathcal{A}}_{n-j+1} \cap \ldots \cap \overline{\mathcal{A}}_n),$$

with $j = 1, 2, \ldots n - 1$. The price of the product with memory is finally obtained by integrating over all possible $c(n-j)$ scenarios, and we obtain

$$A(t, t_n) = v(t_n)\mathbf{Q}(\mathcal{A}_n)c(n)$$
$$+v(t_n) \sum_{j=1}^{n-1} \left(\mathbf{Q}(\overline{\mathcal{A}}_{n-j+1} \cap \ldots \cap \overline{\mathcal{A}}_n) - \mathbf{Q}(\overline{\mathcal{A}}_{n-j} \cap \overline{\mathcal{A}}_{n-j+1} \cap \ldots \cap \overline{\mathcal{A}}_n)\right) c(j).$$

An inspection of the formula allows us to give a representation in terms of replicating portfolios. To see this, consider a partial no-touch option which pays a fixed sum if the event does not take place in the subperiod between t_{n-j} and t_n. The value of this product, that we denote $NT(t_{n-j}, t_n)$, will be:

$$NT(t_{n-j}, t_n) = v(t_n)\mathbf{Q}(\overline{\mathcal{A}}_{n-j} \cap \overline{\mathcal{A}}_{n-j+1} \ldots \cap \overline{\mathcal{A}}_n)$$

and the pricing formula could be rewritten as

$$A(t, t_n) = v(t_n)\mathbf{Q}(\mathcal{A}_n)c(n)$$
$$+ \sum_{j=1}^{n-1} \left(NT(t_{n-j+1}, t_n) - NT(t_{n-j}, t_n)\right) c(j).$$

Notice that the pricing formula is exact if the contract specifies that coupon payments are made at maturity of the bond. In cases in which payment is actually made the first time the event takes place, the above formula can be used to give pricing bounds for the price. In fact, in this case the scenario in which $n-j$ coupons are paid includes many possible scenarios that yield different prices depending on the sequence of the trigger events. Anyway, the price difference is only due to the risk-free discount function $v(t_j)$ and it is thanks to the decreasing shape of such functions that pricing bounds can be recovered. For every scenario of $n-j$ coupons we may distinguish two extreme cases:

- the event takes place at time t_{n-j} for the first time;
- the event takes place at each time up to (and including) t_{n-j}.

It is clear that since the discount function is decreasing the first instance will yield the lowest possible price, and the second the highest possible one. So, the upper price will be

$$A(t, t_n) = v(t_n)\mathbf{Q}(\mathcal{A}_n)c(n)$$

$$+ \sum_{j=1}^{n-1} \frac{v(t_{n-j})}{v(t_n)} \left(NT(t_{n-j+1}, t_n) - NT(t_{n-j}, t_n)\right) c(j)$$

and the lowest one will be

$$A(t, t_n) = v(t_n)\mathbf{Q}(\mathcal{A}_n)c(n)$$

$$+ \sum_{j=1}^{n-1} \frac{V(t_{n-j})}{v(t_n)} \left(NT(t_{n-j+1}, t_n) - NT(t_{n-j}, t_n)\right) c(j),$$

with

$$V(t_{n-j}) = \sum_{k=1}^{n-j} v(t_k).$$

From a casual inspection of the pricing formula and the replicating portfolio, it appears immediately that a dominant role is played by the probability that the trigger event occurs on the last date. This endows the product with a digital-like feature of a European product that pays all the coupons at the final date.

5.4 RISK-NEUTRAL PRICING RESTRICTIONS

Before proceeding, we need to specify the requirements that must be met by marginal distributions and the temporal dependence structure for the prices to be consistent with the no-arbitrage assumption. It is well known that this amounts to requiring that the pricing be carried out under a probability measure such that the forward prices of assets are martingale. This would call for restrictions on the copula functions representing the dynamics of asset prices. It is quite difficult to impose this restriction on the standard DNO approach that specifies the dependence structure of the levels of the process. The problem could instead be addressed with success in a framework that takes into account the dependence structure of increments, such as the one presented in Chapter 3. The solution would be particularly easy under the assumption of independent increments of the logarithm of prices. In this case all one would need to do is normalize the levels by

$$S_j^i = \frac{e^{X_j^i}}{\mathbb{E}[e^{X_j^i}]},$$

where we remember that X_j^i is the logarithm of price S_j^i. In fact, if $\mathcal{F}_t^i$ is the filtration generated by X_t^i, using the independence of increments hypothesis we have

$$\mathbb{E}[S_j^i|\mathcal{F}_k^i] = \mathbb{E}\left[\frac{e^{X_j^i}}{\mathbb{E}[e^{X_j^i}]}|\mathcal{F}_k^i\right]$$

$$= \frac{1}{\mathbb{E}[e^{X_j^i - X_k^i}]\mathbb{E}[e^{X_k^i}]}\mathbb{E}[e^{X_j^i - X_k^i}e^{X_k^i}|\mathcal{F}_k^i]$$

$$= \frac{e^{X_k^i}}{\mathbb{E}[e^{X_j^i - X_k^i}]\mathbb{E}[e^{X_k^i}]}\mathbb{E}[e^{X_j^i - X_k^i}|\mathcal{F}_k^i]$$

$$= \frac{e^{X_k^i}}{\mathbb{E}[e^{X_j^i - X_k^i}]\mathbb{E}[e^{X_k^i}]}\mathbb{E}[e^{X_j^i - X_k^i}]$$

$$= S_k^i.$$

Here we restrict the analysis to this assumption, that is the usual choice in all the literature, and we refer the reader to Cherubini *et al.* (2010a) for the general case of dependent increments. As for the structure of the process, notice that assuming independent increments has stringent implications for the distribution of prices and their dependence structure. Actually, everything is completely defined starting from the sequence of the distributions of log-increments $F_{Y_j^i}$, with $Y_j^i = X_j^i - X_{j-1}^i$. The distribution of log-prices is actually given by the convolution of increments:

$$F_{X_j^i}(x) = \int_0^1 F_{Y_j^i}(x - F_{X_{j-1}^i}^{-1}(w))dw. \qquad (5.4)$$

and the temporal dependence structure is represented by copulas

$$C_{X_{j-1}^i, X_j^i}(u, v) = \int_0^u F_{Y_j^i}(F_{X_j^i}^{-1}(v) - F_{X_{j-1}^i}^{-1}(w))dw. \qquad (5.5)$$

5.5 TIME-CHANGED BROWNIAN COPULAS

Besides the example of α-stable processes, that by definition are closed under convolution, it is very difficult to extend the pricing approach above, based on independent increments, to more general models of market dynamics. An important result that allows us to overcome this problem is the so-called *time-change technique*, that we have already introduced in Chapter 3. This idea is widely used in mathematical finance, and rests on a fundamental result in probability theory due to Monroe (1978), which states that every semi-martingale process can be converted into a Brownian motion process with time change. This finding is actually a generalization of an earlier result, due to Dambins (1965) and Dubins and Schwarz (1965), that only applied to martingale processes. The new time pace at which the process is sampled is a random variable, which is also known as a *stochastic clock*, and which actually represents the volatility and the level of activity of the market. When many new pieces of information reach the market, volatility increases and this is reflected in a faster pace of time. On the contrary, when market activity is flat or the exchange is closed, time slows down, or even stops. This way, departures from Gaussian behavior of innovations are moved from the dynamics of prices to a sort of second-order level, which is the dynamics of the clock measuring the speed of what is called *economic time*.

If we follow this idea in our application of copula functions to the pricing of equity derivatives, it turns out that every pricing problem can be traced back to the Brownian copula, once the proper time-change clock $h(t, \omega)$ has been chosen. This also makes it simple to impose the no-arbitrage requirement, as shown previously.

In fact, for this kind of model, it turns out to be very easy to construct martingales. Consider a stochastic process $(X_t)_{t \geq 0}$ with independent increments and an independent stochastic process

$(\tau)_{t\geq 0}$ representing the clock. For any given trajectory $\tau_t(\omega)$, we know that the stochastic process

$$\frac{e^{X_{\tau(\omega)}}}{\mathbb{E}\left[e^{X_{\tau(\omega)}}\right]}$$

is a martingale; that is, the conditional expectation of the increments worth zero. Now, integrating this conditional expectation with respect to all admissible clock's trajectories, we again get zero, and, as a consequence, the time-changed process is a martingale. More formally, the following result holds.

Proposition 5.5.1. *Let $(X_t)_{t\geq 0}$ be a stochastic process with independent increments and $(\tau_t)_{t\geq 0}$ be an increasing stochastic process with $\tau_0 = 0$ independent of $(X_t)_{t\geq 0}$. Then, if we set $Y_t = \frac{e^{X_t}}{\mathbb{E}[e^{X_t}]}$,*

$$S_t = Y_{\tau_t}$$

is a martingale.

Proof. Let

$$
\begin{aligned}
A &= \{S_s = x_n, \tau_s = t_n, S_{s_{n-1}} = x_{n-1}, \tau_{s_{n-1}} = t_{n-1}, \ldots, S_{s_1} = x_1, \tau_{s_1} = t_1\} \\
&= \{Y_{\tau_s} = x_n, \tau_s = t_n, Y_{\tau_{s_{n-1}}} = x_{n-1}, \tau_{s_{n-1}} = t_{n-1}, \ldots, Y_{\tau_{s_1}} = x_1, \tau_{s_1} = t_1\} \\
&= \{Y_{t_n} = x_n, \tau_s = t_n, Y_{t_{n-1}} = x_{n-1}, \tau_{s_{n-1}} = t_{n-1}, \ldots, Y_{t_1} = x_1, \tau_{s_1} = t_1\},
\end{aligned}
$$

then

$$
\begin{aligned}
\mathbb{E}[S_t - S_s \mid A] &= \mathbb{E}\left[Y_{\tau_t} - Y_{\tau_s} \mid A\right] \\
&= \int_{t_n}^{+\infty} \mathbb{E}\left[Y_v - Y_{t_n} \mid \tau_t = v, A\right] d\mathbb{P}(\tau_t \leq v \mid A) \\
&= \int_{t_n}^{+\infty} \mathbb{E}\left[Y_v - Y_{t_n}\right] d\mathbb{P}(\tau_t \leq v \mid A) = 0.
\end{aligned}
$$

This implies that $(S_t)_{t\geq 0}$ is a martingale with respect to the filtration generated by the pair $(S_t, \tau_t)_{t\geq 0}$. But, as a consequence, it is also a martingale with respect to the filtration generated by $(S_t)_{t\geq 0}$, the latter being smaller. □

Application of standard time-change techniques has already been described in Chapter 3. Here we show that the semi-parametric approach based on copulas can be pushed one step further, that is to model the stochastic clock itself. This may be done by another recursion, leading to the specification of a term structure for *variance swaps*.

Example 5.5.1. *Markovian time change. We are able to construct a dependent increments time-change process τ_t by using the C-convolution with an appropriate distribution of increments, $\Delta\tau_t$. In particular, we need a distribution with a positive support to guarantee strictly increasing trajectories. We have, in fact, a Markovian time-change process whose distribution F_{τ_t} may be recovered by*

$$F_{\tau_t}(t) = \int_0^1 D_1 C_{\tau_{t-1},\Delta\tau_t}(w, F_{\Delta\tau_t}(t - F_{\tau_{t-1}}(w)))dw,$$

where $C_{\tau_{t-1},\Delta\tau_t}$ is the copula function describing the dependence structure. Figure 5.3 displays a number of simulated trajectories from F_{τ_t} in the case where the copula $C_{\tau_{t-1},\Delta\tau_t}$ is Clayton

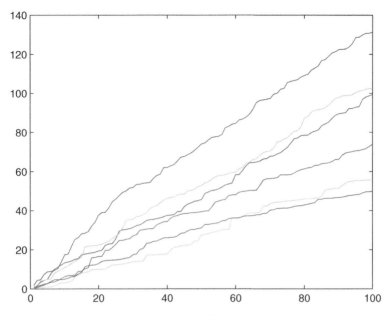

Figure 5.3 Simulation of trajectories of a Markovian time change with increments exponentially distributed.

with $\theta = 0.222$ (consistent with a Kendall's τ equal to 0.1) and it is time-invariant and the increments have an exponential distribution with mean equal to 0.5.

5.6 VARIANCE SWAPS

Variance swaps are contracts that allow us to invest in or hedge against changes in volatility of the market. The payoff is given by the variance actually measured over an investment period, that is referred to as realized variance. In our notation, the contract is specified as

$$VS_j^i = 252 \frac{1}{j-1} \sum_{k=1}^{j} \left(Y_k^i\right)^2, \tag{5.6}$$

where VS_j^i denotes the value of the protection provided against variance. The payoff at expiration, time t_j, is equal to

$$VS^i(t_j) - (VF_j^i)^2, \tag{5.7}$$

with VF_j denoting the fair value of volatility, or fair strike. Fair value means that the net value of the contract at the origin is zero. By convention, this strike is computed as the integral of the prices of *strangles* with exercise at time t_j. We remember that strangles are portfolios of put and call options, with the strike price of the latter higher than those of the former. Formally, we have

$$VF_j^{i\,2} = \frac{2}{t_j} \left(\sum_{K_p \leq F_j^i} \frac{\Delta K_p}{K_p^2} put(K_p, t_j) + \sum_{K_p > F_j^i} \frac{\Delta K_p}{K_p^2} call(K_p, t_j) \right), \tag{5.8}$$

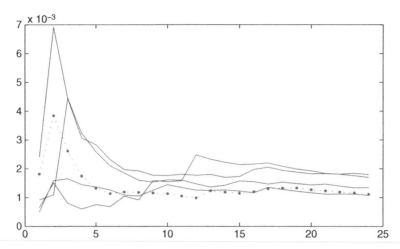

Figure 5.4 Term structure of variance swaps in the case of Clayton copula with $\tau = 0.1$ (dashed line: NIKKEI225).

where K_p is a sequence of strike prices, $\Delta K_p = K_{p+1} - K_p$, and $F_j^i = S_j^i/v(t_j)$ is the forward price of asset i for delivery at time t_j. This value is motivated on the assumption that asset S^i follows a continuous diffusion process.

Using the C-convolution technique it is then possible to design a recursion to yield the evolution of the distribution of realized variance VS_j^i. In fact, given the distribution of the squared log-returns $(Y_j^i)^2$ and the dependence structure between them and VS_{j-1}^i, one can compute

$$F_{VS_j^i}(x) = \frac{j-2}{j-1} \int_0^1 D_1 C\left(w, F_{(Y_j^i)^2}(x - F_{VS_{j-1}^i}^{-1}(w).\right) dw, \tag{5.9}$$

This provides a semi-parametric algorithm to generate the term structure of variance swaps that are typically used for time-change techniques.

Example 5.6.1. *Here we show the term structure of variance swaps of a process with*

- *squared innovations modeled by a Gamma distribution with parameters calibrated on empirical weekly squared returns (NIKKEI225),*
- *dependence structure of squared innovations and integrated variance modeled with a Clayton copula with different values of the parameter.*

Figure 5.4 shows a number of trajectories of the term structure of variance swaps generated by our semi-parametric algorithm.

5.7   SEMI-PARAMETRIC PRICING OF PATH-DEPENDENT DERIVATIVES

We are now ready to gather all the results collected up to this point to generate the term structure of option prices, that is the temporal consistency of option prices for different maturities, and to use it to price path-dependent options. This is typically done by a three-stage process:

- specification of the dynamics of quadratic variation, that represents the stochastic clock;
- evaluation, either in closed form or by simulation, of the path-dependent process conditional on each stochastic clock scenario;
- integration of the price across all possible stochastic clock scenarios.

5.8   THE MULTIVARIATE PRICING SETTING

We now extend the analysis to the case of basket equity derivatives. We assume that each logarithm of price X^i follows a dynamic process specified as

$$X^i_j = X^i_{j-1} + \eta^i_j,$$

and we specify the spatial dependence structure through a copula function linking the innovations η^i_j. This model is called the *semi-parametric copula-based multivariate dynamic model*, or *SCOMDY*, and was first proposed by Chen and Fan (2006). Formally, the dependence structure is given by

$$\mathbb{P}(\eta^1_j \le x^1, \ldots, \eta^m_j \le x^m) = C_j(u^1, \ldots, u^m).$$

An important issue is now how to extend the restrictions due to the risk-neutral valuation principle to this multivariate setting. Notice that in principle we face a problem with dimension $m \cdot n$, where m is the number of assets (the spatial dimension) and n is the number of periods (the temporal dimension). A first restriction comes from the Markovian structure, which implies a representation

$$C_{\mathbf{X}_1,\ldots,\mathbf{X}_n} = C_{\mathbf{X}_1\mathbf{X}_2} \star^m \cdots \star^m C_{\mathbf{X}_{n-1}\mathbf{X}_n},$$

where $\mathbf{X}_j$ denotes the m-dimensional vector of log-prices at time t_j and $\star^m$ is the m-fold $\star$ product operator, defined in Chapter 3.

The issue now is how to extend the martingale pricing principle to higher dimensions. The strategy will be the same as that in the one-dimensional case. First, we will work out a price recursion under the assumption of independent increments. Then, we will assume a model in which, under a suitable random time change, the log-price increments dynamics can be considered independent. Finally, we will integrate across possible time changes. Even though the basic pricing strategy is the same, application to a multivariate setting requires further specifications, concerning, for example, what we mean by independent increments. Before doing that, however, we must first go through a digression on the topic of extension of the martingale pricing principles to higher dimensions.

5.9   H-CONDITION AND GRANGER CAUSALITY

The topic of the possibility of extending the martingale restriction to higher dimensions is very technical. We give here a brief discussion limited to the basic intuition behind the topic. We start by observing that if we are able to impose the martingale restriction on the multivariate system, the restriction also holds for each and every univariate process of the system. The opposite is not true. It may be the case that the martingale restriction holds at the univariate level but is not preserved in the multivariate setting. The insight is that when we include other assets in the set of possible trading strategies, some of them may give rise to arbitrage opportunities. A further insight is that this may actually happen if the new assets help to provide a hedge against states of the world that were not *attainable* before inclusion of these assets. We then require that this is not possible: that is, that arbitrage opportunities are not available even if more assets are included in the trading strategies.

Let us now see how to express these concepts formally. Taking into account more assets means increasing the set of information available. Let us denote by $\mathcal{F}_j$ the set of information available at time t_j encompassing the history of all the available asset prices, and $\mathcal{F}_j^i$ the information set implied in the price history of asset S^i. The problem is to ensure that if S^i is a martingale under filtration $\mathcal{F}_j^i$ it remains a martingale with respect to filtration $\mathcal{F}_j$. In the literature this is called the H-condition. It may be shown that a sufficient condition for this result to hold is the following:

$$\mathbb{P}(X_{j+1}^i \leq x|\mathcal{F}_j) = \mathbb{P}(X_{j+1}^i \leq x|\mathcal{F}_j^i). \tag{5.10}$$

Actually, this reminds us of a concept which is well known in econometrics, and that is called *no-Granger causality* after the name of the Nobel prize winner Clive J. Granger. In econometrics it is usual to find a definition in terms of expected values,

$$\mathbb{E}[X_{k+1}^i|\mathcal{F}_k] = E[X_{k+1}^i|\mathcal{F}_k^i], \tag{5.11}$$

that is clearly implied by (5.10) (see also Florens and Fougere (1991)). It is also clear that if $E[X_{k+1}^i|\mathcal{F}_k^i] = X_k^i$, the no-Granger causality requirement, as defined in equation (5.11), actually implies the H-condition. The same result clearly holds if we start from the stronger assumption in equation (5.10).

Things become more complicated when dealing with Markov processes. Thanks to the results introduced in Chapter 3 concerning the representation of unidimensional and multidimensional Markov processes through suitable $\star$ products, we are interested in working with processes that are Markov both from a unidimensional and a multidimensional perspective. The main problem is that a stochastic process that is Markov with respect to a given filtration is not in general Markov with respect to a larger filtration. But this fact is guaranteed by the no-Granger assumption. In fact,

$$\mathbb{P}(X_{k+1}^i \leq x|\mathcal{F}_k) = \mathbb{P}(X_{k+1}^i \leq x|\mathcal{F}_k^i) = \mathbb{P}(X_{k+1}^i \leq x|X_k^i).$$

Now, a result due to Brèmaud and Yor (1978) states that if X^i is a Markov process with respect to both filtrations $\mathcal{F}_k^i$ and $\mathcal{F}_k$ and it is furthermore a martingale with respect to the filtration $\mathcal{F}_k^i$, it turns out to be a martingale with respect to the enlarged filtration $\mathcal{F}$ as well. All these facts indicate that the no-Granger causality hypothesis guarantees that the martingale assumption extends from the natural filtration generated by each process to the filtration generated by the whole multidimensional process. See Cherubini *et al.* (2010) for more details.

5.10 MULTIVARIATE PRICING RECURSION

We may now go back to the problem of recovering the term structure of prices following the same strategy as in the univariate case, that is addressing the case with independent increments first, and then extending the analysis to the case with time change. In order to do this, we must extend the concept of independent increments to the multivariate setting. This is not straightforward, because for example assuming that each process has independent increments is no longer sufficient. The straight extension to the multivariate setting requires the concept of *vector independence*, defined below.

Definition 5.10.1. We say that a multidimensional Markov process $X = (X^1, \ldots, X^m)$ has *vector* independent increments if

$$\mathbb{P}(\Delta X_j^1 \leq x_1, \ldots, \Delta X_j^m \leq x_m | X_{j-1}^1, \ldots, X_{j-1}^m) = \mathbb{P}(\Delta X_j^1 \leq x_1, \ldots, \Delta X_j^m \leq x_m).$$

This condition is fulfilled for all *multivariate Lévy processes* and it is restrictive, just like the condition of independent increments is in the univariate setting (see Cherubini *et al.* (2010) for proof).

In the case of the SCOMDY model, this hypothesis implies

$$C_{X_j^1, \ldots, X_j^m} \left(z^1, \ldots, z^m\right) = \int_0^1 \cdots \int_0^1 C_j \left(F_{\eta_j^1}\left(F_{X_j^1}^{-1}(z^1) - F_{X_{j-1}^1}^{-1}\left(\omega^1\right)\right), \ldots, \right.$$
$$\left. F_{\eta_j^m}\left(F_{X_j^m}^{-1}(z^m) - F_{X_{j-1}^m}^{-1}\left(\omega^m\right)\right)\right) dC_{X_{j-1}^1, \ldots, X_{j-1}^m}^j \left(\omega^1, \ldots, \omega^m\right).$$

Notice that this may actually be interpreted as a *multivariate convolution*, or a copula of convolutions. In fact, we may verify that

$$C_{X_j^1, \ldots, X_j^m}(1, \ldots, z^i, \ldots, 1) = z^i = \int_0^1 F_{\eta_j^i}(F_{X_j^i}^{-1}(u) - F_{X_{j-1}^i}^{-1}(\omega^i)) d\omega^i$$

and we obtain the standard univariate convolution. Just like in the univariate case, the analysis can be extended by using the change of time technique. This time, however, the transformation must take into account the no-Granger causality constraint in order to be consistent with risk-neutral valuation. For this reason, the assumption of vector independence cannot be dropped *tout court*, but can only be weakened to a concept that we could call *Granger independence*, according to the definition below.

Definition 5.10.2. We say that a set of Markov processes $X = (X^1, \ldots, X^m)$ have *Granger* independent increments if for any k,

$$\mathbb{P}(\Delta X_k^i \leq x | X_{j-1}^1, \ldots, X_{j-1}^m) = \mathbb{P}(\Delta X_k^i \leq x).$$

Let $T = (T^1, \ldots, T^m)$ be a multidimensional stochastic process which is increasing and such that $T_0 = (T_0^1, \ldots, T_0^m) = (0, \ldots, 0)$ representing the stochastic clocks set. We assume that T is a multidimensional Markov process with respect to its natural filtration.

Proposition 5.10.1. *Let $X = (X^1, \ldots, X^m)$ be a multidimensional Granger-independent Markov process. Let $T = (T^1, \ldots, T^m)$ as defined above, independent of X. If $e^{X_j^i}$ is a martingale with respect to its natural filtration $\mathcal{F}_j^i$, then the time-changed stochastic process $S_j^i = e^{X_{T_j^i}^i}$ is a martingale with respect to the filtration $\mathcal{F}_i^{X_T}$ generated by the whole time-changed stochastic process $S_j = (e^{X_{T_j^1}^1}, \ldots, e^{X_{T_j^m}^m})$.*

The proof is very similar to that of the univariate case, but it requires some further technicalities. We refer the interested reader to Cherubini *et al.* (2010) for details.

The Luciano–Schoutens Model

An approach similar to the one above was applied in continuous time to the case of a multidimensional Brownian motion time changed by a one-dimensional stochastic clock in Luciano and Schoutens (2006). They consider a multidimensional Brownian motion $(W_t^{(1)}, \ldots, W_t^{(n)})$ and a geometrical Brownian motion

$$\hat{A}_t^{(i)} = A_0^{(i)} \exp \left(\theta_i t + \sigma_i W_t^{(i)} \right)$$

time changed by a stochastic clock modeled through a Gamma process G_t. We recall that a Gamma process is a subordinator, i.e., a Lévy process with increasing trajectories such that the characteristic function of G_t is

$$\phi_{G_t}(u) = \left(1 - \frac{iu}{a} \right)^{-a}$$

and its density function

$$f(x) = \frac{a^a}{\Gamma(a)} x^{a-1} \exp(-xa), \quad x \geq 0, \quad a > 0.$$

This is the particular case of the general Gamma process (see Appendix B), in which the two parameters characterizing this model are assumed to be equal to a. The ith component of the time-changed multidimensional process is given by

$$A_t^{(i)} = A_0^{(i)} \exp \left(\theta_i G_t + \sigma_i W_{G_t}^{(i)} \right).$$

Now, it is well known that a Brownian motion with drift, time changed by a Gamma process, leads to the variance Gamma process, which is a Lévy process as well (see Appendix B). In this case it is known (see among others Cherubini *et al.* (2010)) that the model can be made risk neutral just by shifting the mean value. Hence, Luciano and Schoutens assume risk dynamics of type

$$A_t^{(i)} = A_0^{(i)} \exp((r - q_i)t + X_t^{(i)} + \omega_i t),$$

where $X_t^{(i)} = \theta_i G_t + \sigma_i W_{G_t}^{(i)}$, r is the constant interest rate, and q_i is the dividend yield continuously paid. The risk-neutral drift rate for the asset is $r - q_i$, and thus to have $\mathbb{E}[A_t^{(i)}] = A_0^{(i)} \exp((r - q_i)t)$ we have to set

$$\omega_i = a \log \left(1 - \frac{1}{2a}\sigma_i^2 - \frac{\theta_i}{a} \right).$$

The common stochastic clock clearly induces dependence among the assets. Luciano and Schoutens study this dependence structure by analyzing the copula linking the process at time $t = 1$. For the sake of simplicity we assume $i = 1, 2$. Let

$$Z_t^{(i)} = \log \left(\frac{A_t^{(i)}}{A_0^{(i)}} \right)$$

and denote by $F(z_1, z_2)$ the joint distribution of $(Z_1^{(1)}, Z_1^{(2)})$, while $F^{(i)}(z_i)$ are the marginal distributions. Setting $m_i = r - q_i + \omega_i$, we have that with respect to the risk-neutral measure

$$\mathbb{P}(Z_1^{(i)} \leq z_i | G_1 = x) = F^{(i)}(z_i | x) = \Phi \left(\frac{z_i - m_i - \theta_i x}{\sigma_i \sqrt{x}} \right),$$

where Φ is the c.d.f. of the standard normal distribution. It is straightforward to get

$$F^{(i)}(z_i) = \int_0^{+\infty} \Phi \left(\frac{z_i - m_i - \theta_i x}{\sigma_i \sqrt{x}} \right) \frac{a^a}{\Gamma(a)} x^{a-1} \exp(-xa)\, dx.$$

Using conditional independence, we obtain

$$F(z_1, z_2) = \int_0^{+\infty} \prod_{i=1}^{2} \Phi\left(\frac{z_i - m_i - \theta_i x}{\sigma_i \sqrt{x}}\right) \frac{a^a}{\Gamma(a)} x^{a-1} \exp(-xa)\, dx.$$

Notice that this distribution only depends on the same parameters as the marginal ones.

Thanks to the Sklar theorem there exists a copula C^A that represents the dependence structure of the vector $(Z_1^{(1)}, Z_1^{(2)})$ given by

$$C^A(u, v) = F\left((F^{(1)})^{-1}(u), (F^{(2)})^{-1}(v)\right).$$

The analysis of the contour plots of C^A varying with a that measures the level of economic activity shows that a decrease in a implies an increase in marginal variance and kurtosis as well as concordance. Moreover, decreasing a seems to increase the lower tail dependence. The increase in concordance and lower tail dependence when a decreases, can be explained as follows: the stochastic clock has a fixed average and variance proportional to $\frac{1}{a}$. Decreasing a then increases the impact of the stochastic time change: this not only causes the processes to depart more from continuity and boosts their jumps at the marginal level, but also causes them to become more dependent, since they are affected by the same stochastic pressure, but the latter is more powerful.

Luciano and Schoutens (2006) use the above model to calibrate the joint behavior of equities or equity indices. The fact stressed above that the model parameters are the same as those of the marginal distributions, allows us to calibrate the market only using data on univariate derivative products. For more details on the analysis of contour plots and calibration results, we refer the reader to the cited paper.

5.11 HEDGING MULTIVARIATE EQUITY DERIVATIVES

Once the term structure of prices is obtained, one can recover sensitivities and hedging policies by the standard total differentiation approach. For the sake of illustration, we stick to hedging of multivariate digital options using univariate digital. This allows us to reconnect our analysis to the standard approach, based on continuous time models, based on *Delta* and *cross-Gamma* replication. For the sake of simplicity we stick to digital products, which represent the pricing kernel of all other multivariate products. We denote by $MDC(t_i, t_j)$ the price of a digital call for maturity t_j. We get

$$\Delta MDC(t, t_j) \approx \sum_{i=1}^{m} \frac{\partial C_{X_j^1, \ldots, X_j^m}(F_j^1(B^1), \ldots, F_j^m(B^m))}{\partial F_i^j(B^i)} \Delta F_i^j(B^i),$$

with $F_j^i(B^i) = \mathbb{P}(S_j^i > B^i)$, and remember that this can also be written in terms of conditional probabilities:

$$\Delta MDC(t, t_j) \approx \sum_{i}^{m} \mathbb{P}(S_j^1 > B^1, \ldots, S_j^m > B^m | S^i = B^i) \Delta F_i^j(B^i).$$

Notice that since the copula $C_{X_j^1, \ldots, X_j^m}(F_j^1(B^1), \ldots, F_j^m(B^m))$ is a multivariate convolution, our approach determines dynamics for the *Delta*, and the hedging strategy is not static like in standard copula function applications.

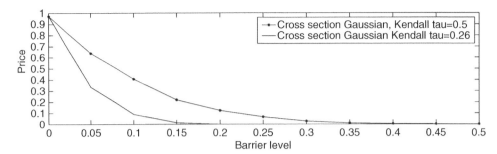

Figure 5.5 Digital barrier prices with respect to the barrier levels for different cross-sectional dependence.

Example 5.11.1. *We consider a digital barrier option (down and out call) on the exchange rates euro/yuan and euro/dollar whose price is:*

$$DBar_t = \mathbb{E}(\mathbf{1}_{\{EU/YUAN_s > H, EU/DOLLAR_s > H, \forall s \in M\}} e^{-r(T-t)}),$$

where M is the set of monitoring dates, H is the barrier, and T is the maturity of the option. We first select a copula $C_{1,2(s)}$ representing the cross-sectional dependence and linking the levels of the univariate processes $(EU/YUAN_t)$ and $(EU/DOLLAR_t)$ at each time. In this example, from the historical data, we compute the empirical copula function and estimate a Kendall's τ equal to 0.26 for the Gaussian copula. Then we model the univariate dynamics choosing a Clayton copula function with Kendall's τ of 0.8 to represent the temporal dependence between the levels and by a dynamic conditional sampling procedure, we generate the univariate dynamics of the exchange rates.

We assume a maturity of one year and a monthly monitoring of the barrier setting at different levels, and we compute the price and the hedging portfolio. At each monitoring date we evaluate the Delta for the two underlyings, corresponding to a conditional probability, i.e. to a partial derivative of the cross-sectional copula function at the same time.

In Figure 5.5 we can see that the digital barrier price decreases for increasing barrier levels and that the curve is shifted up for higher cross-sectional dependence parameters.

In Figure 5.6 we can see that the digital barrier price explains an increasing trend for increasing cross-sectional dependence parameter and that the curve is shifted up for lower barrier levels.

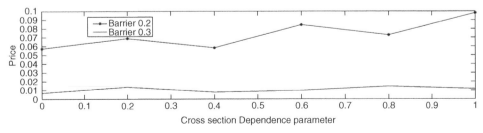

Figure 5.6 Digital barrier prices with respect to the cross-sectional dependence for different barrier levels.

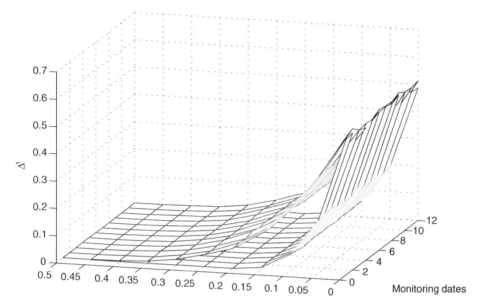

Figure 5.7 Δ^1 Surface with respect to the time and the barrier for different cross-sectional dependence (Kendall's τ 0.26 and 0.5).

Here, Δ^1 and Δ^2 will denote the dynamic replicating portfolio (the static case was studied in Section 2.3). In Figure 5.7 we can see the surface of Δ^1, i.e., with respect to the monitoring dates and the barrier for different levels of cross-sectional dependence. We can observe that Δ^1 decreases for higher barrier levels, in line with the behavior of the price, and that the surface shifts up for higher levels of the cross-sectional dependence parameter.

In Figure 5.8 we can see the histogram of Δ^1 and Δ^2 at time $t = 1$ for barrier 0.3, for different levels of cross-sectional dependence. We can observe that Δ^1 and Δ^2 tend to increase for higher cross-sectional dependence parameters.

In Figure 5.9 we can see the histogram of Δ^1 at different times for barrier 0.2, for different levels of cross-sectional dependence. We can observe the same behavior of Δ^1 at all times, i.e., it tends to increase for higher cross-sectional dependence parameters.

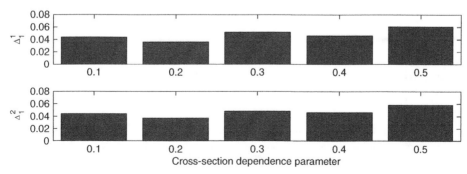

Figure 5.8 Histogram of Δ^1 and Δ^2 at time $t = 1$ for barrier 0.3 and for different cross-sectional dependence.

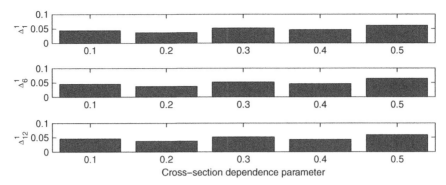

Figure 5.9 Histogram of Δ^1 at different times, for barrier 0.2 and for different cross-sectional dependence.

In Figure 5.10 we can see the histogram of Δ_1^1 for different barriers and for different levels of cross-sectional dependence. We can observe that Δ_1^1 tends to increase for higher cross-sectional dependence parameters and decreases for higher barrier levels.

5.12 CORRELATION SWAPS

Assume two variables X^j and X^k generated by the SCOMDY model:

$$X_i^j = X_{i-1}^j + \epsilon_i,$$
$$X_i^k = X_{i-1}^k + \eta_i,$$

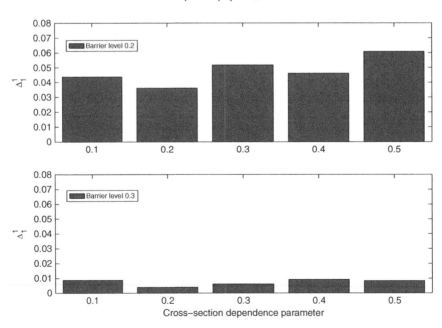

Figure 5.10 Histogram of Δ^1 at time $t = 1$ for barrier 0.2 and 0.3, and for different cross-sectional dependence.

where $(\epsilon_i)_i$ and $(\eta_i)_i$ are sequences of independent r.v.s with marginal distributions F_{ϵ_i} and F_{η_i}. Assume that ϵ_i and η_i are correlated and let ρ_i be their correlation coefficient. We denote by $\mu_{\epsilon,i}$ and $\mu_{\eta,i}$ their mean and by $\sigma_{\epsilon,i}$ and $\sigma_{\eta,i}$ their volatility.

By independence we have

$$\mathbb{E}[X_i^j] = \mu_{X_0^j} + \Sigma_{m=1}^i \mu_{\epsilon,m}, \quad Var(X_i^j) := \sigma_{X_i^j}^2 = \sigma_{X_0^j}^2 + \Sigma_{m=1}^i \sigma_{\epsilon,m}^2,$$

$$\mathbb{E}[X_i^k] = \mu_{X_0^k} + \Sigma_{m=1}^i \mu_{\eta,m}, \quad Var(X_i^k) := \sigma_{X_i^k}^2 = \sigma_{X_0^k}^2 + \Sigma_{m=1}^i \sigma_{\eta,m}^2.$$

We compute the correlation coefficient between log-prices X_i^j and X_i^k. We have

$$\rho_{X_i^j,X_i^k} = \frac{\mathbb{E}[X_i^j X_i^k] - \mathbb{E}[X_i^j]\mathbb{E}[X_i^k]}{\sigma_{X_i^j}\sigma_{X_i^k}}.$$

Since

$$\mathbb{E}[X_i^j X_i^k] = \mathbb{E}[X_{i-1}^j X_{i-1}^k] + \mathbb{E}[X_{i-1}^j]\mathbb{E}[\eta_i] + \mathbb{E}[X_{i-1}^k]\mathbb{E}[\varepsilon_i] + \mathbb{E}[\varepsilon_i \eta_i],$$

and

$$\mathbb{E}[\varepsilon_i \eta_i] = \rho_i \sigma_{\varepsilon,i}\sigma_{\eta,i} + \mu_{\varepsilon,i}\mu_{\eta,i},$$

after some algebra

$$\rho_{X_i^j,X_i^k} = \frac{\mathbb{E}[X_{i-1}^j X_{i-1}^k] - \mathbb{E}[X_{i-1}^j]\mathbb{E}[X_{i-1}^k] + \rho_i \sigma_{\varepsilon,i}\sigma_{\eta,i}}{\sigma_{X_i^j}\sigma_{X_i^k}}$$

$$= \frac{\mathbb{E}[X_{i-1}^j X_{i-1}^k] - \mathbb{E}[X_{i-1}^j]\mathbb{E}[X_{i-1}^k]}{\sigma_{X_{i-1}^j}\sigma_{X_{i-1}^k}}\frac{\sigma_{X_{i-1}^j}\sigma_{X_{i-1}^k}}{\sigma_{X_i^j}\sigma_{X_i^k}} + \frac{\rho_i \sigma_{\varepsilon,i}\sigma_{\eta,i}}{\sigma_{X_i^j}\sigma_{X_i^k}}$$

$$= \rho_{X_{i-1}^j,X_{i-1}^k} K_i + H_i,$$

where $K_i = \frac{\sigma_{X_{i-1}^j}\sigma_{X_{i-1}^k}}{\sigma_{X_i^j}\sigma_{X_i^k}}$ and $H_i = \frac{\rho_i \sigma_{\varepsilon,i}\sigma_{\eta,i}}{\sigma_{X_i^j}\sigma_{X_i^k}}$.

Example 5.12.1. Stationary case. *As an example, we can consider the stationary case. Suppose that $\rho_i = \rho$, $\sigma_{\varepsilon,i} = \sigma_\varepsilon$, and $\sigma_{\eta,i} = \sigma_\eta$ for all i. In this framework, $\sigma_{X_i^j}^2 = \sigma_{X_0^j}^2 + i\sigma_\varepsilon^2$ and $\sigma_{X_i^k}^2 = \sigma_{X_0^k}^2 + i\sigma_\eta^2$. By iterating $\rho_{X_i^j,X_i^k}$, we obtain*

$$\rho_{X_i^j,X_i^k} = \rho_{X_0^j,X_0^k}\left(\prod_{\ell=1}^i K_\ell\right) + \sum_{\ell=1}^{i-1} H_\ell \left(\prod_{\ell'=\ell+1}^i K_{\ell'}\right) + H_i.$$

But it's easy to verify that

$$\prod_{\ell=1}^i K_\ell = \frac{\sigma_{X_0^j}\sigma_{X_0^k}}{\sigma_{X_i^j}\sigma_{X_i^k}}$$

and

$$H_\ell\left(\prod_{\ell'=\ell+1}^i K_{\ell'}\right) = H_i = \frac{\rho\sigma_\varepsilon\sigma_\eta}{\sigma_{X_i^j}\sigma_{X_i^k}}$$

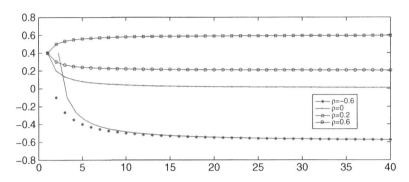

Figure 5.11 Term structure of the correlation for the simulated SCOMDY model with daily parameters when the initial correlation is equal to 0.4.

for $\ell = 1, \ldots, i - 1$, for which

$$\rho_{X_i^j, X_i^k} = \rho_{X_0^j, X_0^k} \frac{\sigma_{X_0^j} \sigma_{X_0^k}}{\sigma_{X_i^j} \sigma_{X_i^k}} + i H_i.$$

Now

$$\lim_{i \to \infty} \frac{\sigma_{X_0^j} \sigma_{X_0^k}}{\sigma_{X_i^j} \sigma_{X_i^k}} = 0,$$

$$\lim_{i \to \infty} i H_i = \lim_{i \to \infty} \frac{i \rho \sigma_\varepsilon \sigma_\eta}{\sigma_{X_i^j} \sigma_{X_i^k}} = \rho,$$

and

$$\lim_{i \to \infty} \rho_{X_i^j, X_i^k} = \rho.$$

Figure 5.11 shows the simulated behavior of $\rho_{X_i^j, X_i^k}$ when the initial correlation is equal to 0.4 and Figure 5.12 when the error volatilities are annualized to 0.2.

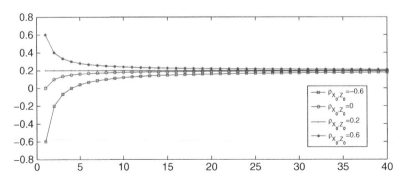

Figure 5.12 Term structure of the correlation for the simulated SCOMDY model with daily parameters when the correlation between the increments is equal to 0.2.

5.13   THE TERM STRUCTURE OF MULTIVARIATE EQUITY DERIVATIVES

In this section we show how to obtain the term structure of prices of multivariate derivatives in a SCOMDY market model with temporally independent, or conditionally independent, increments with assigned spatial dependence structure of the increments. In order to stress the way in which the dependence of increments cumulates into that of the variables, we stick to the vector independent model. We will address the main kinds of products, ranging from multivariate digital derivatives to rainbow options and basket/spread options.

As for the recovery of information, notice that an important implication of our approach is that we can obtain arbitrage-free prices by mixing calibration of the marginal distributions from options data and estimation of the cross-sectional dependence structure of the increments from historical time series. More explicitly, the procedure calls for:

- estimation of the dependence structure of increments from time series;
- fitting of the smile for each asset involved;
- simulation of the dependence structure of prices at different time horizons.

For the sake of illustration, we report here the pricing surfaces of the main multivariate products, under two different scenarios concerning the dependence structure of return innovations. For each product, we also report the multivariate term structure of prices consistent with the no-arbitrage requirement. Of course, the pricing surface is not unique, since the market is incomplete, but it provides an important reference point for pricing under the realized dependence.

For the sake of simplicity, in our simulation design we stick to the bivariate case and to the hypothesis that $C_i(u, v)$ does not depend on i, that is: $C_i(u, v) = C(u, v)$. Based on these assumptions, we obtain two simulated trajectories $(X_i^j)_{0 \leq i \leq n}$ and $(X_i^k)_{0 \leq i \leq n}$ using standard Monte Carlo simulation techniques for copula functions. The marginals are assumed to be Gaussian, and in order to ensure that the prices $e^{X_i^j}$ and $e^{X_i^k}$ be martingales, we assume that

$$\varepsilon_i \sim N\left(-\frac{\sigma_\varepsilon^2}{2}, \sigma_\varepsilon^2\right)$$

and

$$\eta_i \sim N\left(-\frac{\sigma_\eta^2}{2}, \sigma_\eta^2\right).$$

for $i = 0, \ldots, n$. In fact, in this case (and analogously for $e^{X_i^k}$),

$$\mathbb{E}[e^{X_i^j} | \mathcal{F}_{i-1}] = \mathbb{E}[e^{X_{i-1}^j + \varepsilon_i} | \mathcal{F}_{i-1}] = e^{X_{i-1}^j} \mathbb{E}[e^{\varepsilon_i}] = e^{X_{i-1}^j},$$

since $\mathbb{E}[e^{\varepsilon_i}] = e^{-\frac{\sigma_\varepsilon^2}{2} + \frac{\sigma_\varepsilon^2}{2}} = 1$.

As for the dependence structure of the increments, in the base scenario, return innovations in the two markets are assumed to be independent. In the alternative scenario, they are assumed to be dependent, and the dependence structure is assumed to be represented by a Clayton copula with 40% Kendall τ dependence.

Before getting into the details of each product, we report in Figure 5.13 the main innovation of our approach with respect to the copula technique that is currently in use. In standard applications, one would have estimated the dependence structure between the *levels* of the

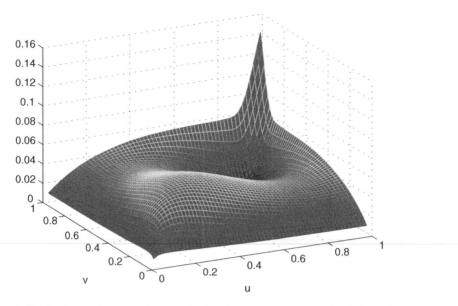

Figure 5.13 Pricing kernel difference between the Clayton copula with 40% Kendall τ and the copula functions obtained with price increments linked to a Clayton copula with 40% Kendall τ.

two assets, and, given the marginals, obtained the price. In our approach one would have estimated the dependence structure between the *increments* of the assets, and then, assuming independent increments, would have recovered the dependence structure by simulation. Only the latter approach provides arbitrage-free prices. Figure 5.13 describes the difference in the probability distribution obtained under the two approaches. In other terms, the figure depicts the difference in the pricing kernel between a price made using copulas on the levels and an arbitrage-free copula.

5.13.1 Altiplanos

We start the analysis with multivariate digital products, called Altiplanos: these products give a fixed amount of payoff if the prices of all assets in a basket are affected by some event (say none of them are below a given threshold on a given date). Pricing Altiplanos was the first, most straightforward application of copulas to equity derivatives. In fact, due to the digital payoff, their price amounts to the computation of a copula function. However, since in Altiplano notes multivariate digital derivatives are used to determine the coupon payoffs on a stream of dates $\{t_1, t_2, \ldots, t_n\}$, the pricing of this kind of security quite naturally calls for temporal consistency of prices of the multivariate derivatives across different maturities. This temporal consistency feature is even more evident in *barrier Altiplanos*, in which the coupon is paid if all the assets are above a given threshold at a set of fixed dates, rather than on the payment date (for a comparison, see Figure 5.14). Differently from the research surveyed above, however, we focus here on such temporal consistency.

In Figures 5.15 and 5.16 we report the term structure of the prices of European Altiplanos with the same contract features and underlying assets but different dependence structure in the

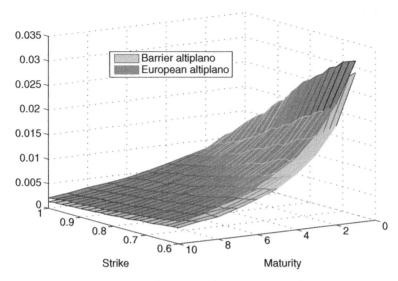

Figure 5.14 Arbitrage-free pricing surface. European vs barrier Altiplanos. Clayton copula with Kendall τ equal to 40%.

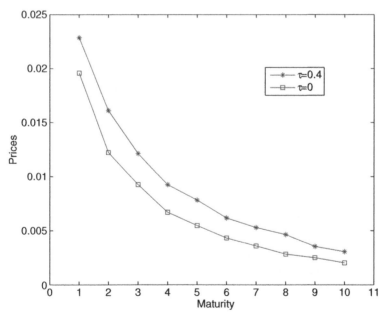

Figure 5.15 Arbitrage-free term structure of European Altiplanos. Product copula vs Clayton copula with Kendall τ equal to 40%.

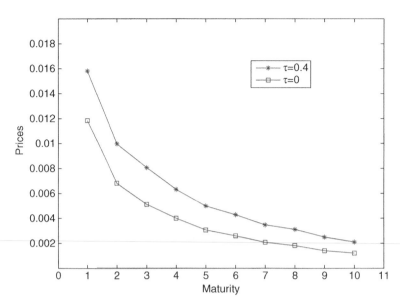

Figure 5.16 Arbitrage-free term structure of barrier Altiplanos. Product copula vs Clayton copula with Kendall τ equal to 40%.

bivariate case. Apart from Monte Carlo errors, the whole term structure appears to be shifted up (down) by a parallel amount by an increase (decrease) in the dependence structure.

5.13.2   Everest

We now extend the analysis to rainbow products. In multivariate equity notes, call on the minimum of assets is typically used to provide the coupon payments over a guaranteed minimum return. These products are known as Everest. The coupon is paid according to the rule

$$\max\left(\min\left(S_T^1/S_0^1, S_T^2/S_0^2, \ldots, S_T^m/S_0^m\right), 1+k\right),$$

where k is the guaranteed return. This is obviously equivalent to a bond paying a coupon equal to $k\%$ for sure, plus a call on the minimum return to the assets, with strike equal to $1 + k$:

$$\max\left(\min\left(S_T^1/S_0^1, S_T^2/S_0^2, \ldots, S_T^m/S_0^m\right) - (1+k), 0\right).$$

The product is actually the integral of multivariate digital options. In Figure 5.17 we report the term structure of a call on the minimum options in the bivariate case. The term structure for this product is not monotone in general. Moreover, not only is the product long correlation as expected, but the term structure is also made steeper by an increase in correlation.

5.13.3   Spread Options

The last class of multivariate product we address is spread options. The payoff is

$$\max\left(S_T^1/S_0^1 - S_T^2/S_0 - K, 0\right).$$

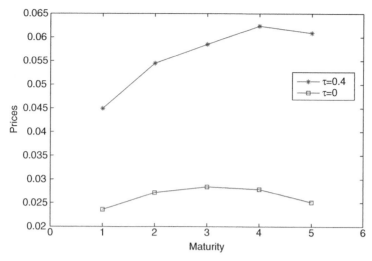

Figure 5.17 Arbitrage-free term structure of call on the min options. Product copula vs Clayton copula with Kendall τ equal to 40%.

Notice that this is nothing but a specific instance of the basket option, that is an option on the sum of two assets, in the case in which the second asset is taken with negative sign. To this purpose, it is worth remembering that there is a specific relationship between the copula function to be used between two variables and monotonic transformations of them. Namely, in the case at hand, if C is the copula linking S_1 and S_2, the copula linking S_1 and $-S_2$

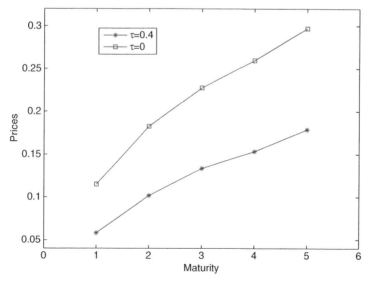

Figure 5.18 Arbitrage-free term structure of spread options. Product copula vs Clayton copula with Kendall τ equal to 40%.

(and any other strictly decreasing transformation of S_2) has to be

$$\widetilde{C} \equiv u - C(u, 1 - v).$$

Notice that the product obviously turns out to be short correlation. In Figure 5.18 we also report the sensitivity of the whole term structure to changes in correlation. Opposite to what happens for options that are long correlation, the effect of an increase in correlation is to flatten out the term structure of the product.

6

Multivariate Credit Products

Credit was the first field of finance to see copula function applications. The reason is that multivariate credit products are intrinsically non-Gaussian, since the possibility of events of default implies a huge downside risk that cannot be captured by standard elliptical distributions. Typically, this asymmetry persists even if a credit portfolio is widely diversified, and this is due to the fact that default events may be highly dependent on a common systematic factor, and this commonality may also get more relevant in periods of crisis. The recent global credit crisis started in 2008 is the most clear evidence of this regularity. Moreover, in applications in which the set of credit exposures is limited, when one tries to simulate default scenarios the relevant variable is the time at which default of an obligor takes place, and time cannot be modeled like any other variable, at least for the trivial (even though maybe disturbing) reason that it must be defined on the positive line. It is then no surprise that this tool was among the first applied to the study of dependence in credit exposures. The first application goes back to the beginning of the century, and is due to Li (2000). Excellent reviews of copula applications to finance may be found in McNeil *et al.* (2005), Burtschell *et al.* (2007), and the book by Schönbucher (2003).

6.1 CREDIT TRANSFER FINANCE

Credit risk has always been the most formidable source of risk. Even sticking to the last hundred years, the two worst crises, that of 1929 and the one that we are still living through, were inherently crises of credit. It does not come as a surprise that financial innovation in the last decade of the last century has focused on products to allocate credit risk among institutions. Despite the fact that these products are blamed as one of the reasons for the crisis, they remain an indispensable tool for efficient management of credit risk in portfolios.

Multivariate credit products have developed for several reasons, on both the supply and demand sides. As for supply, financial institutions have created instruments to transfer and manage credit exposures with more flexibility. The reason for this flexibility was not only directed to the virtuous goal of more efficient implementation of credit management decisions, but also to the less honorable search for *regulatory arbitrage* opportunities. In something of a paradox, the credit risk regulation launched at the international level with the so-called Basel I agreements at the end of the 1980s spurred the banks to transfer credit exposures to the market in an attempt to retain on their balance sheet those that were most profitable, and so most risky and illiquid. Another reason for resorting to credit risk transfer policies was the search for funding and the creation of fee income in what is now known as the *originate-to-distribute* business model. On the demand side, the success of credit transfer products was also due to the existence of regulation. On the one hand, investors subject to regulation – like mutual and pension funds and insurance companies – were in a desperate search for diversification, and the possibility of investing in products sensitive to new risk factors, in which they were not allowed to invest directly, was an important opportunity to achieve the task. On the other hand, alternative investment managers – such as hedge and private equity funds – were provided with an opportunity to manage the credit risk of particularly risky, and so profitable, exposures. Notice that this need to discriminate the output of the credit risk transfer process according to the different appetite for risk of the investors requires us to differentiate the securities offered to the final investors in terms of *seniority*. This is accomplished by applying a technique called *tranching*. More risk-averse investors are offered products with higher credit protection, and remain exposed to credit events that are far in the right tail of the distribution. On the other hand, investors with a high appetite for risk acquire positions that leave them exposed to the first credit losses occurring in the portfolio. As we will see, it is this feature of the credit transfer process that calls into play the degree of dependence among credit events in the pool that is transferred.

6.1.1 Univariate Credit Transfer Products

When transferring risks the usual method that comes to mind is that of a swap. Just like interest rate risk can be switched by entering an interest rate swap, credit risk may be transferred by underwriting suitable swap transactions. In the practice of credit risk transfer, we can distinguish two classes of swaps, depending on whether the underlying is a specific asset or an event. As for the first class, typical deals are *asset swaps* or *total return swaps*. The second class is represented by the most famous and liquid market for the transfer of credit risk, *credit default swaps* (*CDS*). The former class includes derivative contracts written on assets instead of credit events, and for this reason these contracts enable us to transfer credit risk inasmuch as it is recorded in prices, and it may be somewhat garbled by other factors. The example below makes the point clear.

Example 6.1.1. *Asset swap spread. An asset swap is a package of a bond and a swap. With respect to the standard swap contract, the asset swap includes an* upfront *payment (that is a payment at the origin of the contract) equal to the difference between the value of the bond and par. Intuitively, if the value of the bond is below par, the part contributing the asset would pay the difference between par and the value of the bond. If instead the price of the bond is above par, then the part contributing the swap would pay the difference. For the rest, the swap contract would require floating payments plus a spread, against payment of the coupons of the asset. The spread is determined to equate the present value of payments. We then have*

$$spread = (coupon - swap\ rate) + \frac{1 - DP}{\sum_{i=1}^{n} v(t, t_i)},$$

where DP is the price of the bond, and $v(t, t_i)$ is the discount factor extracted from the swap curve. Notice that if the market price of the bond were consistent with the discount factor, the part of the spread due to the upfront payment would be exactly the same amount as the difference between the coupon and the swap rate, with opposite sign, and the spread would be zero.

For example, the price of the Italian BTP (Italian long-term notes) with maturity February 2034 and coupon equal to 4% was 90.83% on September 30, 2010. Considering that on the same date the swap rate for the same maturity was 2.91%, the bond should have traded at a price above par. In fact, computing the asset swap spread one would obtain 1.62%.

In the example above, the 1.62% value may be due to the credit risk of the Republic of Italy, but also to the specific features of the bond on which the asset swap spread is calculated. Differently, the credit default swap is not affected by particular features of a single bond, because it is a contract in which insurance is paid if a credit event involving the Republic of Italy takes place, no matter on which bond the default event occurs.

Example 6.1.2. *Credit default swaps. A credit default swap is an insurance contract in which the party providing protection stands ready to pay the loss on a given notional amount the first time that a* credit event *involves a* reference entity, *called the* name, *within a given time horizon. The set of credit events triggering payment of protection is defined, according to the ISDA standard, among the following:*

- *bankruptcy,*
- *obligation acceleration,*
- *obligation default,*
- *failure to pay,*
- *moratorium/repudiation,*
- *restructuring.*

In case the credit event takes place, payment of the protection can take place either by physical delivery, *that is by accepting delivery of a bond issued by the* name *against payment of the nominal value, or by* cash settlement, *that is by cash payment of the loss, according to expert judgements. The party buying protection usually pays a fixed premium on a quarterly or semi-annual basis, until the end of the protection period or the time the event takes place, whatever comes earlier. At the end of September 2010, the annual insurance premium of the CDS on Republic of Italy was around 2%.*

Quotes for purchase and sale of protection are offered for different time horizons, even though the most liquid market typically corresponds to the five-year maturity. Notice that the CDS enables us to enact a credit transfer process to the general public of final investors, as shown in the following example.

Example 6.1.3. *Equity-linked note. Consider the following structure:*

- *Build a vehicle, that is a new corporate entity.*
- *The vehicle sells protection to a CDS dealer on a given name, say FIAT, and receives the payment of a premium on a running basis.*
- *The vehicle issues bonds to the general public and invests the proceeds at the risk-free rate.*

The flow of payments is as follows:

- *Before maturity, the vehicle pays to the final investors the revenues from the investment in the risk-free rate plus the premium earned from the sale of protection.*
- *In case of default, the vehicle pays the loss on the CDS deal and impairs the collateral by the same amount, so that the final investor would receive risk-free rate payments on an amount corresponding to the recovery rate.*

6.1.2 Multivariate Credit Transfer Products

Consider now the problem of transferring the credit risk of an entire portfolio of exposures. The first idea that comes to mind would be to buy insurance on each exposure separately. A problem with this may be that insuring everything may cost too much. As an alternative, one could think of buying insurance only on a portion of the exposures. For example, one could buy insurance on the first x defaults of the basket of exposures. These products are actually called *basket credit derivatives* and the first-to-default case analyzed in Example 6.1.2 is part of this class. If we denote by m the number of exposures of the basket, LGD the loss in case of default (by assumption the same for all the exposures) and Y the number of defaults occurred, the first-x-to-default product would give the payoff

$$LGD \, Y \, \mathbf{1}_{\{Y \leq x\}} + x \, LGD \, \mathbf{1}_{\{Y > x\}}.$$

Hence the value of this product is

$$FTD(x) = v(t, T) \left(LGD \sum_{i=1}^{x} i \, \mathbb{Q}(i) + x \, LGD \sum_{i=x+1}^{m} \mathbb{Q}(i) \right),$$

where $\mathbb{Q}(i)$ denotes the risk-neutral probability of observing i default events, and $v(t, T)$ denote the risk-free discount factor. This way, the institution holding the exposures may find someone with more appetite for risk willing to take over the riskiest part of the portfolio. The institution would then find itself facing only large credit crises causing more than x defaults: in other words, it would mainly face *systemic* shocks.

The same principle above is at the basis of the securitization business and the so-called *collateralized debt obligations* (*CDO*) or, more generally, *asset-backed securities* (*ABS*). To start, consider a standard, so-called *cash* CDO. The purpose of the deal is to sell a basket of assets, namely in our case a set of debt instruments, to the general public (mainly institutional investors), for reasons that may carry advantage both to the seller and the buyers. The deal is also referred to as *securitization* because it is bound to change into marketable securities whatever kind of asset the *originator* is willing to dispose of. The formal structure of a CDO deal closely resembles that of the equity-linked note in Example 6.1.3. On the selling side, the originator transfers a set of assets to an intermediary, called a special purpose vehicle (SPV). The SPV collects funds to pay for the assets by issuing securities. These securities are differentiated in a set of so-called *tranches*, defined by different degrees of seniority of debt. By seniority we mean a ranking according to which, in front of default events on the assets, the *senior tranches* are affected after the *junior tranches* are depleted. So, the income accruing from the assets sold from the originator to the SPV are passed through from the SPV to the final investors. Losses incurred on the assets are also passed through, but they affect, and eventually may fully erode, junior tranches before the senior ones even begin to be affected. The process by which losses on the asset side are transferred to the tranches is called *waterfall*. The first tranche to be affected by losses is the so-called *equity tranche* (in standard corporate finance, in fact, equity is the residual claim). Mezzanine is the tranche affected as soon as the equity tranche has been swept away. On the opposite side of the spectrum, senior and super-senior *tranches* are made much safer, because only massive losses can erode the principal invested in these *tranches*: as in the basket equity derivative case above, senior tranches are exposed to *systemic* shocks. To see how the two techniques are related, consider a random variable L representing losses on a portfolio of exposures. The payoff of the *equity tranche*, absorbing up to L_a losses, will be

$$(L_a - L)\mathbf{1}_{\{L \leq L_a\}},$$

whose value is

$$Equity(t) = v(t, T) \int_0^{L_a} (L_a - L)f(w)dw,$$

where $f(.)$ is the density of losses. This is nothing but a short position in the *first-x-to-default* product above. It is immediate to see that the value of this product at time t is that of a put option

$$Equity(t) = v(t, T)\mathbb{E}_t[\max(L_a - L, 0)].$$

The same technology of options can then be applied to *tranches*. Consider for example a spread between two equity tranches: one buys insurance on losses up to L_b and sells losses on the same pool up to L_a. The value would be that of a put option spread, namely

$$T(t; L_a, L_b) = v(t, T)[\mathbb{E}_t[\max(L_b - L, 0)] - \mathbb{E}_t[\max(L_a - L, 0)]],$$

which would give the value of a tranche absorbing losses from level L_a up to L_b. In the jargon, L_a is called the *attachment* of the tranche and L_b is the *detachment*. Notice that equity tranches are defined by setting the attachment equal to zero, namely $T(.; 0, L_i)$. Furthermore, it is easy to see that buying insurance against all losses above L_a would give

$$T(t : L_a, \bar{L}) = v(t, T)[\mathbb{E}_t[\max(\bar{L} - L, 0)] - \mathbb{E}_t[\max(L_a - L, 0)]],$$

where $\bar{L}$ is the maximum amount of losses. We could also use put–call parity to obtain

$$T(t : L_a, \bar{L}) = v(t, T)(\bar{L} - L_a) - v(t, T)\mathbb{E}_t[\max(L - L_a, 0)]$$

and the value of a senior tranche corresponds to a short position in a call option on losses. Notice that all tranches are affected in the same direction by an increase of expected losses on the pool of assets, while senior and equity tranches are affected with opposite sign by an increase in correlation among the losses. Namely, an upsurge of correlation increases the value of equity at the expense of senior tranches.

Example 6.1.4. *Tranche hedging.* *Typically the* delta *of the options on losses is used to hedge positions on tranches. So, a delta value of 20 on an equity tranche means that an exposure on this tranche can be hedged by buying protection against an exposure 20 times as much as the position in the tranche. If this hedging strategy is too expensive, one could consider hedging using other tranches. So, if the mezzanine tranche of the same CDO has a hedge of 5, one could hedge exposure to the equity tranche by buying protection on 4 times as much exposure on the mezzanine tranche* (mezz-equity hedging)*. Of course, these hedging strategies work as a first-order approximation, and may break down if correlation moves abruptly. Moreover, the hedge ratio is obviously contingent on the CDO pricing model that one uses. A reference case is the sudden drop in correlation that the market experienced around the first weekend of May 2005. Then, the change in correlation caused equity and mezzanine tranches to move in opposite directions, and lots of hedge funds following the mezz-equity policy experienced major losses.*

As for a taxonomy of CDO deals, a structure that has made the most of the market over the last decade is that of *synthetic CDOs*. By these contracts, one can avoid the actual sale of the assets and use credit derivatives to provide protection against losses on the assets themselves. Differently from the cash CDO, on the asset side the SPV is just selling protection, typically by means of CDS. As the CDSs are worth zero at origin, the funds raised in exchange for the issuance of bonds have to be invested in default-free collateral. The interest rate risk of this investment is typically hedged by an interest rate swap. The proceedings from interest payments and CDS spreads are passed over to final investors. If some default happens, the SPV pays for the losses by reducing the amount of collateral and the principal of the tranches accordingly. Of course, the reduction of the principal is structured according to the tranching and waterfall rules. Synthetic CDOs, in which on the liability side the SPV raises funds, are called *funded*. On the contrary, unfunded CDOs are structures in which both the asset and the liability sides of the SPV are represented by CDS: in this case, typically the SPV as a corporate entity is not even used in the deal.

6.2 CREDIT INFORMATION: EQUITY VS CDS

Before taking care of the dependence structure of credit risk exposures, it is important to say a few words on the main data sources from which information about the credit risk of each exposure can be extracted. Of course, if no information is available for the individual obligor, a route that can be taken is to search similarities with other default cases. This approach calls for a statistical analysis of so-called *scoring* models, that uses tools of multivariate statistical analysis. We will see some of these below.

A more significant source of information may come from market data if the liabilities of the obligor are traded. For a corporate obligor whose equity is listed on the market, the most informative information may be found in the price of common stock. The idea rests on the so-called structural approach to credit risk, that we will discuss below, but the intuition is straightforward. Since equity-holders are residual claimants, in case of default the value of equity is zero. Then, any model of the probability that the price of common stock may reach zero is a model of the probability of default. We leave aside here more complex phenomena of strategic debt servicing that may imply that the barrier is not actually zero: anyway, even accounting for this, the value of common stock in the event of default drops very close to zero, because no recovery rate is expected on equity. The most well-known example of this approach is Moody's KMV model.

Example 6.2.1. *Moody's KMV model. Building on the idea that equity is a call option (having positive value only if the market value of an asset is higher than that of liabilities), KMV extract a measure of credit risk from equity data. Differently from standard option pricing theory, however, the underlying asset of the option, that is the value of the firm, is not observed. The volatility of this value is not observed either, so that the value of equity is not sufficient to recover the probability that the value of this call option will reach zero. KMV uses a second equation involving volatility of common stock prices. The value of the firm V and its volatility σ_V are then used in a single measure of* distance-to-default *(DD):*

$$DD = \frac{\ln(V/B) + \left(\mu - 0.5\sigma_V^2\right)}{\sigma_V},$$

where B is the amount of debt on a one-year horizon, and μ is the expected profit rate of the firm. Distance-to-default data are then used to determine empirical frequencies of defaults over the one-year horizon.

For large firms, financial institutions and countries, there is another source of information that has taken over the role of the most relevant indicator of credit risk. This is the market of credit default swaps illustrated above. The reason why this market became so important is that it is the only market, besides equity, in which the *name* of the company is traded, rather than a single bond issued by it. In other words, the whole debt of the obligor is the determinant of the CDS value. This endows this market with the same role that futures markets have for commodities. Everyone who is interested in hedging a bond issued by a name can buy protection by going long a CDS contract. On the other side of the deal, everybody who wants to take a positive position on the credit standing of the name can write a CDS contract, without being involved in buying bonds whose return may be garbled by liquidity problems. Furthermore, CDSs are quoted for different maturities and from the term structure of spreads it is possible to construct the entire term structure of default risk of the *name*. This procedure is called *bootstrapping* of default probability.

Example 6.2.2. *Bootstrapping of default probability.* *For the sake of simplicity, assume payments are made every year and, in case the credit event takes place, payment of protection is made at the end of the year. Denote by s_n the CDS premium to be paid every year for protection lasting n years. Then, since CDS contracts must be worth zero at origin, we have*

$$v(t, t_1)s_1 = v(t, t_1)(1 - \mathbb{Q}(1))LGD,$$

where $\mathbb{Q}(k)$ denotes the probability of survival of the name *k years and beyond. Then, the implied survival probability in one year is given by*

$$\mathbb{Q}(1) = 1 - \frac{s_1}{LGD}.$$

For the two-year contract we have instead

$$v(t, t_1)s_2 + v(t, t_2)\mathbb{Q}(1)s_2 = (v(t, t_1)(1 - \mathbb{Q}(1)) + v(t, t_2)(\mathbb{Q}(1) - \mathbb{Q}(2)))LGD.$$

Since the $\mathbb{Q}(1)$ value is known from the one-year contract, we can solve the equality for $\mathbb{Q}(2)$. Then, we may iterate the procedure to get $\mathbb{Q}(3)$ and so on. In general, we would have

$$\mathbb{Q}(k) = \mathbb{Q}(k-1)\left(1 - \frac{s_k}{LGD}\right) - \frac{s_k - s_{k-1}}{v(t, t_k)LGD}\sum_{i=1}^{k-1} v(t, t_i)\mathbb{Q}(i-1).$$

Once the term structure of survival probabilities is recovered, one can either determine the term structure of default probability in each year or the cumulative probability of default. In Figure 6.1 we report the conditional one-year probability of default of the Republic of Italy.

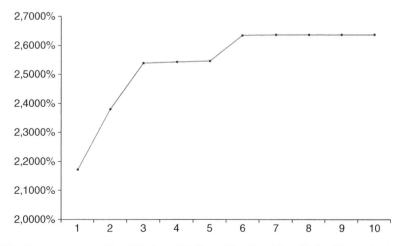

Figure 6.1   Term structure of the default probability of the Republic of Italy: 10-year horizon.

Since 2003 a new market was created to extend the possibility of hedging and investing in the credit risk of portfolios. These products consisted of standardized unfunded CDOs written on a standard set of names (a credit index) with standard maturities, attachment, and detachment points. Currently, the most famous among these products are the *iTraxx* and *CDX* indices, which are based on the most representative 125 *names* for different markets. iTraxx are specialized in the European market, but indices are also available for the Australian and Asian markets, while CDX mainly refer to the US market. Curiously, CDX and iTraxx have different detachment points, except for the equity tranche. Being synthetic unfunded, the value is quoted in terms of running basis premia, except for the equity tranche that is quoted *upfront* plus 500 bp running premium. These products may be used to hedge tranches of specific CDOs (so-called *bespoke CDOs*) once a hedge ratio is suitably selected. So, we may say that just like CDSs play the role of futures markets for single-name credit risk, CDX and iTraxx are meant to represent a sort of futures market for hedging portfolio credit risk. Following the same idea, other credit indices have been launched, such as ABX indices, which are meant to represent a benchmark for large credit portfolios with retail customer credit and mortgages as collateral, and Sovx, indices of sovereign risks. Differently from the CDX and iTraxx indices, a market for tranches on these indices has not developed yet, at least to the best of our knowledge.

6.3 STRUCTURAL MODELS

We now introduce the main credit risk models, with special focus on how the different approaches suggest different dependence structures and multivariate models. The first set of models is the so-called *structural approach* to credit risk. This approach is based on the idea that corporate liabilities, starting from plain equity and debt, are actually derivatives, or portfolios of derivatives, written on the value of the firm (that is, the value of its assets). The first idea that comes to mind when one is in front of options is to apply Black and Scholes. This seminal model, that started all the option pricing literature, is based on the Gaussian distribution. Building on this, we will see that this approach quite naturally leads to the Gaussian copula for multivariate credit exposures.

6.3.1 Univariate Model: Credit Risk as a Put Option

We will describe the structural approach to single-name credit risk in a nutshell, referring to the seminal papers by Merton (1974) and Black and Cox (1976). Assume a bond is issued to fund an entrepreneurial project that will give its payoff at time T. The payoff is denoted $V(T)$. Assume this project is funded using a zero-coupon bond whose face value is B and with maturity matching the end of the project T. At maturity, the value of debt will be

$$D(T, T) = \min(B, V(T)).$$

In order to isolate the non-linearity due to default risk, notice that we have

$$D(T, T) = B - \max(B - V(T), 0).$$

Default risk is then measured as a short position in a put option written on the value of assets for a strike price equal to the face value of debt. By the same token, using the decomposition

$$D(T, T) = V(T) - \max(V(T) - B, 0),$$

it is easy to see that what is left after repayment of debt, that is equity, is a call option with the same underlying, strike price, and exercise dates. If one assumes that the value of the firm follows a geometric Brownian motion, its value is given by the risk-neutral expectation of a log-normal variable. The probability that its value would be less than the amount of debt B is then

$$\mathbb{P}(V(T) \leq B | V(0)) = \mathbb{P}(\ln(V(T)) \leq \ln(B) | V(0)) \sim \Phi \left(-\frac{\ln(V(0)/B) + \left(\mu - 0.5\sigma_V^2 \right) T}{\sigma_V \sqrt{T}} \right),$$

where $\Phi(.)$ is the standard normal distribution, μ is the instantaneous percentage drift, that under the risk-neutral measure is equal to the instantaneous risk-free rate r. Notice that the argument of the probability distribution is actually the *distance-to-default* in the Moody's KMV model above (which uses the objective measure). Essentially, the survival probability may be seen as a function of the ratio of debt to the value of the firm, and of the volatility of assets. If we measure debt by discounting its face value with the risk-free rate, we get the variable used in the seminal Merton (1974) paper, called the *quasi-debt-to-firm-value* ratio (or *quasi-leverage*) $d = \exp(-rT)B/V(0)$, and the probability of default is actually the complement:

$$\mathbb{P}(V(T) \leq B | V(0)) \sim \Phi(-DD(d; \sigma_V)),$$

where $DD(.)$ is the function determining the *distance-to-default*. Many extensions have been provided to this early idea. One of the first and most interesting is that of allowing for default prior to the end of the project and the maturity of the bond. This extension was first studied by Black and Cox (1976), who introduced *covenants* in the debt contract. Covenants are trigger rules written in the contracts, so that default occurs the first time the value of the firm reaches one of these triggers. This idea corresponds to extending the option-based approach above to barrier options.

Example 6.3.1. *Black–Cox approach.* *The Black–Cox approach transforms the survival probability of the firm to a condition that depends on the path of the value of assets. We can then apply the representation presented in Chapter 5 concerning barrier options. The default probability can then be expressed with reference to the running minimum of the value of assets, $m(T)$, instead of the level at maturity T:*

$$\mathbb{P}(\tau \leq T) = \mathbb{P}(m(T) \leq B),$$

where τ denotes the time of default.

6.3.2 Multivariate Model: Gaussian Copula

If we extend the original structural Merton approach to a multivariate setting, we immediately obtain the Gaussian copula, which was severely put under question in the discussion following the recent credit crisis. Assume we have a set of firms V_i, for which

$$\mathbb{P}(V_i(T) \leq B_i | V(0)) \sim \Phi \left(-DD(d_i; \sigma_{V_i}) \right) = u_i,$$

where u_i is in $[0, 1]$. The natural extension of this model to recover the joint survival distribution of a set of m names immediately yields

$$\mathbb{P}(V_1(T) \leq B_1, \ldots, V_m(T) \leq B_m | V_1(0), \ldots, V_m(0)) = \Phi_m \left(\Phi^{-1}(u_1), \ldots, \Phi^{-1}(u_m); \mathbf{R} \right),$$

where $\Phi_m(.)$ denotes the m-dimensional Gaussian distribution with correlation matrix $\mathbf{R}$. The choice of this copula was put into question, particularly because correlation is kept constant over all the distribution and the copula displays no tail dependence. These shortcomings could actually be overcome by making correlation stochastic and changing the probability law from Gaussian to some other elliptical distribution with tail dependence, such as *Student's t*.

Besides these extensions, there is an issue concerning the specification of the correlation matrix from data, particularly in cases in which the copula has particularly large dimensions. A typical shortcut that is often used in statistics is to reduce the dimension of the problem by data compression techniques, such as factor models. This idea gives rise to the concept of *factor copula*. We stick here to the usual case of a single factor, which is typically used in the market. The assumption is that there exists a standard normal credit driver determining default of each obligor k as a function of a common factor y, which is also assumed to be standardized and normally distributed, and a white-noise disturbance ϵ_i, independent of y. The relationship is assumed linear, and written as

$$v_i = \rho_i y + \sqrt{1 - \rho_i^2} \epsilon_i.$$

The conditional probability of default is then

$$\mathbb{P}(v_i \le x_i | y) = \mathbb{P}\left(\epsilon_i \le \frac{x_i - \rho_i y}{\sqrt{1 - \rho_i^2}}\right) = \Phi\left(\frac{\Phi^{-1}(p_i) - \rho_i y}{\sqrt{1 - \rho_i^2}}\right), \tag{6.1}$$

where $p_i = \Phi(x_i)$ (that is, the marginal default probability of the ith exposure).

By conditional independence, the joint default probability can be recovered by the product of conditional probabilities and integration over the common factor:

$$\mathbb{P}(V_1(T) \le B_1, \ldots, V_m(T) \le B_m) = \int_{-\infty}^{+\infty} \prod_{i=1}^{m} \Phi\left(\frac{\Phi^{-1}(p_i) - \rho_i y}{\sqrt{1 - \rho_i^2}}\right) f(y) dy,$$

where $f(y)$ is the density of the common factor, that is assumed normal.

Implied Correlation

The Gaussian copula is also used as an alternative quotation device with respect to premium, that is *implied correlation*. Like implied volatility in option pricing, implied correlation is that level of correlation (assuming correlation to be the same across all of the names) which produces, using the standard pricing formula, the prices observed in the market. So, parallel to what happens in option pricing, the standard pricing model for tranches is also based on the assumption of a Gaussian dependence structure, and uses the Gaussian copula model.

What is strange, and breaks down the similarity between implied correlation and implied volatility, is that, at least at the beginning, the implied correlation was applied to all the tranches, including the intermediate ones. Remembering that intermediate tranches are like option spreads, it is easy to see that this is incorrect. Call spread prices may be (and usually are) non-monotone in volatility, just like intermediate tranches are generally non-monotone in correlation. The implied correlation concept can instead be correctly applied to equity tranches or senior tranches, that correspond to positions in options and may be proved to be monotone in correlation. For this reason, after some years in which implied correlation had been applied to all the tranches, a new concept was introduced, and it was called *base correlation* to distinguish it from the past practice. *Base correlation* is the implied correlation of equity tranches. So, for example, the base correlation for the CDX index refers to the correlation of the 0–3% equity tranche, the 0–7% equity tranche, and so on. The price of the mezzanine tranche is then obtained by arbitrage, as specified above, by computing the difference between the two values.

Example 6.3.2. *Correlation skew. Just like in option pricing applications, one can design smiles or skews in correlation. The idea is to recover base correlations for all detachment values, in such a way as to fit the prices of all tranches. So, the algorithm would recall other applications of the bootstrapping principle and would sound something like the following:*

1. *Evaluate the base correlation of the first equity tranche.*
2. *Use the intermediate tranche to update the equity tranche.*
3. *Compute the base correlation of the new equity tranche.*
4. *If there are no more tranches exit, otherwise go back to step 2.*

6.3.3   Large Portfolio Model: Vasicek Formula

As a further shortcut that is used for portfolios of particularly high dimensions, it is typically assumed that the portfolio is *homogeneous*, meaning that

- all exposures have the same default probability $p_i = p$;
- all exposure have the same correlation with the common factor: $\rho_i = \rho \in [0, 1]$.

An interesting question is the asymptotic distribution of the joint probability of default in this model. Vasicek (1991) showed that this distribution is given by

$$F(q) = \mathbb{P}(L \le q) = \Phi\left(\frac{\sqrt{1 - \rho^2}\Phi^{-1}(q) - \Phi^{-1}(p)}{\rho}\right),$$

and the corresponding density is given by

$$f(q) = \sqrt{\frac{1 - \rho^2}{\rho^2}} \exp\left(0.5(\Phi^{-1}(q))^2 - \frac{1}{2\rho^2}\left(\Phi^{-1}(p) - \sqrt{1 - \rho^2}\Phi^{-1}(q)\right)^2\right),$$

where L is now the percentage of defaults in the asymptotic portfolio.

The idea of the proof is the following: let L_i be the gross loss of the ith loan, so that $L_i = 1$ if the ith name defaults and $L_i = 0$ otherwise. When the common factor is fixed, the conditional probability of loss on any name is

$$p(y) = \mathbb{P}(L_i = 1 | Y = y) = \Phi\left(\frac{\Phi^{-1}(p) - y\rho}{\sqrt{1 - \rho^2}}\right)$$

where, in the last equality, we have used (6.1). Conditionally on the value of the common factor, the variables L_i are independent equally distributed random variables with finite variance. The law of large numbers implies that the average loss $L = \frac{\sum_{i=1}^{n} L_i}{n}$ converges to $p(y)$. Then, one has

$$\mathbb{P}(L \le q) = \mathbb{P}(p(y) \le q) = \mathbb{P}(y > p^{-1}(q)) = \Phi(-p^{-1}(q))$$

and the result follows by substituting equation (6.1).

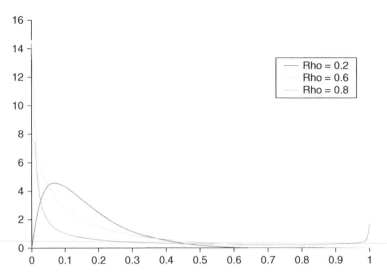

Figure 6.2 Density function of the Vasicek distribution.

The shape of the density function is reported in Figure 6.2. We see that the distribution is unimodal for $\rho > 0.5$ and bimodal otherwise. This approximation is largely used and provides a quite good approximation, except for very low correlation, and less than 20 exposures.

6.4 INTENSITY-BASED MODELS

Even though they represent the most elegant and informative approach to credit risk analysis, structural models face serious limits when compared with real market data. The main problem is that they generate a credit spread term structure that is too low with respect to the values observed in the market. This undervaluation is even more relevant for short maturities, due to the fact that the value of the assets is modeled as a diffusive process and is assumed to be observed at each point in time. A point that is even harder to address is how to model the process and outcome of a credit event, and the recovery rate that results. Actually there exists a rich stream of literature addressing these issues and amending the approach. Going over these models is beyond the scope of this book, so that we follow the typical response that in practical applications was given to these limitations. The answer was to introduce maximum flexibility in the model and to drop the economic content typical of structural models in favor of an approach with pure statistical content. The basic idea was to search for the most tractable and effective model to represent the two variables that define credit risk: (i) the probability of a default event; (ii) the recovery rate in case of default (or its complement, the loss given default). Notice that in the structural approach both these variables were endogenously determined by the unique state variable, namely the value of the firm. Now that both variables can be modeled independently, the first thing that comes to mind is to keep the loss-given-default figure fixed (or anyway independent from the event of default), and to model the credit event by the most straightforward process, the Poisson process. Since the Poisson process is characterized by a parameter, called intensity, models of this class are also called *intensity-based* (they are also referred to as *reduced-form* models, borrowing the term from the econometric literature).

6.4.1 Univariate Model: Poisson and Cox Processes

Poisson processes model the arrival of independent discrete events. Then, the survival probability of a firm, or of a portfolio of firms, is given by the probability that no credit event would take place in a period of time $[0, T]$. Formally, this is given by

$$\mathbb{P}(\tau > T) = \exp(-\lambda T),$$

where τ is the time of default and λ is the intensity parameter of the Poisson process. Notice that this representation is extremely suggestive for specialists in term structure theory (where intensity corresponds to the instantaneous interest rate). So, just like structural models provide option-like applications to credit risk, the reduced-form approach allows us to apply all the models that were designed to describe term structure dynamics. Actually, if one assumes zero recovery rate and independence of credit risk and interest rate risk, one obtains

$$D(t, T) = v(t, T)\mathbb{P}(\tau > T) = \exp(-(r + \lambda)T),$$

where r represents the interest rate computed with continuous compounding. It is easy to see that λ immediately reports a measure of the credit spread implied by the price of a defaultable bond. The approach can easily be extended to allow for dependence between interest rate and credit risk (see Duffie and Singleton, 1999), but this is beyond the scope of this book. What is instead important here is how to extend this approach to a multivariate setting.

6.4.2 Multivariate Model: Marshall–Olkin Copula

Mai and Scherer (2009a,b) provide a multivariate default model that has the extendable Marshall–Olkin copula as its own dependence structure and admits a mathematical viable large homogeneous portfolio approximation. Notice that if the credit exposures in the portfolios were independent, then the multivariate extension of the intensity-based model would be immediate. Actually, the probability of observing zero credit events in a T-year time span would be

$$\mathbb{P}(\tau_1 > T, \ldots, \tau_m > T) = \exp(-(\lambda_1 + \ldots + \lambda_m)T),$$

and the intensity of the portfolio would merely be the sum of individual (marginal) intensities. The question is how to extend this idea to allow for dependence among the default events. Marshall and Olkin (1967a,b) provided a straightforward answer to this problem. The catch is to include new Poisson processes in the model, independent of the others, causing failure of more than one element in the system. These joint shocks then induce dependence in the failures of several components in the system.

To explain the model, we start from a bivariate case. We then have three Poisson processes with intensities λ_1, λ_2, and λ_{12}. The first two intensities refer to idiosyncratic shocks, while the last one refers to a shock that is fatal to both the components of the system (in our case, both the exposures). Corresponding to these shocks there are three variables denoting the waiting time of these shocks, that we could call z_1, z_2, and z_{12} respectively. It is clear to see that the default time of each exposure will be the minimum of the idiosyncratic and the common shock, namely: $\tau_1 = \min(z_1, z_{12})$ and $\tau_2 = \min(z_2, z_{12})$. Then, since the Poisson processes are independent, we have

$$\mathbb{P}(\tau_1 > T_1, \tau_2 > T_2) = \exp(-\lambda_1 T_1 - \lambda_2 T_2 - \lambda_{12} \max(T_1, T_2)).$$

In order to rewrite this joint distribution as a copula function, notice that the marginal probability of default of each of the two exposures is

$$\mathbb{P}(\tau_i > T_i) = \exp(-(\lambda_i + \lambda_{12})T_i) = \exp(-(\hat{\lambda}_i T_i)),$$

where $\hat{\lambda}_i$ denotes the marginal default intensity of obligor $i = 1, 2$. We now rewrite the joint distribution as a function of marginal default probabilities:

$$\mathbb{P}(\tau_1 > T_1, \tau_2 > T_2) = \exp(-\hat{\lambda}_1 T_1 - \hat{\lambda}_2 T_2 + \lambda_{12} \min(T_1, T_2)),$$

where we have used the identity $\max(T_1, T_2) = T_1 + T_2 - \min(T_1, T_2)$. What remains to be done to obtain a copula is to transform the marginals to variables with uniform distribution, but we already know that $u_i = \exp(-\hat{\lambda}_i T_i)$ is a univariate marginal probability. Then, it only remains to observe that we may write $\exp(\lambda_{12} T_i) = u_i^{-\alpha_i}$, with $\alpha_i = \lambda_{12}/\hat{\lambda}_i$. Collecting everything we have

$$C(u_1, u_2) = u_1 u_2 \min \left(u_1^{-\alpha_1}, u_2^{-\alpha_2} \right) = \min \left(u_1^{1-\alpha_1} u_2, u_1 u_2^{1-\alpha_2} \right), \qquad (6.2)$$

which defines the *Marshall–Olkin copula*.

A couple of points are worth noticing about the economic interpretation of the parameters of this copula model. First, notice that the parameters α_i that define the copula are actually the ratio of the intensity of the common shock and the marginal intensity of the credit event. An obligor with higher α_i values is more sensitive to the common shock. Second, since the parameters α_i are generally different from one obligor to the other, this implies that the joint probability of default is actually driven by the marginal probability of the obligor with the highest α_i parameter. In other words, the Marshall–Olkin copula is *non-exchangeable* and changes in the marginal default probability of different obligors have different effects on the joint probability of default, as the example below illustrates.

Example 6.4.1. *Consider two obligors with $\alpha_1 = 0.95$ and $\alpha_2 = 0.05$. Say they both have marginal probability of default of 5%. The joint probability of default turns out to be 0.29%. Now see what happens if the marginal probability of default of obligor 1 doubles: the joint probability doubles and we have 0.58%. If instead the default probability of the second obligor doubles, the joint probability becomes 0.56%.*

From the technical viewpoint, the Marshall–Olkin copula has the interesting property of having both an absolutely continuous and a singular component. The singular component is concentrated on the curve $u_1^{\alpha_1} = u_2^{\alpha_2}$ and the probability mass is

$$\mathbb{P}\left(U_1^{\alpha_1} = U_2^{\alpha_2} \right) = \frac{\alpha_1 \alpha_2}{\alpha_1 + \alpha_2 - \alpha_1 \alpha_2},$$

where $U_i, i = 1, 2$ are uniform random variables. The economic meaning is that if we simulate pairs of default times from a Marshall–Olkin copula we have a probability of extracting the same times equal to the measure of the singular component. Another important peculiarity of this copula is that both the *Pearson correlation* and the *Kendall's τ* parameters are equal to the measure of the singular component,

$$\rho = \tau = \frac{\alpha_1 \alpha_2}{\alpha_1 + \alpha_2 - \alpha_1 \alpha_2}.$$

The Spearman's ρ parameter is instead given by

$$\rho_S = \frac{3\alpha_1\alpha_2}{2\alpha_1 + 2\alpha_2 - \alpha_1\alpha_2}.$$

6.4.3 Homogeneous Model: Cuadras Augé Copula

Extending the Marshall–Olkin distribution to higher dimensions does not pose conceptual problems. Nevertheless, it raises huge practical problems for the specification of the dependence structure. Even if we consider a case with three obligors, we see that the dimension of the problem increases dramatically: actually we should take into account the idiosyncratic shocks, the shocks common to every couple, and the shock fatal to all the three obligors. The complexity of the problem then increases very quickly. For this reason, in applications it is usual to find restriction to two extreme cases: one could allow for bivariate shocks only, or one could consider a unique common shock affecting all the obligors. The latter solution is actually particularly suggestive because it reminds us of the factor copula model that we met in the structural approach. In this specification the marginal default intensity of each obligor $i = 1, 2, \ldots, m$ may be decomposed as

$$\hat{\lambda}_i = \lambda_i + \bar{\lambda},$$

where $\bar{\lambda}$ denotes the intensity of the common shock. Just like in the Gaussian factor model, the entire dependence structure of the system can be reconstructed knowing the intensity of the common shock. For every pair of names, the correlation of default times will be

$$\rho_{ij} = \frac{\bar{\lambda}}{\lambda_i + \lambda_j + \bar{\lambda}},$$

for all $i, j = 1, 2, \ldots, m, i \neq j$.

Just like for the Gaussian model, when we would like to extend the analysis to a large portfolio, it would be natural to require, at least as an approximation, that the portfolio of exposures be homogeneous, meaning to have the same default probability for every obligor and the same dependence structure for all pairs. As for the latter point, this particular specification transforms the Marshall–Olkin copula in an exchangeable distribution and for this reason it is called an *exchangeable Marshall–Olkin* or *Cuadras–Augé* copula from the authors that first proposed it. In this case, we have $\alpha_i = \alpha$ for all i and the copula becomes a geometric mixture of the cases of perfect positive dependence and independence:

$$C_\alpha(u_i, u_j) = [\min(u_i, u_j)]^\alpha (u_i u_j)^{1-\alpha}, \tag{6.3}$$

for all pairs $i, j = 1, 2, \ldots, m$. If we also add the restriction that there is a unique common shock in the system, we also necessarily obtain $u_i = u_j = u$ and the portfolio of credit exposures is homogeneous.

A Cuadras–Augé Filter

Consider the following problem. We want to represent a multivariate Marshall–Olkin set of credit exposures by a homogeneous portfolio as close as possible to the original one. Say that by the term "as close as possible" we mean that the homogeneous portfolio must have the same default probability and the same correlation as the average default probability and correlation of the original portfolio. We show that this can be accomplished very easily by filtering the average default probability of the original portfolio by a suitable function of the average correlation. Baglioni and Cherubini (2010) called this technique the *Cuadras–Augé filter*. To see how the filter is derived, consider that the problem is to replace the Pearson correlation statistics ρ_{ij} by a single correlation figure ρ. We remember that in a general Marshall–Olkin system the correlation is given by

$$\rho_{ij} = \frac{\alpha_1 \alpha_2}{\alpha_1 + \alpha_2 - \alpha_1 \alpha_2} = \frac{\lambda_{ij}}{\hat{\lambda}_i + \hat{\lambda}_j - \lambda_{ij}},$$

where $\hat{\lambda}_i$ denotes the marginal intensity. In the homogeneous system ($\hat{\lambda}_i = \hat{\lambda}$, $\lambda_{i,j} = \hat{\lambda}$ for all i, j) that we want to recover, we will then have

$$\rho = \frac{\bar{\lambda}}{2\hat{\lambda} - \bar{\lambda}}.$$

A straightforward choice for the homogeneous marginal intensity $\hat{\lambda}$ would be to take the average of marginal intensities

$$\hat{\lambda} = \sum_{i=1}^{m} \frac{\hat{\lambda}_i}{m}.$$

Now the problem is to recover the common intensity $\bar{\lambda}$ in such a way as to match the average (defined in some suitable sense) correlation. It turns out that in the case at hand it is well suited to work with the *harmonic mean*. We remember that the harmonic mean is defined as

$$H = \frac{k}{\sum_{i=1}^{k} \frac{1}{X_i}},$$

for a set of k positive variables X_i. Let us now work with the average value of the inverse of ρ_{ij} in the m-dimensional correlation matrix:

$$\frac{1}{H} = \frac{2}{m(m-1)} \sum_{i=1}^{m-1} \sum_{j=i+1}^{m} \frac{1}{\rho_{ij}}. \tag{6.4}$$

Notice that the sum includes $m(m-1)/2$ terms, that is the number of elements above the diagonal of the correlation matrix. We now rewrite each correlation as a function of marginal intensities and common intensity:

$$\frac{1}{H} = \frac{2}{m(m-1)} \sum_{i=1}^{m-1} \sum_{j=i+1}^{m} \left(\frac{\hat{\lambda}_i + \hat{\lambda}_j}{\lambda_{ij}} - 1 \right). \tag{6.5}$$

If we substitute $\bar{\lambda}$ for λ_{ij} and consider the number of elements in the sum, we obtain

$$\frac{1}{H} = \frac{2}{m(m-1)} \sum_{i=1}^{m-1} \sum_{j=i+1}^{m} \left(\frac{\hat{\lambda}_i + \hat{\lambda}_j}{\bar{\lambda}} \right) - 1. \tag{6.6}$$

Now, since in the double sum each marginal intensity $\hat{\lambda}_i$ appears $m-1$ times, we have

$$\frac{1}{H} = \frac{2}{m} \sum_{i=1}^{m} \left(\frac{\hat{\lambda}_i}{\bar{\lambda}} \right) - 1 = 2\frac{\hat{\lambda}}{\bar{\lambda}} - 1, \tag{6.7}$$

where we substituted the average marginal intensity. Rearranging, we obtain

$$\bar{\lambda} = \frac{2}{1 + \frac{1}{H}}\hat{\lambda} \tag{6.8}$$

and we have a filter that extracts an intensity representing the common components of the marginal intensities. Notice that in general this is not an estimator of the intensity of the common shock, but can only be considered as a synthetic representation of the portfolio, just like in the Gaussian factor copula. Only if it happens that the data are actually generated by a Marshall–Olkin model with a unique common shock, is the filter actually an estimator of the intensity of the common shock. Notice that the same can be done using rank correlation instead of correlation, in which case we use the relationship

$$\rho_{s,ij} = \frac{3\lambda_{ij}}{2(\hat{\lambda}_i + \hat{\lambda}_j) - \lambda_{ij}},$$

where $\rho_{s,ij}$ denotes the Spearman's ρ rank correlation between exposure i and j. After some computation we obtain

$$\bar{\lambda} = \frac{4}{3}\frac{1}{\frac{1}{3} + \frac{1}{H_s}}\hat{\lambda},$$

where H_s is the harmonic mean of rank correlations. In Figure 6.3. we report an example of application of the filter to a sample of eight German banks from January 2008 to September 2010. The filter was computed from rank correlation estimates on a rolling window of one year.

Figure 6.3   An application of the Cuadras–Augé filter to a set of German banks.

6.5 FRAILTY MODELS

Let us see how introducing a slight modification in a reduced-form approach may pave the way for application of one of the main families of copula functions, Archimedean copulas. The idea is borrowed from actuarial and reliability applications, and is known as a *frailty model*. The frailty is an unobserved random variable that multiplies the hazard rate (or intensity) of an event. At the basis of the application there is an algorithm due to Marshall and Olkin for the simulation of Archimedean copulas.

Example 6.5.1. *(Marshall–Olkin algorithm, 1988)*

1. *Draw m independent variables with uniform distribution $U_1, \ldots, U_m$.*
2. *Draw a positive Y with distribution $G(.)$ and denote by $\phi^{-1}(.)$ the corresponding Laplace transform.*
3. *Compute*

$$X_i = \phi^{-1}\left(-\frac{1}{Y}\ln U_i\right).$$

The dependence structure of the variables X_i is represented by an Archimedean copula with generator $\phi(.)$.

In the above algorithm, Y is the frailty. The application to credit that we show below is due to Schönbucher (2003).

6.5.1 Multivariate Model: Archimedean Copulas

Let us start to check how the Archimedean dependence structure is impounded in the model. Each variable X_i is actually generated in such a way that the conditional distribution with respect to Y is

$$\mathbb{P}(X_i \leq x_i|Y) = \exp(-\phi(x_i)Y).$$

It is easy to see that Archimedean dependence is generated by using the law of iterated expectations and conditional independence:

$$\mathbb{P}(\mathbf{X} \leq \mathbf{x}) = \mathbb{E}\left[\prod_{i=1}^{m}\mathbb{P}(X_i \leq x_i|Y)\right] = \mathbb{E}\left[\prod_{i=1}^{m}\exp(-\phi(x_i)Y)\right] = \mathbb{E}\left[\exp\left(-\sum_{i=1}^{m}\phi(x_i)\right)Y\right].$$

Now remember the definition of Laplace transform and the fact that Y was actually generated from a distribution having Laplace transform $\phi^{-1}(.)$:

$$\mathbb{E}\left[\exp(-sY)\right] = \phi^{-1}(s).$$

We then have

$$\mathbb{E}\left[\exp\left(-\sum_{i=1}^{m}\phi(x_i)\right)Y\right] = \phi^{-1}\left(\sum_{i=1}^{m}\phi(x_i)\right)$$

and we know that by hypothesis, $\phi(.)$ is a generator of Archimedean copula functions.

In the Schönbucher model, the default of obligor i takes place whenever $X_i \leq p_i$, where p_i denotes the probability of default of obligor i over the investment horizon T. The conditional default probability is then

$$\mathbb{P}(X_i \leq p_i | Y = y) = \exp(-y\phi(p_i)). \tag{6.9}$$

6.5.2 Large Portfolio Model: Schönbucher Formula

Just like we did for the Vasicek formula, we can assume a homogeneous portfolio and recover an asymptotic approximation if such portfolio is very large. The key point is again the law of large numbers that for any realization of the mixing variable Y ensures that the percentage of defaults L in the large portfolio is $p(Y)$. The result follows by applying the conditional default probability in equation (6.9):

$$\mathbb{P}(L \leq q) = \mathbb{P}(p(Y) \leq q) = \mathbb{P}(\exp(-Y\phi(p)) \leq q) = \mathbb{P}\left(Y > -\frac{\ln q}{\phi(p)}\right) = 1 - G\left(\frac{\ln q}{\phi(p)}\right),$$

$$\tag{6.10}$$

where we remember that $G(.)$ is the distribution function of the mixing variable Y. The density is

$$f(q) = \frac{1}{q\phi(p)} g\left(-\frac{\ln q}{\phi(p)}\right),$$

where $g(.)$ is the density of the mixing variable, if it exists.

Example 6.5.2. *We remember here the distributions of the mixing variable leading to the most well-known components of the family of Archimedean copulas:*

- **Clayton.** *The mixing variable Y is Gamma with parameter $1/\theta$ (where θ is the parameter representing dependence). In this case, everything is available in closed form.*
- **Gumbel.** *The mixing variable has α-stable distribution with parameter $\alpha = 1/\theta$. Here, the form of the density is not known in closed form, but can be recovered by Laplace inversion from the known Laplace transform.*
- **Frank.** *The mixing variable in this case is discrete and has the distribution of logarithmic series, with $\alpha = 1 - \exp(-\theta)$.*

6.6 GRANULARITY ADJUSTMENT

How large is a large portfolio? This question sounds like a paradox of ancient Hellenistic philosophy. It was called the *sorites*, from the Greek term soros, which means "heap." How many grains of sand give rise to a heap? Our question is very similar. How many exposures constitute a well-diversified portfolio, that by the law of large numbers I could represent by a large homogeneous portfolio? Setting a cutoff point is impossible, like it is impossible to set a precise number of grains of sand to constitute a heap. What can be done is to measure how far we are from a well-diversified portfolio. This is typically done by adjusting the asymptotic distribution using a *granularity adjustment*. The idea exploits a result from Gouriéroux *et al.* (2000) and is based on a second-order Taylor expansion of the percentiles of losses around the fully diversified portfolios (Martin and Wilde, 2003).

The intuitive idea is that asymptotically the percentile of losses tends to the conditional probability of losses measured on the corresponding quantile of the conditioning variable, plus a second term that depends on the dimension of the portfolio. More formally, we denote by $\alpha(L)$ the percentile of losses, $\mu(L|Y)$ and $\sigma^2(L|Y)$ the mean and variance of losses conditional on the percentile of the factor, $h(Y)$ the density of the factor, and m_α its α-quantile. The Taylor expansion yields

$$\alpha(L) \cong \mu(L|Y = \mu_\alpha) - \frac{1}{2}\frac{\partial}{\partial Y}\left[\frac{\sigma^2(L|Y)\ln h(Y)}{\partial \mu(L|Y)/\partial Y}\right]. \tag{6.11}$$

Emmer and Tasche (2003) and Rau-Breedow (2002) provide a measure of granularity adjustment in the Gaussian model, while Gordy (2003, 2004) develops the same approach in a CreditRisk + framework, that is based on the frailty approach developed above.

6.7 CREDIT PORTFOLIO ANALYSIS

Up to now, we have been how to back out the distribution of losses from a set of exposures. Throughout all the models we have seen that when the number of obligors increases, specifying the risk of each exposure and the dependence on all of the others very soon becomes unmanageable, and one has to resort to asymptotic approximations, under the assumption that the portfolio is homogeneous. Granularity adjustment may improve the approximation, but the assumption of homogeneous exposures remains. This hypothesis may be too strong for many credit portfolios, which may include exposures that are different for geographical reasons, industrial sectors, type of obligor, and the like. For these reasons, before applying large portfolio approximations, it is important to assess the actual degree of homogeneity of the portfolio, and in case the test is not passed, to break the portfolio in subsets that may be considered homogeneous. The credit risk in these groups of exposures may be first analyzed in isolation by applying the large-portfolio approximation techniques, and then the dependence structure with other portions of the portfolio may be measured focusing on average cases in representation of each group. While doing this, we are actually imposing a *hierarchical structure* on the dependence specification of the portfolio. To borrow a concept from econometrics, homogeneous approximation may be used as *within estimators*, inside the groups, while standard low-dimensional copula models could be used as *between estimators*. An obvious question is how to split the portfolio into homogeneous groups. This is more art than science, and one should use information on the different credit risk drivers that may be present in the portfolio. Nevertheless, science can help: below we illustrate methods that can be used to partition the portfolio. These tools are called *unsupervised clustering* techniques in statistics, and may be used either to automatically generate partitions of the portfolio or to corroborate prior information that one may have.

6.7.1 Semi-unsupervised Cluster Analysis: *K-means*

K-means clustering is a partitioning method, i.e., the function *K*-means partitions the data into *K* mutually exclusive clusters and returns a vector of indices pointing to the clusters which have been assigned to each observation.

K-means treats each observation in a data set as an object having a location in space. It finds a partition in which objects within each cluster are as close as possible to each other, and as far as possible from objects in other clusters. Each cluster in the partition is defined by its member objects and by its *centroid*, i.e., the point to which the sum of distances from all objects in that cluster is minimal. The procedure uses an iterative algorithm that minimizes the sum of distances from each object to its cluster centroid, over all clusters. This algorithm moves objects between clusters until the sum cannot be decreased any further. The result is a set of clusters that are as compact and well-separated as possible.

K-means is a semi-unsupervised method since it requires choosing a distance metric and a number of clusters. Here we give a bird's-eye review of solutions available for this choice.

Choice of the Metrics

Metric selection is necessary to establish a rule for assigning patterns to the domain of a particular cluster. As expected, the most popular is the Euclidean distance, which is one of the most common so-called Minkowski distance metrics. With the Euclidean distance, the K-means algorithm tends to detect spherical clusters, and this may not always be a good choice *a priori*. Su and Chou (2001) proposed a new non-metric measure, based on the idea of symmetry, which can be used to group a given data set into a set of clusters of different geometrical structures. In Wang *et al.* (2006), a new data-dependent dissimilarity measure was designed, called the density-sensitive distance metric, which has the property of elongating the distance among points in different high-density regions and then can reflect the characters of data clustering. The density-sensitive K-means algorithm can identify non-convex clustering structures, thus generalizing the application area of the original K-means algorithm. In the same spirit, Xing *et al.* (2002) determine the metric as a parameterized Mahalanobis distance which solves a convex-optimization problem for a given data set.

Other contributions have focused more on the performance of a metric in a randomized data set, arguing that in the case of absence of noise of any kind, virtually all metrics are equivalent. In Francois *et al.* (2005) it is suggested to choose the metric according to the shape of the noise that is assumed on the data. While it is known that the Euclidean metric is optimal in the presence of white Gaussian noise, other types of noise may require other metrics.

Choice of the Number of Groups

The choice of the number of groups is a critical problem which involves the consideration of the trade-off between the complexity of the problem in terms of computing time and the precision of the method. Several papers address the problem of clusters number selection by using a K-means approach that exploits local changes of internal validity indices to split or merge clusters. In Pelleg and Moore (2000) this issue is addressed by developing a dynamic K-means variant known as X-means that performs bisecting splits of clusters where the resulting clustering solution is addressed by the Bayesian Information Criterion (BIC). Here, the bisecting K-means algorithm splits the data samples into two disjoint clusters and if the split increases the overall fitness measured by internal validity indices, the cluster is split and the bisecting K-means continues recursively. If none of the bisecting steps of each existing cluster leads to an improved result, the algorithm stops and returns the current clusters as result. In Muhr and Granitzer (2009) a comparison is proposed between an X-means method with several internal multivariate validity indices that measure cluster quality based on the total scatter of between-cluster and within-cluster sum of squares, and the Calinski and Harabasz index (Calinski and Harabasz, 1974), the Hartigan index (Hartigan, 1985), and the Krzanowski and Lai index (Krzanowski and Lai, 1985).

6.7.2   Unsupervised Cluster Analysis: Kohonen Self-organizing Maps

Self-Organizing Maps (SOM) are a type of artificial neural network with the capability of mapping high-dimensional continuously distributed input patterns into an ordered array of competing output units so as to capture the global structure of the input space through the use of local adaptation rules. This neural network produces a low-dimensional, discretized representation of the input space of the training samples, called a map. The model was first described as an artificial neural network by Kohonen (1982) and is sometimes called the Kohonen map. A self-organizing map consists of components called nodes or neurons; associated with each node is a weight vector of the same dimension as the input data vectors and a position in the map space. The self-organizing map describes a mapping from a higher-dimensional input space to a lower-dimensional map space. The procedure for placing a vector from the data space onto the map is to find the node with the closest weight vector to the vector taken from the data space and to assign the map coordinates of this node to the vector itself.

More formally, the Kohonen map is a two-layer, feed-forward network such that every output unit has incoming connection from every input unit. Let K be the number of units in the input layer and M the number of units in the output layer. For each output unit j, the weights of the corresponding input layer connections form a vector w_j of dimension K which can therefore be plotted in the same space as the input data. Output units compete so that, given an input pattern x, only one input unit $j(x)$ will become activated. This competition is based on the similarity between the input vector and each weight vector measured on Euclidean distance $\|w_j - x\|$. Thus, the active output unit is:

$$j(x) = \arg \min_{j \in \{1,\dots,M\}} \|w_j - x\|.$$

In the learning phase, the weight vector of the active output is updated so that it becomes closer to the input pattern by an amount proportional to a learning rate α. Even if in the conventional competitive learning model only the winning neuron should update its weight, lateral interactions between the output units are required to develop the self-organization. Kohonen proposed modulating the weight adaption according to a Gaussian neighborhood function $\Lambda : (j, r) \in \{1, \dots, M\}^2 \to [0, 1]$, such that

$$\Lambda(j, j(x)) = \exp\left(-\frac{\|j - j(x)\|^2}{2\sigma^2}\right),$$

where $\|j - j(x)\|$ is the distance between any output unit j and the activated one $j(x)$ given a certain metric defined over the input layer; σ is the size of the neighborhood around an output unit, i.e., how far should an output unit be from the activated one so as not to be affected by it. The learning rule is

$$w_y \leftarrow w_j + \alpha \Lambda(j, j(x))(x - w_j).$$

The output layer metric induces a topology. Provided a good match between the output layer topology and the topology of the input data exists, the network will develop into a topology-preserving map: the best-matching output units $j(x_1)$ and $j(x_2)$ corresponding to two input vectors x_1 and x_2 that are close in the input space will also be close in the output layer metric. Kohonen (1982) suggested shrinking the neighborhood size as the algorithm proceeds, in order to accelerate the convergence of the expected squared quantization error to a local minimum. In the Gaussian choice, that is done by slowly decreasing its width $\sigma(t) \to 0$ as $t \to \infty$.

Many variants of SOM are proposed in the literature; Mohebi and Sap (2009) proposed an optimized clustering algorithm based on SOM which employs the rough set theory and the simulated annealing as a general technique for optimization problems (Pawlak, 1982; Munakata, 2008). The idea is to implement a two-level clustering algorithm: the first level is to train the data by the SOM neural network and the second level is a rough set-based incremental clustering approach (Asharaf *et al.*, 2006), which will be applied on the output of SOM. The optimal number of clusters can be found by rough set theory, which groups the given neurons into overlapping clusters. Then the overlapped neurons will be assigned to the true clusters they belong to, by applying the simulated annealing algorithm which optimizes the clustering operation[1] by assigning the nearest cluster center to the overlapped neurons. A simulated annealing algorithm has been adopted to optimize the uncertainty that comes from the clustering operations in order to be as precise as possible.

6.7.3   (Semi-)unsupervised Cluster Analysis: Hierarchical Correlation Model

Hierarchical clustering groups data by creating a cluster tree or *dendrogram* that is a multilevel hierarchy where clusters at one level are joined as clusters at the next level. The first step consists of finding the similarity or dissimilarity between every pair of objects in the data set by computing a dissimilarity matrix. There are many ways to calculate this distance information, for example through the Euclidean distance between the objects. Once the proximity between objects in the data set has been computed, the distance information is used to determine the proximity of every object to each other: the procedure links pairs of objects that are close together into binary clusters, then the newly formed clusters are grouped into larger clusters until a hierarchical tree is formed. The cluster tree is easily understood when graphically represented as a dendrogram, where the horizontal axis reports the indices of the objects in the original data set and the links between objects are represented as upside-down U-shaped lines whose height indicates the distance. The cluster tree is determined as a function of a distance measure which allows us to link objects at some level; the height of the link represents the distance between two clusters that contain those objects (called the *cophenetic distance* between two objects). To verify that the distances (i.e., the heights) in the tree reflect the original distances, it is possible to compare the *cophenetics distances* with the original distance between data: the stronger the correlation between the linking and the distance between data, the better the performance of the clustering method. The *cophenetic correlation* allows us to compare the results of clustering using different distance and linkage methods and to choose those best suited to represent the original distances.

The last step of the hierarchical method consists of creating a partition of the data by taking into account the clustering tree. We can create these clusters by detecting natural grouping in the hierarchical tree or by cutting off the hierarchical tree at an arbitrary point; the natural cluster divisions in a data set are reached by comparing the height of each link in the cluster tree with the heights of neighboring links below in the tree (Figure 6.4). A link with approximately the same height as the links below it is said to be *consistent*, whereas a link whose height differs noticeably from the heights of the links below it is said to be *inconsistent*. The inconsistent links indicate the border of a natural division in a data set. The procedure uses a quantitative measure of inconsistency, i.e. an *inconsistency coefficient*, to determine where to partition the

---

[1] The optimization phase corresponds to a minimization of the energy function that sums the distances between each neuron i and each cluster center j, $F = \sum_i \sum_j d_{ij}$.

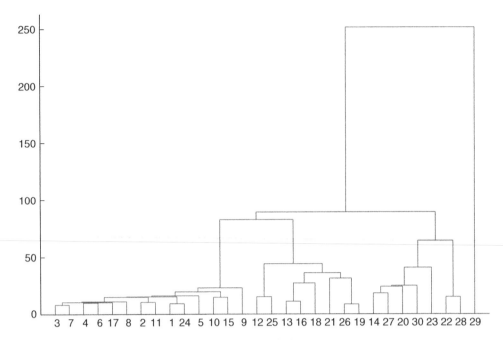

Figure 6.4   An example of a dendrogram of a hierarchical tree.

data set into clusters: each link in the cluster hierarchy is compared with the adjacent links that are less than two levels below it in the cluster hierarchy, determining the so-called *depth* of the comparison. The leaf nodes, that are the objects at the bottom of the cluster tree, have no further objects below them and then clearly their inconsistency coefficients are zero. This way, the inconsistency coefficients allow us to naturally divide the data into distinct, well-separated clusters identified in the corresponding dendrogram by densely packed objects in certain areas and not in others. Instead of a natural partition of data, it is possible to specify the desired number of clusters, which corresponds to applying a cutoff value to the inconsistency coefficient.

6.8   DYNAMIC ANALYSIS OF CREDIT RISK PORTFOLIOS

Up to this point, we have illustrated how to evaluate the credit risk of a portfolio for losses in a unit of time. Now we finalize our analysis by addressing the dynamic behavior of such losses. So, we assume we have estimates of the distribution of losses L_i where i denotes time on a schedule $\{t_1, t_2, \ldots, t_n\}$. Our task is to evaluate the dynamics of the cumulated losses, that we denote X_i. The outcome will be the entire term structure of the distribution of losses. Knowing this term structure is a crucial piece of information for designing credit transfer policies. In fact, if the risk on these losses were to be transferred to some counterpart, we would be able to evaluate different premia for the different time horizons of the insurance. Transferring the same risk for five years would cost less than insuring it for seven or ten years, and the difference would be given not only by the distribution of losses that will arise more than five years from now, but also by the dependence between those losses and the cumulated losses in the first five years.

Assume we have a sequence of credit losses L_i and one of cumulated losses X_i. We want to compute the distribution of

$$X_i = X_{i-1} + L_i,$$

knowing the distribution of L_i and the dependence structure with X_{i-1}. The problem is nothing but computing a C-convolution (see Chapter 3), that is the main feature of this book and that we report here for this special problem:

$$F_{X_i}(t) = F_{X_{i-1}+L_i}(t) = \int_0^1 D_1 C_{X_{i-1},L_i}\left(w, F_{L_i}\left(t - F_{X_{i-1}}^{-1}(w)\right)\right) dw, \qquad (6.12)$$

with $F_{X_1} = F_{L_1}$.

Here we describe the algorithm for the propagation of the probability distribution of losses based on the C-convolution approach.

Algorithm   Propagation of losses:
1. Start with the distribution of losses in the first period and set $X_1 = L_1$.
2. Numerically compute the integral in the C-convolution equation (6.12) yielding F_{X_2}.
3. Go back to step 2 and use F_{X_2} and F_{L_2} to compute F_{X_3}.
4. Iterate until $i < n$.

The result of the procedure would be a matrix representing the quantiles of the distribution for every time horizon up to t_n.

The Term Structure of CDX

We conclude this chapter with an application to the evaluation of the term structure of a CDX contract. In this example we put together the main concepts we have discussed throughout the

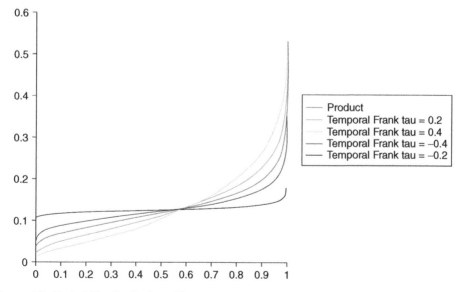

Figure 6.5   Probability distribution of losses: 10-year horizon, different temporal dependence values.

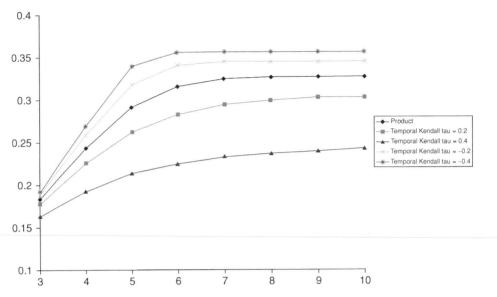

Figure 6.6   The term structure of equity tranches, different temporal dependence values.

book both concerning spatial and temporal dependence of losses. We remember that CDXs are synthetic unfunded CDOs. The collateral is made by a credit index of the main 125 *names* representing the American economy. We use data of the CDS of each of these 125 names for maturities from 1 to 10 years. The data refers to the series number 8, and was collected on July 3, 2007. We report here a macro-algorithm of the evaluation procedure.

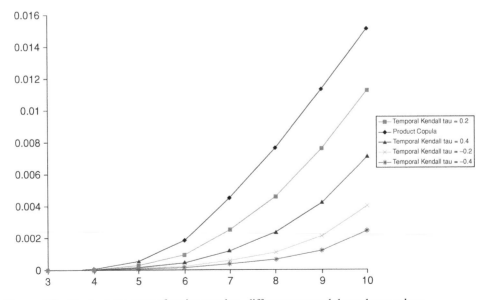

Figure 6.7   The term structure of senior tranches, different temporal dependence values.

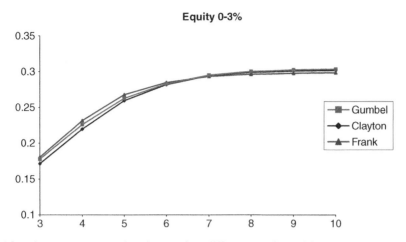

Figure 6.8   The term structure of equity tranches, different copula models.

Algorithm   Term structure of CDX:
1. Apply the bootstrapping procedure to each CDS of the index and recover individual probabilities of default year-by-year.
2. Compute the average probability of default year-by-year.
3. Select a dependence model for the aggregated losses year-by-year (example, a large portfolio approximation) and compute the distribution of losses in each year.
4. Run the C-convolution algorithm to propagate the distribution of cumulated losses.
5. Store the distributions of cumulated losses year-by-year and compute the value of running premia for every tranche.

In Figures 6.5–6.7 we report the simulation results of application of the algorithm to our data. The Vasicek distribution was used as a large portfolio approximation for the law of losses of each year. Cross-sectional dependence was set equal to 40% for all years. The

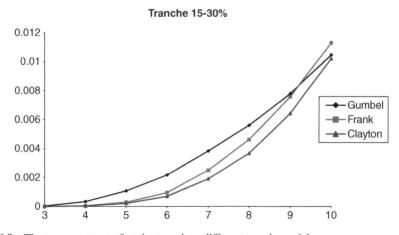

Figure 6.9   The term structure of senior tranches, different copula models.

temporal dependence structure was assumed to be Archimedean, namely the Clayton copula. In Figure 6.5 we report the evolution of the distribution of percentage losses for different levels of temporal correlation.

In Figures 6.6 and 6.7 we report the term structure of the equity tranche and the senior tranche for different levels of temporal dependence. We see that temporal correlation impacts on tranches in the same direction as spatial correlation, even though very often only the latter is taken into account in pricing applications.

Finally, Figures 6.8 and 6.9 compare the effect on prices of the choice of different Archimedean copulas. We see that the difference is not remarkable. Model risk does not seem to be an issue, if we mean choice of a copula function instead of another. It may instead be substantial if the temporal correlation is completely forgotten.

7
Risk Capital Management

Risk management is an intrinsically multivariate concept. Whatever position one can take in the market, as plain as it may be, it implies exposure to several sources of risk. This is due to the high degree of interdependence of markets and risk factors. You enter a plain investment in the equity market and you find yourself exposed to a crisis of credit or a turbulence in the foreign exchange market. More to the point, this dependence is particularly subject to change in times of turbulence. In the standard paradigm that we were taught, there exists a central limit theorem that enables us to collect all minor and independent shocks into a single variable that we would call *noise*, and that would be normally distributed. It seems that in today's markets, and particularly in periods of turbulence, this result is no longer available, and the reason is that these shocks are not independent anymore. Aside from this, dealing with correlation has always been central to the function of the risk manager because it is connected to capital allocation. How much capital you commit to support a set of activities depends crucially on the degree of diversification of the business. Taking as given the risk of every operation, that we could call *marginal risk*, the amount of capital needed to insure the business would depend crucially on how much the losses on the operations are likely to occur at the same time. This is the problem addressed in this chapter. A bird's-eye preliminary analysis of the problem reveals three main issues. Assume you find yourself wanting to commit capital to a firm that is operating two lines of business, *A* and *B*. Even before knowing anything about the tools to be used and other technicalities, what is clear is that you have to address three questions:

- How much capital to devote to the firm as a whole?
- How much capital to devote to each business line *A* and *B*?
- For how long to commit the capital, one day, one month, one year?

Notice that these questions naturally call for the tools we presented in the methodological Chapters 3 and 4. Furthermore, we will see (and this may sound new to the reader who is used to standard treatment of capital allocation) that even in this case the dependence enters the problem with two different meanings. The relevance of spatial dependence is well known to everyone. The question "for how long?" foreshadows that even in capital allocation, temporal dependence may be relevant. In this chapter we will see that there exist examples of applications in which temporal aggregation is the most relevant concept.

7.1 A REVIEW OF VALUE-AT-RISK AND OTHER MEASURES

Beginning in the 1990s, a new measure appeared on the scene of risk management. Initially, it was called *Maximum Probable Loss* (MPL), but soon after it became universally known as *Value-at-Risk*, or VaR. Probability was the key innovation with respect to the previous standard paradigm. Before VaR, risk management was merely an ancillary concept in the process of asset and liability (ALM) management, and the analysis was mainly carried out in terms of sensitivity of assets and liabilities, and the balance of the two, to changes in market prices and rates. The introduction of probability was actually inspired by the method that futures exchanges use to protect their integrity in the event of defaults of the participants to the market. Brokers in these markets are required to post a guarantee deposit, called *margin*, at inception of any position either long or short, and they are required to replenish it whenever this margin, because of impairment from losses, falls short of a lower bound (*margin call*). Of course, setting the right level of margin requirement calls for an analysis of the chance that the margin itself be exhausted in the time period elapsing from one valuation period to the other (*marking-to-market*). In futures markets, this period is one working day. At the beginning of the 1990s, banks began to apply this same technique to their proprietary deals, and this brought probability into the picture. The notion of margin was transferred inside the operation of banks, taking the name of *economic capital*; namely, the amount of capital to be set aside and kept in safe and immediately cashable investments to absorb losses. Moreover, the concept, which in the first stage had been applied to market risk, from which it had actually been borrowed, was later extended to all sources of risk of the financial intermediation activity, covering both credit and operation risk. As of today, the use of this concept has also been extended beyond the financial intermediation business to corporate risk management and communication of risk information to retail investors.

The concept of VaR is then a matter of time and chance, and it is meant to represent the amount of capital that one has to set aside to resist adverse events for a given period of time. The time span is actually conceived as the *unwinding period*, having in mind the standard case in which losses come from changes in the market price of a position or a portfolio when it is dismantled. More generally, this time length is meant to ensure the ability to repair the loss incurred: in the example of operation risk, this could be for example the time needed to restore operation after a break of business continuity. The main ingredient of Value-at-Risk analysis is then a probability distribution of losses over a given period. The VaR measure is simply defined as the quantile of this measure. The quantile of a distribution is defined as

$$q_\alpha(X) = \inf\{x : \mathbb{P}(X \le x) \ge \alpha),$$

where X is a random variable representing the value of a portfolio of assets or exposures to risk factors. Notice that

$$q_\alpha(X) = F_X^{-1}(\alpha),$$

where F_X^{-1} denotes the generalized inverse defined in Chapter 2. Then, the VaR of an exposure X at confidence level α (α usually close to 1) will be defined as

$$VaR_\alpha(X) = q_\alpha(-X) = F_{-X}^{-1}(\alpha).$$

From the definition, we get

$$\begin{aligned} VaR_\alpha(X) = q_\alpha(-X) &= \inf\{x : \mathbb{P}(-X \le x) \ge \alpha) \\ &= \inf\{x : \mathbb{P}(X + x \ge 0) \ge \alpha) \\ &= \inf\{x : \mathbb{P}(X + x < 0) \le 1 - \alpha), \end{aligned} \qquad (7.1)$$

that is $VaR_\alpha(X)$ is the smallest amount of money which, if added to X, keeps the probability of a negative outcome below the level $1 - \alpha$.

It is also immediate from the definition that in the case where F_X is invertible

$$
\begin{aligned}
F_X(-VaR_\alpha(X)) &= \mathbb{P}(X \leq -VaR_\alpha(X)) \\
&= \mathbb{P}(-X \geq VaR_\alpha(X)) \\
&= \mathbb{P}(-X \geq F_{-X}^{-1}(\alpha)) \\
&= \mathbb{P}(F_{-X}(-X) \geq \alpha) = 1 - \alpha
\end{aligned}
\tag{7.2}
$$

or, otherwise written,

$$
VaR_\alpha(X) = -F_X^{-1}(1 - \alpha).
\tag{7.3}
$$

The measure of VaR was the object of much discussion and criticism at the end of the last century. Other kinds of measure have been proposed, and the selection of risk measures has actually become a new branch of theory of its own. In any case, getting into the details of this discussion is immaterial here, where the main feature of the analysis is the probability distribution of profits and losses. Once this distribution is recovered, different measures can be used to describe its shape, but at that point, at least for what we are interested in here, the work has already been done.

VaR Homogeneity

VaR homogeneity is an important property that will be used below to turn the VaR measure into a risk management policy of capital allocation. More precisely, we give here the proof that VaR is homogeneous of degree 1, meaning that

$$
VaR_\alpha(\lambda X) = \lambda VaR_\alpha(X).
\tag{7.4}
$$

VaR inherits this property from the generalized inverse. In fact, if $\lambda > 0$,

$$
\begin{aligned}
F_{\lambda X}^{-1}(\alpha) &= \inf\{t : F_{\lambda X}(t) \geq \alpha\} \\
&= \inf\{t : \mathbb{P}(\lambda X \leq t) \geq \alpha\} \\
&= \inf\{t : \mathbb{P}(X \leq \frac{t}{\lambda}) \geq \alpha\} \\
&= \inf\{t : F_X\left(\frac{t}{\lambda}\right) \geq \alpha\} \\
&= \lambda \inf\{\frac{t}{\lambda} : F_X\left(\frac{t}{\lambda}\right) \geq \alpha\} \\
&= \lambda \inf\{z : F_X(z) \geq \alpha\} \\
&= \lambda F_X^{-1}(\alpha).
\end{aligned}
$$

With $VaR_\alpha(X) = F_{-X}^{-1}(\alpha)$, we get (7.4).

A Tale of Two Bond Markets

Here we introduce two indices that will accompany us in the explanation of capital allocation to a portfolio. The current situation of crisis would suggest a bond portfolio. We selected an equally weighted portfolio of Italian government securities (proxied by the 30-year benchmark BTP index) and corporate bonds, represented by the IBOXX index. If we take the perspective

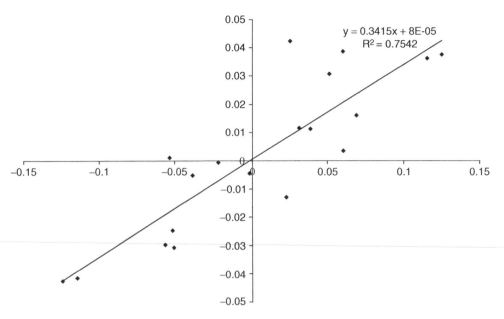

Figure 7.1   Scatter plot of returns on the bond portfolio: six-month investment horizon.

of an institutional or retail investor, instead of a trader, a reasonable investment horizon could be six months. In Figure 7.1, we report the scatter plot of six-month returns on the two indices. The data set spans the period from 2000 to 2010. The figure highlights a high degree of association, which is confirmed by the regression line.

Moving one step forward, we address the problem of specifying the marginal distributions and the dependence structure. As for distributions, they turn out to be almost Gaussian, with higher volatility of the Italian government bond index (about 10%) than the European bond index (about 4%). Our focus is obviously on the specification of dependence. We choose Archimedean copulas, that, as we will see, are endowed with features that make them very useful in capital allocation applications. Since our focus is on risk management, we concentrate on the specification of the lower tail, that is losses. For this reason, it suffices to limit the comparison to the Frank copula, which is radially symmetric, and the Clayton copula, which is skewed to the left and has lower tail dependence. For the selection of the copula function, we use the Genest and McKay procedure based on the Kendall function. Namely, we compute the empirical Kendall function and check which copula provides the closest fit. The empirical Kendall function is reported in Figure 7.2. For reference, in the figure we also report the functions corresponding to perfect dependence and independence, along with those representing the Frank and Clayton copula for Kendall's τ estimated from the data. The evidence of high association gleaned from the regression line in Figure 7.1 is confirmed by a Kendall's τ figure of about 65.30%. Visual inspection shows that the Frank copula provides a better fit of the upper tail, while the Clayton copula is much more accurate for the lower tail. This is quite a typical case, with left skewness and lower tail dependence. Since, in risk management, we are particularly interested in the specification of the lower tail, in this typical case we are led to choose the Clayton copula as the most reliable representation risk.

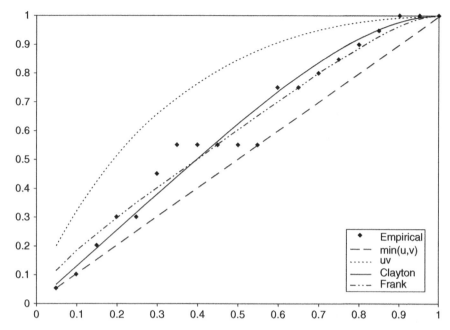

Figure 7.2 Kendall function of six-month returns on Italian and European bonds. Clayton copula parameter 3.758; Frank copula parameter 9.529.

7.2 CAPITAL AGGREGATION AND ALLOCATION

Once the risk measure is estimated for two or more risk factors or business lines, the problem emerges of how to translate this measure into a practical risk management decision, concerning how much capital to set aside or, which is the same, how much insurance to buy against losses. At first sight, it appears immediately that the problem can be partitioned into two questions:

- How much capital to allocate to the overall portfolio, that is to all business lines and risk factors?
- How much capital to allocate to the elements of the portfolio, that is to each business line and risk factor?

The first question is actually about the risk of the sum of all elements in the portfolio, and it is a problem of *capital aggregation*. The second question refers to a trade-off, or sort of budget constraint, in the use of this capital: how much of it should be devoted to each business line or risk factor. What should be computed first? There is no clear answer to that. A natural choice would be to decide first on the size of the cake and then on the size of the slices. Arguments could, however, be raised against this: one could argue that the size of the cake should actually be the result of the appetite for slices, rather than an input. In fact, on the one hand, taking the size of the overall risk as exogenous may either lead to leaving capital unused or to a shortage of capital. It must be remembered that the former event is as negative as the latter: in fact, devoting more capital than needed to the operation of business units amounts to imposing an unnecessary burden of cost on them, which may turn out fatal in a competitive environment.

On the other hand, computing the amount of capital for each business unit first and the overall capital in the second stage may lead to an excessive amount of capital for the portfolio as a whole.

The two different approaches have been given the names of *top-down* versus *bottom-up* techniques. In the *top-down* approach, the capital needed to support all the portfolio is computed first and then it is allocated across the business units. In the *bottom-up* approach, the need for capital is computed for each business unit and then it is aggregated over all the portfolio. Discussing which comes first between capital aggregation and allocation is not in the scope of this book. What counts as far as the computational aspects are concerned (and they are our main concern here) is that aggregation and allocation problems can be collected under a common representation of the problem using different aggregation operators in the computation of the risk measure. Keeping again the VaR given in (7.1) as a reference for these measures, and assuming two exposures X_i, $i = A, B$, we may measure the multivariate risk through a multivariate VaR as

$$VaR_\alpha^f(X_A, X_B) = \{(x_a, x_B) : \mathbb{P}(f(X_a + x_a, X_B + x_b) < 0) = 1 - \alpha\}, \tag{7.5}$$

where $f : \mathbb{R}^2 \to \mathbb{R}$ is an aggregation function. In the above definition as well as in the sequel, we assume that all level curves are well defined. In the simplest form, the function will be linear, representing a portfolio. Other forms of aggregation function, such as $f(x, y) = \max(x, y)$, would lead to representations that will be useful for capital allocation.

Euler Attribution Principle

It is now time to see why the homogeneity property proved above for VaR is important in risk management. The reason is called the *Euler principle*. Throughout the literature it is called an "allocation" principle. Here we use the different term of "attribution" to distinguish the concept from the allocation policy that will be addressed below. The concept of attribution denotes how much of the capital committed to a set of risk sources is accounted for by each of them. If the risk measure is homogeneous of degree one, the Euler principle provides an answer. We report the Euler result here.

Theorem 7.2.1.   *Euler theorem on homogeneous functions.* Let *f* be an *m*-variate function. *The function is homogeneous of degree κ, that is, for all $\lambda > 0$*

$$f(\lambda u) = \lambda^\kappa f(u),$$

if and only if

$$\kappa f(u) = \sum_i^m u_i \frac{\partial f(u)}{\partial u_i}, \tag{7.6}$$

with $u = [u_1, \ldots, u_m]$.

If the aggregation function is homogeneous of degree one (and this is the case for linear combination and the maximum), the same homogeneity property is preserved by the VaR^f of the aggregated risk. In fact, if $\lambda > 0$

$$
\begin{aligned}
VaR_\alpha^f(\lambda X_1, \ldots, \lambda X_m) &= \{(y_1, \ldots, y_m) : \mathbb{P}(f(\lambda X_1 + y_1, \ldots, \lambda X_m + y_m) < 0) = 1 - \alpha\} \\
&= \{(y_1, \ldots, y_m) : \mathbb{P}(\lambda f(X_1 + \frac{y_1}{\lambda}, \ldots, X_m + \frac{y_m}{\lambda}) < 0) = 1 - \alpha\} \\
&= \{(y_1, \ldots, y_m) : \mathbb{P}(f(X_1 + \frac{y_1}{\lambda}, \ldots, X_m + \frac{y_m}{\lambda}) < 0) = 1 - \alpha\} \\
&= \lambda\{(x_1, \ldots, x_m) : \mathbb{P}(f(X_1 + x_1, \ldots, X_m + x_m) < 0) = 1 - \alpha\} \\
&= \lambda VaR^f(X_1, \ldots, X_m).
\end{aligned}
$$

We then have

$$
VaR_\alpha^f(X) = \sum_{i=1}^m X_i \frac{\partial VaR_\alpha^f(X)}{\partial X_i}.
$$

7.2.1   Aggregation: *C*-convolution

In the case in which the aggregation function is the sum

$$
\begin{aligned}
VaR_\alpha^f(X_1, \ldots, X_m) &= \{(x_1, \ldots, x_m) : \mathbb{P}(f(X_1 + x_1, \ldots X_m + x_m) < 0) = 1 - \alpha\} \\
&= \{(x_1, \ldots, x_m) : \mathbb{P}(X_1 + \cdots + X_m + x_1 + \cdots + x_m < 0) = 1 - \alpha\}
\end{aligned}
$$

and, if $X = X_1 + \cdots + X_m$, we have that

$$
(x_1, \ldots, x_m) \in VaR_\alpha^f(X_1, \ldots, X_m) \text{ if and only if } x_1 + \cdots x_m = VaR_\alpha(X).
$$

Since $VaR_\alpha(X)$ is again homogeneous if thought of as a function of $(X_1, \ldots, X_m)$, meaning that, if $\lambda > 0$ and $(Y_1, \ldots, Y_m) = (\lambda X_1, \ldots, \lambda X_m)$, if $Y = Y_1 + \cdots, Y_m$, then

$$
VaR_\alpha(Y) = \lambda VaR_\alpha(X),
$$

thanks to Euler Theorem we have

$$
VaR_\alpha(X) = \sum_{i=1}^m X_i \frac{\partial VaR_\alpha(X)}{\partial X_i}.
$$

The above result means that the VaR of a portfolio of business units can be represented as a linear combination of VaR sensitivities. What remains to be found for this result to be of practical use is a formula for the partial derivatives of the VaR function. This was provided by Gourièroux et al. (2000) and Tasche (1999) in the case where f is linear, under suitable smoothness conditions:

$$
\frac{\partial VaR_\alpha(X)}{\partial X_i} = -\mathbb{E}[X_i | X = -VaR_\alpha(X)].
$$

This means that these partial derivatives are conditional expectations computed at the point at which all the VaR is exhausted.

Consider a typical top-down approach to the measurement of risk and to the allocation of capital to the overall firm and to its business units. Assume for the sake of simplicity that we have two business units denoted A and B. It is immediate to see that the most natural way of addressing this problem is to specify the aggregation function as the sum: $f(x, y) = x + y$.

As a result, the solution to the capital aggregation problem is the C-convolution technique studied throughout this book.

If we consider that the two sources of losses X_A and X_B have marginal distributions F_A and F_B and that their dependence structure is represented by some copula $C(u,v)$, it turns out that the distribution of the overall exposure to loss is given by the C-convolution of the two distributions, namely

$$\int_0^1 D_1 C\left(w, F_B(K - F_A^{-1}(w))\right) dw = F_A \overset{C}{*} F_B(K).$$

The capital to be posted to insure against the losses on the overall position $X = X_A + X_B$ at confidence level α is then, thanks to (7.2),

$$F_A \overset{C}{*} F_B(-VaR_\alpha(X)) = 1 - \alpha.$$

Remember also that

$$C_{X_A,X}(u, v) = \int_0^u D_1 C\left(w, F_B(F_X^{-1}(v) - F_A^{-1}(w))\right) dw.$$

Now, we may analyze the relationship between episodes of failure of the VaR insurance policy. For example, the probability that the capital devoted to losses from X_A will be wiped out given that the overall capital will be exhausted is:

$$\mathbb{P}(X_A \leq -VaR_\alpha(X_A)|X \leq -VaR_\alpha(X))$$
$$= \frac{C_{X_A,X}(F_A(-VaR_\alpha(X_A)), F_X(-VaR_\alpha(X))}{F_X(-VaR_\alpha(X))}$$
$$= \frac{C_{X_A,X}(1 - \alpha, 1 - \alpha)}{1 - \alpha}$$
$$= \frac{1}{1 - \alpha} \int_0^{1-\alpha} D_1 C\left(w, F_B(F_X^{-1}(1 - \alpha) - F_A^{-1}(w))\right) dw$$
$$= \frac{1}{1 - \alpha} \int_0^{1-\alpha} D_1 C\left(w, F_B(-VaR_\alpha(X) - F_A^{-1}(w))\right) dw.$$

Notice that if the two risks are co-monotonic, this conditional probability will be equal to 1, while under independence we have

$$\mathbb{P}(X_A \leq -VaR_\alpha(X_A)|X \leq -VaR_\alpha(X)) = \frac{1}{1 - \alpha} \int_0^{1-\alpha} F_B(-VaR_\alpha(X) - F_A^{-1}(w)) dw.$$

Notice that it is also easy to calculate the risk attribution to each component of the sum. In fact, we remember that

$$\frac{\partial VaR_\alpha(X)}{\partial X_i} = -\mathbb{E}[X_i|X = -VaR_\alpha(X)],$$

that is the expected loss of business i conditional on all the VaR being exhausted. In the C-convolution approach the conditional probability should be given by

$$\mathbb{P}(X_A \leq x_A|X = -VaR_\alpha(X)) = D_2 C_{X_A,X}(F_A(x_A), F_X(-VaR_\alpha(X)))$$
$$= D_2 C_{X_A,X}(F_A(x_A), 1 - \alpha).$$

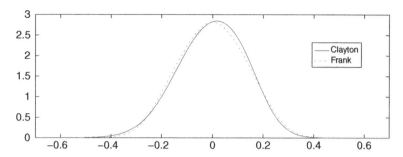

Figure 7.3 Density function of six-month returns on an equally weighted portfolio of Italian and European bonds: Clayton and Frank copulas.

If we assume that C has a density c, and that F_X and F_B have densities f_X and f_B respectively, we get

$$\mathbb{P}(X_A \leq x_A | X = -VaR_\alpha(X))$$
$$= \frac{1}{f_X(-VaR_\alpha(X))} \int_0^{F_{X_A}(x_A)} c\left(w, F_B(-VaR_\alpha(X) - F_A^{-1}(w))\right)$$
$$\times f_B\left(-VaR_\alpha(X) - F_A^{-1}\right)(w)dw$$

and the conditional expectation is simply obtained by integration of the conditional probability.

Capital Allocation in a Bond Portfolio

In Figure 7.3 we plot the density of six-month returns on the equally weighted portfolio of Italian and European bonds described in Section 7.1. We remember that the marginal distributions have volatility 10% and 4% and the Kendall's τ is 65.30%, corresponding to a parameter value of 3.758 for Clayton and 9.529 for Frank copulas. The aggregated VaR figure at 99% confidence level turns out to be equal to 32.40% for the Clayton copula against a value of 29.73% if the Frank copula is used.

7.2.2   Allocation: Level Curves

Consider now the problem of capital allocation between the business units A and B. A reasonable approach could be to commit capital to the two businesses in such a way as to leave unchanged the joint probability that capital allocated to both businesses be wiped out. This could be done by looking for those values of the risky positions that keep constant at α, the joint probability of exhausting both capital endowments. As we are going to see, this amounts to using as aggregation function the maximum, that is $f(x_A, x_B) = \max(x_A, x_B)$ and the level curves of the copula functions. According to the definition of aggregate VaR given in (7.5), we define the value at risk of level α of the two risks X_A and X_B as follows:

$$VaR_\alpha^f(X_A, X_B) = \{(x_a, x_B) : \mathbb{P}(\max(X_a + x_a, X_B + x_b) < 0) = 1 - \alpha\}$$
$$= \{(x_A, x_B) : F_{X_A, X_B}(-x_A, -x_B) = 1 - \alpha\}. \tag{7.7}$$

From the definition it immediately follows that VaR_α^f is the set of all possible pairs of amounts of capital that, added to each risk, guarantees a probability of joint default equal to $1 - \alpha$. From (7.7) we get

$$
\begin{aligned}
VaR_\alpha^f(X_A, X_B) &= \{(x_A, x_B) : F_{X_A, X_B}(-x_A, -x_B) = 1 - \alpha\} \\
&= \{(x_A, x_B) : C_{X_A, X_B}(F_{X_A}(-x_A), F_{X_B}(-x_B)) = 1 - \alpha\} \\
&= \{(-F_{X_A}^{-1}(1 - \alpha_A), -F_{X_B}^{-1}(1 - \alpha_B)) : C_{X_A, X_B}(1 - \alpha_A, 1 - \alpha_B) = 1 - \alpha\} \\
&= \{(VaR_{\alpha_A}(X_A), VaR_{\alpha_B}(X_B)) : C_{X_A, X_B}(1 - \alpha_A, 1 - \alpha_B) = 1 - \alpha\}.
\end{aligned}
\tag{7.8}
$$

The condition

$$
C(1 - \alpha_A, 1 - \alpha_B) = 1 - \alpha
\tag{7.9}
$$

induces a trade-off between the VaR figures of A and B and the capital allocated to those. A particularly simple example to show the approach is given by Archimedean copulas, and it is also an advantage of the use of this class of copulas in capital allocation. Remember that Archimedean copulas are completely defined by a convex and decreasing generator $\phi(x)$, through

$$
C(u, v) = \phi^{-1}(\phi(u) + \phi(v)).
$$

Now it is easy to recover the $1 - \alpha$ level curve and condition (7.9) can be rewritten as

$$
1 - \alpha = \phi^{-1}(\phi(1 - \alpha) - \phi(1 - \alpha_B)).
$$

These level curves are convex and the extreme cases of perfect positive and negative dependence define a triangle in the unit square, with coordinates $1 - \alpha$. All the possible $(1 - \alpha)$-level curves of the class are included in this triangle.

Remark 7.2.1. Since the function max is homogeneous, VaR_α^f is homogeneous. We may ask what this fact induces on level curves. Namely, if we pick a pair of VaR measures on a level curve, the homogeneity property holds for each of them and the level curve does not change with changes of scale of the exposure. In fact, if $\lambda > 0$, $\mu > 0$, then

$$
\begin{aligned}
VaR_\alpha^f(\lambda X, \mu Y) &= \{(VaR_u(\lambda X), VaR_v(\mu Y)) : C_{X,Y}(1 - u, 1 - v) = 1 - \alpha\} \text{ by (7.4)} \\
&= \{(\lambda VaR_u(X), \mu VaR_u(Y)) : C_{X,Y}(1 - u, 1 - v) = 1 - \alpha\} \\
&= \{(VaR_u(X), VaR_u(Y)) \begin{pmatrix} \lambda & 0 \\ 0 & \mu \end{pmatrix} : C_{X,Y}(1 - u, 1 - v) = 1 - \alpha\} \\
&= Var_\alpha^f(X, Y) \begin{pmatrix} \lambda & 0 \\ 0 & \mu \end{pmatrix}.
\end{aligned}
$$

If $\lambda = \mu$ we recover the known homogeneity

$$
VaR_\alpha^f(\lambda X, \lambda Y) = \lambda VaR_\alpha^f(X, Y).
$$

Level Curves of a Portfolio of Bonds

In Figures 7.4 and 7.5. we report the level curves for the Italian and European bond indices described in Section 7.1. We notice that the level curves of the Frank copula are less steep and convex than those of the Clayton copula.

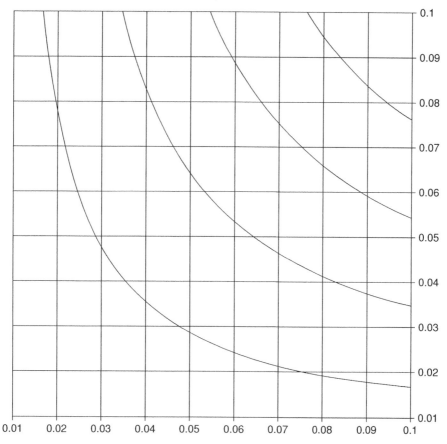

Figure 7.4   Level curves for allocation of capital to Italian and European bonds: Frank copula parameter 9.529.

7.2.3   Allocation with Constraints

We said that the question of what comes first between aggregation and allocation is a matter of taste, and leads to the distinction between the top-down and the bottom-up school. Here we suggest a policy that may be of help to find a compromise between the use of C-convolution and that of level curves. Heuristically, one could partition the process of capital aggregation and allocation as follows:

1. Computation of the capital for the overall portfolio.
2. Allocation of the capital to the business units, setting the capital computed in step 1 as an upper bound.

In a sense, the risk manager would set a global reinsurance for both businesses A and B. The two business lines will use this capital to absorb losses and will be closed when all capital is exhausted. From a technical point of view, this amounts to studying the probability representation of the convolution of two variables whose sum is bounded. This problem was

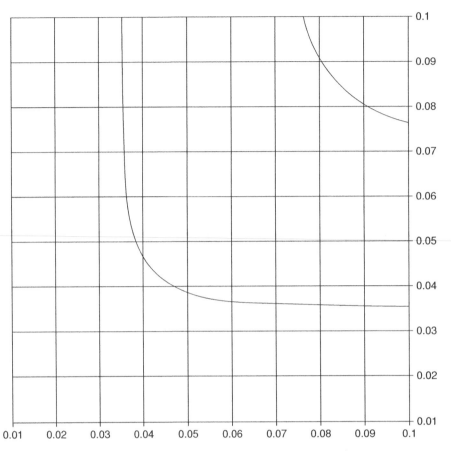

Figure 7.5 Level curves for allocation of capital to Italian and European bonds: Clayton copula 3.758.

studied in Cherubini et al. (2011). Referring to the paper for details, the main idea that is of interest for this application is that this provision, which substantially provides reinsurance to the portfolio, modifies the dependence structure between the risks. It is worth stressing that this has nothing to do with incentives and reactions of the managers of business lines *A* and *B*, but is only due to the existence of the reinsurance policy. It is sort of a Heisenberg paradox of risk management that the sole act of attributing capital to a set of business units modifies the dependence structure among them. The main intuition is quite clear if we take an extreme example. Assume that a very small amount of capital is attributed to the aggregated losses of lines *A* and *B*, so that the constraint will be binding with probability 1. In this case, the dependence structure between the profit and losses in the two businesses will have to be perfectly negative. Then, in case of a loss exhausting the capital from one business line, this will have to be compensated with a good performance of the other business line, or else the other business would also be wiped out. In other words, business lines *A* and *B* will not be allowed to be positively correlated.

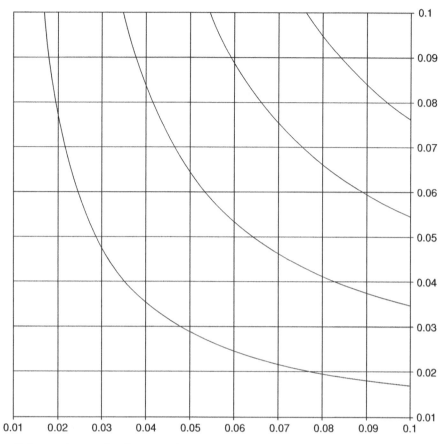

Figure 7.6 Level curves for allocation of capital to Italian and European bonds with constraints: Frank copula parameter 9.529.

Constrained Level Curves of Portfolio Bonds

In Figures 7.6 and 7.7 we report the level curves for the allocation strategy discussed above. Here the idea is that 5% of capital is allocated to both the bond portfolios, meaning that the loss cannot break, either on a stand-alone or on a joint basis, the VaR figure corresponding to a confidence level of 95%. Notice that in both cases the level curves are flattened out, and become negatively sloped straight lines (corresponding to perfect negative dependence) for scenarios in which the constraint is binding.

Finally, Figure 7.8 reports the difference between the copula without and with the constraint. Notice that the joint probability of losses is reduced for all choices with confidence levels lower than 95%.

7.3 RISK MEASUREMENT OF MANAGED PORTFOLIOS

We now tackle an issue that is quite new in the risk management literature, that is the impact of asset management on the distribution of returns. Standard risk measurement theory only allows

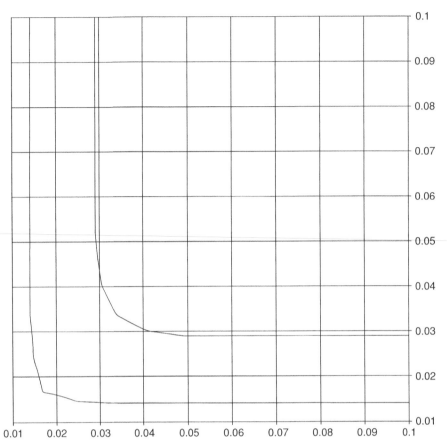

Figure 7.7 Level curves for allocation of capital to Italian and European bonds with Constraints: Clayton copula parameter 3.758.

for movements in the market, while most risks also depend on decisions made by the trader or the asset manager in charge of the investment. Curiously, this is a major area of interest in the asset management literature, and it is known as *performance attribution analysis*, but this has never crossed over to the research field in risk management. We are going to show that accounting for acts of men on top of changes in the market may have a crucial impact on risk measurement and risk management decisions. We will work on a very simple model in which the profit and loss on a managed fund Z over a given investment horizon is given by the sum of changes in the market, called X, and an investment strategy followed by the asset manager Y:

$$Z = X + Y.$$

We are interested in arriving at a risk measure, a VaR measure to keep things simple, for Z. Intuitively, this is again a convolution problem, and the aggregation technique used above may also apply to this case. Moreover, we have the opportunity of interpreting the C-convolution technique with the economic notions of the fund management literature.

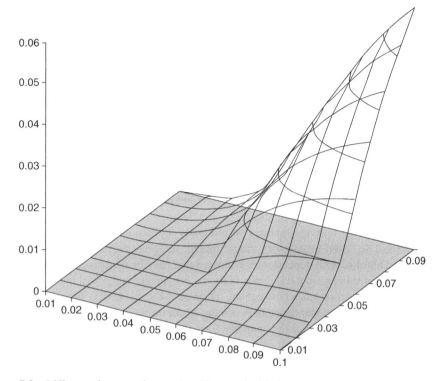

Figure 7.8   Difference between the copula without and with the constraint.

7.3.1   Henriksson–Merton Model

A good way to introduce the problem of the distribution of managed funds is to refer to the well-known Henriksson–Merton model (HM), developed in Merton (1981) for the theoretical part and Henriksson and Merton (1981) for the empirical implementation. Even though the model refers to an investment strategy that is very stylized, it has the advantage of being funded on an equilibrium model and includes all the aspects of the problem. It is then a good starting point for generalizations. The main idea is that the asset manager follows a strategy of moving all its funds to the risky asset (the market) whenever he expects it will outperform the risk-free asset, and back to the risk-free asset whenever he expects that the market will underperform. The assumption is then a sequence of sudden shifts back and forth to the market with no frictions involved and no leverage allowed. This policy leads to a specification of the returns as

$$Y = \delta + \gamma \max(0, -X) + \epsilon,$$

where we have assumed that the risk-free rate pays no interest, δ and γ are parameters, and ϵ is a noise component.

This specification allows us to comment on the impact of an active investment policy on the risk measure of a managed fund. A first look at the return of the investment policy reveals that it is made up of one part independent of the market and one linked to it. In the consolidated jargon of performance analysis, the former part is called *stock picking* while the latter is referred to as *market timing* analysis. Before this jargon came into fashion, in the HM article these asset management activities were called more simply microeconomic and macroeconomic analysis, respectively. A further look at the return from the investment policy reveals that the stock-picking policy has the effect of changing the location (if the asset manager is successful) and increasing the noise in the overall return of the managed fund. The market-timing activity instead induces a dependence structure between the investment strategy and the market. In the HM strategy this association is negative, which may sound counter-intuitive at first sight, but which is fully consistent with the fact that this policy is actually granting a protection against downturns of the market. This provides a portfolio insurance that is represented by a position in a quantity γ of put options. To those who have an advanced knowledge of the HM model, the γ parameter reveals an even more subtle meaning: this parameter is actually an association measure between the forecasts of the asset manager and the actual movements of the market. If the asset manager is completely unable to predict the market, Y is independent of X, while if her predictions are perfect, the investor would enjoy a position in a *protective put option* against all losses. So, even in this very single model, the final distribution of the managed fund Z is the result of several dependence structures that may be conceived nested one in the other. The dependence between the forecast of the asset manager and actual movements of the market determines the dependence between the return on the investment policy and the market, which in turn specifies the dependence between the return on the market and the return on the managed fund. The distribution of the managed fund is then much more than a convolution of two variables. Rather, it is the result of a complex interplay of explicit and hidden variables (such as the forecast of the asset manager).

HM Copula

Sticking to this very simple model, let us try a first generalization. Let us say that we want to extend the formula in full generality. Then, we want it to be applied to any arbitrary distribution X for market returns, and for a very general concept of noise ϵ. All we want to retain of the Henriksson–Merton model is the very specific association between market and investment strategy, indexed by the forecast accuracy parameter γ, that is

$$Y = \delta + \gamma \max(0, -X) + \epsilon.$$

The result will be a copula linking X and Y that we call the Henriksson–Merton copula (or *HM copula*). The best way to represent this copula is of course by simulation, based on the distribution of the market and noise. Nevertheless, we can also compute an analytical formula for the HM copula

Proposition 7.3.1. *Assume $\gamma > 0$ and marginal distributions F_X and F_ϵ assigned. Then*

$$C_{X,Y}(u, v) = G_{X,Y}(F_X^{-1}(u), F_Y^{-1}(v))\mathbf{1}_{\{u > F_X(0)\}} + H_{X,Y}(F_X^{-1}(u), F_Y^{-1}(v))\mathbf{1}_{\{u \leq F_X(0)\}},$$

where

$$G_{X,Y}(x, y) = F_X(x)F_\epsilon(y - \delta) - \int_{-\infty}^{y-\delta} F_X\left(\frac{e - y + \delta}{\gamma}\right)F_\epsilon(de)$$

and

$$H_{X,Y}(x, y) = F_X(x)F_\epsilon(\gamma x + y - \delta) - \int_{-\infty}^{\gamma x + y - \delta} F_X(\frac{e - y + \delta}{\gamma})F_\epsilon(de),$$

where

$$F_Y(y) = F_\epsilon(y - \delta) - \int_{-\infty}^{y - \delta} F_X(\frac{e - y + \delta}{\gamma})F_\epsilon(de). \qquad (7.10)$$

Proof. To construct this copula function we first need the joint distribution between X and Y, say $F_{X,Y}$. We have

$$\begin{aligned}
F_{X,Y}(x, y) &= \mathbb{P}(X \leq x, Y \leq y) \\
&= \mathbb{P}(X \leq x, \delta + \gamma \max\{0, -X\} + \epsilon \leq y) \\
&= \int_{-\infty}^{+\infty} \mathbb{P}(X \leq x, \delta + \gamma \max\{0, -X\} + e \leq y | \epsilon = e)F_\epsilon(de) \\
&= \int_{-\infty}^{+\infty} \mathbb{P}(X \leq x, \max\{0, -X\} \leq \frac{y - e - \delta}{\gamma})F_\epsilon(de) \\
&= \int_{-\infty}^{+\infty} \Big[\mathbb{P}(X \leq x, X \geq \frac{e - y + \delta}{\gamma}, X < 0) \\
&\quad + \mathbb{P}(X \leq x, 0 \leq \frac{y - e - \delta}{\gamma}, X \geq 0) \Big] F_\epsilon(de).
\end{aligned}$$

We have to distinguish two cases: $x > 0$ and $x \leq 0$. In the first case

$$\begin{aligned}
\mathbb{P}(X \leq x, X &\geq \frac{e - y + \delta}{\gamma}, X < 0) + \mathbb{P}(X \leq x, 0 \leq \frac{y - e - \delta}{\gamma}, X \geq 0) \\
&= \mathbb{P}(X \leq 0, X \geq \frac{e - y + \delta}{\gamma}) + \mathbb{P}(0 \leq X \leq x, e \leq y - \delta),
\end{aligned}$$

therefore, if we indicate by $G_{X,Y}$ the joint distribution in this case

$$\begin{aligned}
G_{X,Y}(x, y) &= \int_{-\infty}^{+\infty} \Big[\mathbb{P}(X \leq 0, X \geq \frac{e - y + \delta}{\gamma}) + \mathbb{P}(0 \leq X \leq x, e \leq y - \delta) \Big] F_\epsilon(de) \\
&= \int_{-\infty}^{y - \delta} \mathbb{P}\Big(\frac{e - y + \delta}{\gamma} \leq X \leq 0 \Big) F_\epsilon(de) + (F_X(x) - F_X(0))F_\epsilon(y - \delta) \\
&= F_X(0)F_\epsilon(y - \delta) - \int_{-\infty}^{y - \delta} F_X(\frac{e - y + \delta}{\gamma})F_\epsilon(de) + (F_X(x) - F_X(0))F_\epsilon(y - \delta) \\
&= F_X(x)F_\epsilon(y - \delta) - \int_{-\infty}^{y - \delta} F_X(\frac{e - y + \delta}{\gamma})F_\epsilon(de).
\end{aligned}$$

In the case where $x \leq 0$ we have

$$\begin{aligned}
\mathbb{P}(X \leq x, X &\geq \frac{e - y + \delta}{\gamma}, X < 0) + \mathbb{P}(X \leq x, 0 \leq \frac{y - e - \delta}{\gamma}, X \geq 0) \\
&= \mathbb{P}(X \leq x, X \geq \frac{e - y + \delta}{\gamma}),
\end{aligned}$$

then, the only relevant case is when $\frac{e-y+\delta}{\gamma} \leq x$ and therefore if we indicate by $H_{X,Y}$ the joint distribution in this case

$$H_{X,Y}(x,y) = \int_{-\infty}^{\gamma x + y - \delta} \mathbb{P}(X \leq x, X \geq \frac{e-y+\delta}{\gamma}) F_\epsilon(de)$$

$$= F_X(x) F_\epsilon(\gamma x + y - \delta) - \int_{-\infty}^{\gamma x + y - \delta} F_X(\frac{e-y+\delta}{\gamma}) F_\epsilon(de).$$

Briefly, we can conclude that

$$F_{X,Y}(x,y) = G_{X,Y}(x,y) \mathbf{1}_{\{x>0\}} + H_{X,Y}(x,y) \mathbf{1}_{\{x\leq 0\}}.$$

Now, the copula between X and Y is given by

$$C_{X,Y}(u,v) = F_{X,Y}(F_X^{-1}(u), F_Y^{-1}(v))$$

$$= G_{X,Y}(F_X^{-1}(u), F_Y^{-1}(v)) \mathbf{1}_{\{u>F_X(0)\}} + H_{X,Y}(F_X^{-1}(u), F_Y^{-1}(v)) \mathbf{1}_{\{u\leq F_X(0)\}},$$

or more explicitly

$$C_{X,Y}(u,v) = \left[u F_\epsilon(F_Y^{-1}(v) - \delta) - \int_{-\infty}^{F_Y^{-1}(v)-\delta} F_X(\frac{e - F_Y^{-1}(v) + \delta}{\gamma}) F_\epsilon(de) \right] \mathbf{1}_{\{u>F_X(0)\}}$$

$$+ \left[u F_\epsilon(\gamma F_X^{-1}(u) + F_Y^{-1}(v) - \delta) \right.$$

$$\left. - \int_{-\infty}^{\gamma F_X^{-1}(u)+F_Y^{-1}(v)-\delta} F_X(\frac{e - F_Y^{-1}(v) + \delta}{\gamma}) F_\epsilon(de) \right] \mathbf{1}_{\{u\leq F_X(0)\}}.$$

We call this copula function the Henriksson–Merton copula. It remains to determine the marginal distribution F_Y. We have

$$F_Y(y) = \lim_{x \to +\infty} F_{X,Y}(x,y) = \lim_{x \to +\infty} G_{X,Y}(x,y)$$

$$= F_\epsilon(y - \delta) - \int_{-\infty}^{y-\delta} F_X(\frac{e-y+\delta}{\gamma}) F_\epsilon(de). \qquad \square$$

Given $C_{X,Y}$, F_X and F_Y, we can find the distribution of $Z = X + Y$ by using the C-convolution formula, that is

$$F_Z(z) = \int_0^1 D_1 C_{X,Y}\left(w, F_Y(z - F_X^{-1}(w))\right).$$

Remark 7.3.1. Notice that the HM copula has a peculiarity with respect to other copula functions. While in standard copula functions we are completely free to select any distribution for both the marginals, here the distribution of Y is indirectly recovered from the distribution of noise, see (7.10).

In Figure 7.9 we report the shape of the probability density function of $Z = X + Y$ for different levels of the γ parameter, taking the usual choice of Gaussian distribution for the market and noise. As expected, the higher the parameter, the more effective the portfolio insurance strategy, and the lower the left tail of the distribution.

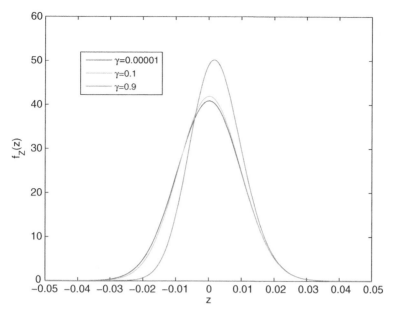

Figure 7.9 Density of the return of a managed fund with different market timing skills of the asset manager: Clayton copula.

Two extreme cases: $\epsilon \to 0$ and $\gamma \to 0$

1. $\epsilon \to 0$.

If $\epsilon \to 0$ then $F_\epsilon(e) \to \mathbf{1}_{[0,+\infty)}(e)$ and, to the limit,

$$F_Y(y) = \left(1 - F_X(\frac{\delta - y}{\gamma})\right) \mathbf{1}_{\{y \geq \delta\}}. \tag{7.11}$$

Notice that $F_Y(\delta) = 1 - F_X(0)$. It follows that for $(u, v) \in \{(u, v) \in [0, 1]^2 : v \geq 1 - F_X(0)\}$

$$C_{X,Y}(u, v) \to \left(u\mathbf{1}_{[0,+\infty)}(F_Y^{-1}(v) - \delta) - F_X(\frac{\delta - F_Y^{-1}(v)}{\gamma})\mathbf{1}_{[0,+\infty)}(F_Y^{-1}(v) - \delta)\right) \mathbf{1}_{\{u > F_X(0)\}}$$

$$+ \left(u\mathbf{1}_{[0,+\infty)}(\gamma F_X^{-1}(u) + F_Y^{-1}(v) - \delta)\right.$$

$$\left. - F_X(\frac{\delta - F_Y^{-1}(v)}{\gamma})\mathbf{1}_{[0,+\infty)}(\gamma F_X^{-1}(u) + F_Y^{-1}(v) - \delta)\right) \mathbf{1}_{\{u \leq F_X(0)\}}$$

$$= \left(u - F_X(\frac{\delta - F_Y^{-1}(v)}{\gamma})\right) \left(\mathbf{1}_{\{u > F_X(0)\}} + \mathbf{1}_{\{u \leq F_X(0), \gamma F_X^{-1}(u) + F_Y^{-1}(v) - \delta \geq 0\}}\right)$$

$$= (u - 1 + v) \left(\mathbf{1}_{\{u > F_X(0)\}} + \mathbf{1}_{\{u \leq F_X(0), v \geq F_Y(\delta - \gamma F_X^{-1}(u))\}}\right)$$

$$= (u - 1 + v)) \left(\mathbf{1}_{\{u > F_X(0),\}} + \mathbf{1}_{\{u \leq F_X(0), v \geq 1 - u\}}\right)$$

$$= (u - 1 + v)\mathbf{1}_{\{u > F_X(0)\}} + \max(u + v - 1, 0)\mathbf{1}_{\{u \leq F_X(0)\}} = W(u, v).$$

2. $\gamma \to 0$.

If $\gamma \to 0$, then, to the limit,

$$F_Y(y) = F_\epsilon(y - \delta) - \int_{-\infty}^{y-\delta} F_X(-\infty) F_\epsilon(de) = F_\epsilon(y - \delta). \tag{7.12}$$

It follows that

$$\begin{aligned}
C_{X,Y}(u, v) \to & \left[u F_\epsilon(F_Y^{-1}(v) - \delta) - \int_{-\infty}^{F_Y^{-1}(v)-\delta} F_X(-\infty) F_\epsilon(de) \right] \mathbf{1}_{\{u > F_X(0)\}} \\
& + \left[u F_\epsilon(F_Y^{-1}(v) - \delta) - \int_{-\infty}^{F_Y^{-1}(v)-\delta} F_X(-\infty) F_\epsilon(de) \right] \mathbf{1}_{\{u \leq F_X(0)\}} \\
= & \ u F_\epsilon(F_Y^{-1}(v) - \delta) \text{ by } (7.12) \\
= & \ uv.
\end{aligned}$$

7.3.2 Semi-parametric Analysis of Managed Funds

The discussion above paves the way to a general representation of managed fund distributions. It is clear that the investment policy above is particularly *ad hoc*. One can, for example, think that a more cautious asset manager would set a strictly positive level above which to switch from the risk-free rate to the market. We could also conceive that he would turn to leverage if his forecast of the market is particularly bullish (and if the regulation allows him to do so). In cases in which no specific reference portfolio (*benchmark*) is declared, we may also have problems understanding which is the actual market X that the manager is forecasting (there exists a branch of the fund management literature, called *style analysis*, which is entirely devoted to this). Nevertheless, from the HM model, stylized as it may be, we may take home the general conclusion that if there is a market-timing activity in the asset management process, that must result in a dependence structure between the return on the investment strategy and the market.

Based on this argument, a general representation of a managed portfolio may be based on three elements:

- A return from a reference market.
- A return from an asset management strategy.
- The dependence between the two.

In Figure 7.10 we report the value of the corresponding Kendall's τ statistic, as a function of different values of the γ parameter. Notice that the γ parameter in the Merton model is a measure of concordance between the forecast of the asset manager and market movements. So, the figure shows how this concordance measure translates into a concordance measure (the Kendall τ) between the return on the investment strategy and the market. Since the market-timing strategy implies a position in *protective* put options, it is not surprising that association is negative. The more effectively an asset manager is able to forecast future market movements, the more effective the protective put position is.

Remark 7.3.2. Notice that the Kendall function remains exactly the same in as far as $\gamma < 1$. In the case $\gamma = 1$ the Kendall function remains the same above $t = 0.5$ with 0.5 mass concentrated at zero.

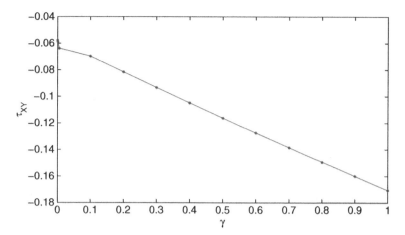

Figure 7.10   HM-Copula: Kendall's τ statistic as a function of different values of the γ parameter.

7.3.3   Market-neutral Investments

A very interesting specific case of managed fund risk analysis is that of market-neutral funds. It is well known that hedge funds should, by definition, be endowed with this feature, even though this has been questioned recently. Market neutrality can easily be used to extract the return on the investment strategy of the hedge fund manager. That is, to find the probability distribution of $Y = Z - X$ given the dependence structure between Z and X and the distribution of X and Y. The general formulas are

$$F_Y(t) = \int_0^1 D_1 C_{Z,-X}(w, F_{-X}(t - F_Z^{-1}(w))dw = 1 - \int_0^1 D_1 C_{Z,X}(w, F_X(F_Z^{-1}(w) - t))dw$$

and

$$C_{X,Y}(u, v) = u - \int_0^u D_1 C_{Z,X}(w, F_X(F_Z^{-1}(w) - F_Y^{-1}(v))dw,$$

In the present case of market neutrality. as a matter of fact, one would simply have to set $C_{Z,X} = uv$ and obtain

$$F_Y(t) = 1 - \int_0^1 F_X(F_Z^{-1}(w) - t)dw$$

for the stock-picking strategy and

$$C_{X,Y}(u, v) = u - \int_0^u F_X(F_Z^{-1}(w) - F_Y^{-1}(v))dw$$

for the market-timing activity.

By the same approach, it is also quite straightforward to describe the distribution of the so-called *core satellite* strategies. In these, a small investment in market-neutral funds is added to a position in mutual funds in order to add some stock-picking return and to decrease correlation in the portfolio. Namely, if a hedge fund H is added to the fund Z, the distribution of the return would be given by

$$F_{\overline{Z}}(t) = \int_0^1 F_H(t - F_Z^{-1}(w))dw, \tag{7.13}$$

where $\overline{Z} = Z + H$ denotes the *core satellite* portfolio. The gain in diversification can also be gauged, and this is done by measuring

$$C_{Z,\overline{Z}}(u, v) = \int_0^u F_H(F_{\overline{Z}}^{-1}(v) - F_Z^{-1}(w))dw. \tag{7.14}$$

7.4 TEMPORAL AGGREGATION OF RISK MEASURES

Like in all the other applications surveyed in this book, also in the case of risk management and capital allocation there exists a temporal dimension. Even though this problem is seldom addressed, the question arises inevitably in many applications. It arises in standard risk capital aggregation and allocation, when for example we are required to come up with an aggregate measure of different kinds of risk. Take, for example, the problem of aggregating market risk with credit and operational risk. When one has to compare and aggregate these measures, one finds that typically (but it would be better to say always) market risk is computed at much higher frequency, typically one to ten days, than the other risks, which are typically measured on a monthly basis. Then, before aggregating these measures one has to change the reference time period over which they are computed: either one reports credit and operational risk to a higher frequency, or, which is preferable, temporally aggregates market risk to measure it on the same frequency as that of the other risks. This raises a temporal aggregation question: given a market risk measure at daily frequency, how much will the measure be for a month or a year?

The same temporal aggregation question arises, and in this case it is the key feature of the problem, in cases in which risk measurement is applied to investments in mutual funds, hedge funds and the like, and in all applications in which the risk measure is used to communicate the risk of an investment plan to retail investors. Differently from traders, these investors have longer time horizons, so that it may make a difference whether they intend to lock in their money in a given investment for a month or a number of years. In some cases, such as closed-end funds, this lock-in period is even required by the regulations and is mandatory. Moreover, also for institutional investors, considering longer investment periods than the trading horizon may be a way of addressing the problem of *liquidity risk*. Actually, when international regulatory bodies required a ten-day *unwinding period* for the computation of the VaR figure, that was to ensure enough time to close an open position in the face of adverse market movements: so, it was motivated by liquidity concerns. Actually, ten days can be an extremely long time for a very liquid futures markets, while it may be too short to allow the unwinding of positions in individual securities even in liquid markets. As an example, consider the well-known phenomenon of *seasoned* bonds versus *on-the-run* bonds. More to the point, whether this length of time is too long or too short for this and that market may change significantly from one period to another: markets tend to become illiquid mostly when liquidity is needed, that is in periods of stress.

There should be no further need to justify the question of how long to commit capital to a given position. So, it may come as a surprise that very little information can be found in the risk management literature. More to the point, the main solution, called the *square-root formula*, is too simple to represent even a mild attempt to address the problem. We will see that the C-convolution studied in this book may be of great help to address this topic.

7.4.1 The Square-root Formula

Consider an investor who wants to decide how much risk he is taking when subscribing to a closed-end fund from which he can withdraw his money after three years or more, with respect to the risk he would take by investing in an open-end mutual fund from which he can get his money back in two days. Of course, the two investments will be deeply different for the asset class to which they refer, and that is the very reason why regulators have imposed different requirements concerning the duration of the investment period. Being sure that the investor will not be able to withdraw the money for a long period of time makes it possible for the asset manager to invest in long-term projects with high profit perspectives. These cannot be funded by open-end fund managers because they invest under the pressure of earning a competitive return at least on a semester basis, or the time in which the returns of their funds are disclosed to the investors. So, this short-term problem motivates the managed part of the fund to be very different in the two cases. Nevertheless, both funds will be exposed to the same common risk factor of the general evolution of the market. The open-end fund investor will be able to react to adverse changes in the market and the economy as a whole more promptly than the closed-end one, which is locked in for years. The question then arises, and it is crucial to the selection of the investment, how to compare the market risk of the two investments.

As of now, the standard answer would be the square-root formula, meaning that the VaR figure for a unit of time should be multiplied by the square-root of the time length

$$VaR_{T,\alpha} = \sqrt{T} VaR_{1,\alpha}. \tag{7.15}$$

In the case above, three years amounts to about 750 business days. Compared with the two working days of the open-end fund investment, we obtain that the market risk of the closed-end fund investment should be about 19 times higher. Actually, the square-root formula is based on very stringent assumptions:

- serial independence of profit and losses,
- normal distribution of losses.

It is difficult to assume that both assumptions hold for every market. In fact, while it is reasonable to assume that for short horizons market movements are serially uncorrelated, there is overwhelming evidence that they are not Gaussian. On the contrary, while over longer horizons the assumption of normality can be accepted for many markets, there is evidence against independence of market changes. For example, there is a vast stream of literature on what is called market *predictability*: so, if over a long time horizon there is some mean-reversion tendency, the square-root rule above would actually end up overestimating the market risk of the closed-end fund investor.

7.4.2 Temporal Aggregation by *C*-convolution

The C-convolution approach is well suited to address the problems above. For the sake of simplicity, assume the distribution of profit and losses per unit of time is given by F_Y and the temporal dependence between the returns cumulated at any time t_{n-i} and the increment between t_{n-i} and time t_n is given by $C(u,v)$. Then, the α VaR figure at horizon t_n will be given by solving

$$\int_0^1 D_1 C \left(w, F_Y(-VaR_{n,\alpha} - H_{n-1}(w))\right) dw = H_n(-VaR_{n,\alpha}) = 1 - \alpha,$$

where H_j denotes the cumulated profit and losses up to time t_n.

Under this approach, one can easily relax the assumptions supporting the square-root formula. We may, for example, retain the assumption of serial independence of profits and losses, while dropping the assumption of Gaussian returns. In this case, we would have

$$\int_0^1 F_Y(-VaR_{n,\alpha} - H_{n-1}(w)) dw = H_n(-VaR_{n,\alpha}) = 1 - \alpha.$$

Working out the temporal aggregation in general would require a numerical algorithm, so that the simplicity of the square-root formula will be lost. Exceptions are cases in which the distributions of profit and losses in the time unit happen to be closed under C-convolution. Within the realm of serially independent profits and losses, this is the case of α-stable distributions. So, if $F = \Phi_\alpha$, with $0 < \alpha \leq 2$ a parameter denoting heavy tails, the square-root formula would naturally be extended to

$$VaR_{T,\alpha} = VaR_{1,\alpha} T^{1/\alpha}. \tag{7.16}$$

The square-root formula would then only hold for the case $\alpha = 2$ and would in general yield higher numbers for fat-tailed distributions. Losses would be aggregated linearly for $\alpha = 1$, the case in which the first moment of the distribution is not defined.

In the more general case we can also drop the assumption of serial independence of profits and losses and allow for $C(u, v) \neq uv$. Even in this case, while a recursive algorithm of numerical integration will be the rule, we may have shortcuts selecting the copula from families which are closed under convolution. An example is given by elliptical copulas.

Finally, to summarize this section it may be useful to observe that the same aggregation law may be generated by very different convolution models. The simplest, and most extreme, example of this is linear aggregation. We saw that taking serially independent profits and losses with α-stable distribution for $\alpha = 1$ yields linear aggregation. The same happens if one takes perfect serial dependence of profits and losses, $C(u, v) = \min(u, v)$.

Market Risk of an Investment in a Private Equity Fund

We now give an example of the relevance of temporal dependence of market profits and losses in the evaluation of risk capital. Consider an investment in a private equity fund. A private equity fund is a closed-end fund investing in the capital of unlisted firms. An example is venture capital, which assists the growth of new firms with participation in their equity for a long period of time, until the firm goes public, that is files for listing status on a stock exchange. Investments in private equity funds are locked in for a minimum period of time, typically required to be three years. The question for the investor is then how much capital to commit to absorb the risk of losses that may be brought about by changes in the equity market over this long period of time. To address this question, Figure 7.11 reports a statistical analysis of the dependence of increments in the dynamics of the Eurostoxx equity index. The analysis is performed on data from 2000 to 2010 with the usual Kendall function technique.

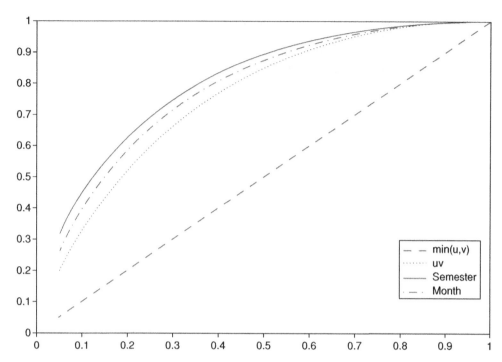

Figure 7.11 Incremental dependence of the European stock index (Eurostoxx): monthly and semi-annual increments, Clayton copula.

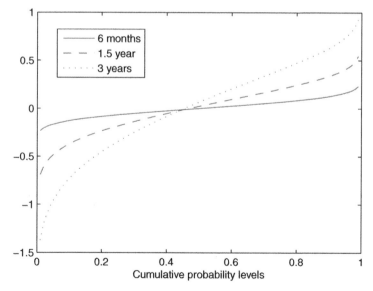

Figure 7.12 VaR at different time horizons for Clayton copula function with $\tau = 0.6$.

We report the dependence of both monthly and semi-annual increments. Notice that in both cases there is evidence of mean reversion, and the evidence is stronger for the longer time horizon. In fact, semi-annual increments report strong negative dependence, with a Kendall's τ score of -21.10%. The corresponding Clayton copula parameter is -0.348. The risk capital requirement for arbitrary investment horizons can now be computed by C-convolution. The law for the amount of capital required for different time horizons is reported in Figure 7.12. In comparison with the square-root law, we see that the amount of capital required is substantially lower. In particular, for the three-year horizon of our private equity investor it turns out to be equal to 5.89 times as much as the six-month requirement, against 2.45 times what one would obtain under the square-root law.

8

Frontier Issues

We have now finished our journey through the use of copula functions to represent the dynamics of multivariate systems. Throughout the whole book the analysis has been carried out in a discrete time setting. Even though at some point we have assumed a continuous time change machine to sample discrete time observations endowed with features that would make the model more tractable, the heart of analysis was carried out in a discrete time setting. This raises a natural question. What about the dependence of processes defined in continuous time? And what is the relationship to our approach? As of today, we do not have an answer to the latter question, and all we can do is provide a review of answers that were given to the former starting from the specification of the model in continuous time. The question of exploring convergence of our model (but also of the DNO approach) to a continuous time setting is particularly intriguing. One is stimulated by the curiosity of discovering which processes would be found in continuous time, and whether they are only those that have been studied until now, or whether some others can be found. So, the eagerness is the same as that of a worker digging a tunnel, expecting to reach people who are digging from the opposite side of the mountain, and the spirit of discovery is the same as that of someone expecting to find some weird new creature in the process. We leave this question and consider the future research of this new decade. Here we describe the tools and work of people who are digging from the other side of the mountain. They work directly with processes defined in continuous time and design tools to link those processes into multivariate ones. The idea is then to represent dependence among stochastic processes so that the laws of individual components of a multivariate process agree with some prescribed laws. Therefore, with an abuse of terminology we could refer to the models reviewed in this chapter as *semi-martingale copulas* and *Markov copulas*, that must not be confused with the representation of Markov processes with copulas, which has been the subject of this book.

8.1 LÉVY COPULAS

Copulas give a characterization of the dependence structure of random vectors, given the marginals, and allow us to construct a multidimensional distribution with specified marginal distributions. In the same spirit, Lévy copulas provide a way to construct multivariate Lévy processes with given one-dimensional marginals.

Given a Lévy process X, the Lévy copula describes only the dependence structure of the purely discontinuous part of X, since the Brownian part of the process is characterized by its covariance matrix and since the Brownian part is independent of the jump part (see Appendix B for a survey of the main properties of Lévy processes). Here we will present formal definitions and main results following the seminal approach of Tankov (2003), and Kallsen and Tankov (2006), dropping all proofs; the interested reader is referred to the cited literature.

Parallel to what happens with standard copulas, we have first to lay out the concept of volume and the d-increasing property. It is to be noticed that, contrary to what happens with standard copulas, our task is to couple measures taking the whole real line as support. So we have:

d-increasing property   Let $\overline{\mathbb{R}} := (-\infty, +\infty]$. A function $F : \overline{\mathbb{R}}^d \to \mathbb{R}$ is d-increasing if $F(u_1, \ldots, u_d) \neq \infty$ for $(u_1, \ldots, u_d) \neq \infty$ and

$$\sum_{c \in \{a_1, b_1\} \times \ldots \times \{a_d, b_d\}} (-1)^{N(c)} F(c) \geq 0, \qquad (8.1)$$

for any $-\infty < a_i \leq b_i \leq \infty$ and $N(c) := \sharp\{k : c_k = a_k\}$.

We design a property that should match the uniform margin property in standard copulas. The one-dimensional ith margin of F is now given by

$$F^{\{i\}}(x) = \lim_{c \to -\infty} (F(+\infty, \ldots, +\infty, x, +\infty, \ldots, +\infty) - F(c, \ldots, c, x, c, \ldots, c)). \qquad (8.2)$$

We are now ready to write the definition of a Lévy copula with properties similar to those of the ordinary copula.

Lévy copula   A function $F : \overline{\mathbb{R}}^d \to \overline{\mathbb{R}}$ is called a Lévy copula if

1. $F(u_1, \ldots, u_d) = 0$ if $u_i = 0$ for at least one $i \in \{1, \ldots, d\}$.
2. F is d-increasing.
3. $F^{\{i\}}(u) = u$ for any $i \in \{1, \ldots, d\}, u \in \mathbb{R}$.

In order to use Lévy copulas to investigate dependence between the components of a multi-variate Lévy process, we must define a tail integral. This concept will replace the role of multivariate distributions in the realm of ordinary copulas. As in standard copulas the dependence function is recovered from the multivariate distribution, in Lévy copulas the dependence structure will be recovered from the multivariate specification of the Lévy measure. We remind the reader that intuitively a Lévy measure summarizes the intensity and the size of jumps.

Tail integral   Let X be an $\mathbb{R}^d$-valued Lévy process with Lévy measure ν. The tail integral of X is the function $U : (\mathbb{R} \backslash \{0\})^n \to \mathbb{R}$ defined by

$$U(x_1, \ldots, x_d) := \prod_{i=1}^{d} sgn(x_i) \nu \left(\prod_{j=1}^{d} \mathcal{I}(x_j) \right), \qquad (8.3)$$

where for $x \in \mathbb{R}$, we define

$$\mathcal{I}(x) := \begin{cases} (x, \infty), & x \geq 0 \\ (-\infty, x], & x < 0. \end{cases}$$

Notice that, since the Lévy measure ν is a positive measure, the sign function allows us to distinguish between positive and negative jumps.

To determine the Lévy measure uniquely it is not sufficient to know the tail integral since it does not give any information about mass on coordinate axes. We remind the reader that by definition, the Lévy measure in dimension one does not include the origin and the mass is concentrated on the rest of the real line. When we extend the definition to a multivariate setting, dropping the origin for measure i would entail dropping the whole coordinate for the multivariate measure. As shown in Kallsen and Tankov (2006), Lemma 3.5, the Lévy measure is uniquely determined by the set of all marginal tail integrals, that is we collect the multivariate Lévy measure for all possible subprocesses with dimension lower than d.

***I*-marginal tail integral**   The I-marginal tail integral, U^I, for non-empty set $I \subset \{1, .., d\}$ is the tail integral of the Lévy process $(X_i)_{i \in I}$ or equivalently the tail integral of the I-marginal of the Lévy measure ν^I defined by

$$\nu^I(A) := \nu(x \in \mathbb{R}^d : (x_i)_{i \in I} \in A \setminus \{0\}), \quad A \in \mathcal{B}(R^{|I|}).$$

In order to finally recover the equivalent result of the Sklar theorem for Lévy processes, we have to define the tail integral that enables us to compute the Lévy measures for marginals of all dimensions lower than d.

***I*-margin of F**   Let F be a d-increasing function. For any non-empty index set $I \subset \{1, \ldots, d\}$ the I-margin of F is the function $F^I : \overline{\mathbb{R}}^I \to \overline{\mathbb{R}}$ defined by

$$F^I((u_i)_{i \in I}) := \lim_{k \to \infty} \sum_{(u_j)_{j \in I^c} \in \{-k, \infty\}^{I^c}} F(u_1, \ldots, u_d) \prod_{j \in I^c} \text{sgn}(u_j), \quad (8.4)$$

where $I^c := \{1, \ldots, d\} \setminus I$.

We are now ready to write the counterpart of Sklar's theorem for Lévy measures.

Theorem 8.1.1.   (see Kallsen and Tankov (2006))

(i) *Let $X = (X_1, \ldots, X_d)$ be an $\mathbb{R}^d$-valued Lévy process. Then there exists a Lévy copula F such that the tail integrals of X satisfy*

$$U^I((x_i)_{i \in I}) = F^I(((U_i(x_i)))_{i \in I}),$$

for any non-empty set $I \subset \{1, \ldots, d\}$ and any $(x_i)_{i \in I} \in (\mathbb{R} \setminus \{0\})^I$. The Lévy copula F is unique on $\prod_{i=1}^d \overline{RanU_i}$.

(ii) *Let F be a d-dimensional Lévy copula and $U_i, i = 1, \ldots, d$ be the tail integrals of real-valued Lévy processes. Then there exists an R^d-valued Lévy process whose components have tail integrals $U_i, i = 1, \ldots, d$ and whose marginal tail integral satisfies condition (i) for any non-empty $I \subset \{1, \ldots, d\}$ and any $(x_i)_{i \in I} \in (\mathbb{R} \setminus \{0\})^I$. The Lévy measure ν of X is uniquely determined by F and $U_i, i = 1, \ldots, d$.*

Example 8.1.1.   *In Kallsen and Tankov (2006) it is shown that a pure jump Lévy process has independent coordinates if and only if its Lévy copula is given by the following formula:*

$$F_\perp(x_1, \ldots, x_d) := \sum_{i=1}^d x_i \prod_{j \neq i} \mathbf{1}_{\{\infty\}}(x_j).$$

Notice that the formula implies that the Lévy processes never jump together so that all the probability mass is concentrated on the coordinate axes.

Example 8.1.2.   *For the characterization of complete dependence we recall that a subset S of $\mathbb{R}^d$ is called* ordered *if, for any two vectors $u, v \in S$, either $u_k \leq v_k, k = 1, \ldots, d$ or $u_k \geq v_k, k = 1, \ldots, d$. Similarly, S is called* strictly ordered *if, for any two different vectors $u, v \in S$, either $u_k < v_k, k = 1, \ldots, d$ or $u_k > v_k, k = 1, \ldots, d$.*

In Kallsen and Tankov (2006) it is shown that if X is an $\mathbb{R}^d$-valued Lévy process whose Lévy measure is supported by an ordered set $S \subset K$ where $K := \{x \in \mathbb{R}^d : sgn(x_1) = \ldots = sgn(x_d)\}$, then the complete dependence Lévy copula, given by

$$F_K(x_1, \ldots, x_d) := \min(|x_1|, \ldots, |x_d|)\mathbf{1}_K(x_1, \ldots, x_d) \prod_{i=1}^{d} sgn(x_i),$$

is a Lévy copula of X. Conversely, if F_K is a Lévy copula of X, then the Lévy measure of X is supported by an ordered subset of K. If, in addition, the tail integral U_i of X^i is continuous and satisfies $\lim_{x \to 0} U_i(x) = \infty, i = 1, \ldots, d$, then the jumps of X are completely dependent, i.e. there exists a strictly ordered subset S of K such that $v(\mathbb{R}^d \backslash S) = 0$.

Example 8.1.3. *Kallsen and Tankov (2006) introduced the Archimedean Lévy copula by defining*

$$F(x_1, \ldots, x_d) := \varphi\left(\prod_{i=1}^{d} \widetilde{\varphi}(u_i)\right),$$

where $\varphi : [-1, 1] \to [-\infty, \infty]$ is a strictly increasing continuous function with $\varphi(-1) = -\infty, \varphi(0) = 0, \varphi(1) = \infty$, and having derivatives up to order d on the intervals $(-1, 0)$ and $(0, 1)$, satisfying

$$\frac{\partial^d \varphi(e^x)}{\partial x^d} \geq 0, \quad \frac{\partial^d \varphi(-e^x)}{\partial x^d} \leq 0, \quad x \in (-\infty, 0)$$

and where $\widetilde{\varphi}(u) := 2^{d-2}(\varphi(u) - \varphi(-u))$.

8.2 PARETO COPULAS

While the copula of a multivariate distribution is the distribution transformed so that one-dimensional marginal distributions are uniform, Klüppelberg and Resnick (2008) suggest a different transformation which yields standard Pareto distributions for the marginal laws. Here we briefly summarize the results presented in Klüppelberg and Resnick (2008), referring the interested reader to the original paper for details and related literature. We will limit our discussion to the task that this branch of related literature shares with our approach, concerning the aggregation of risk. Actually, the reason why it is wise to switch from uniform to Pareto marginals is to study the asymptotic behavior of aggregated risks.

Given a random vector $(X^{(1)}, \ldots, X^{(d)})$, we define

$$(P_1, \ldots, P_d) = \left(\frac{1}{1 - F^{(i)}(X^{(i)})}, i = 1, \ldots, d\right),$$

where $F^{(i)}$ is the marginal cumulative distribution function of $X^{(i)}$. Note that P_j is standard Pareto distributed:

$$\mathbb{P}(P_i > x) = x^{-1}, \quad x \geq 1, \quad i = 1, \ldots, d.$$

Pareto copula   Suppose $(X^{(1)}, \ldots, X^{(d)})$ has joint distribution F with continuous marginals. We call the distribution ψ of $(P_1, \ldots, P_d)$ a Pareto copula.

We now show how to use Pareto copulas for risk aggregation. We first need the definition of the domain of attraction that is well known in extreme value theory.

Definition 8.2.1. Let $\{\mathbf{X}, \mathbf{X}_n, n \geq 1\}$ be i.i.d. random vectors with common distribution F. Then $\mathbf{X}$ is in a multivariate maximal domain of attraction if there exists

$$\mathbf{b}(t) = \left(b^{(1)}(t), \dots, b^{(d)}(t)\right) \in \mathbb{R}^d, \quad \mathbf{a}(t) = \left(a^{(1)}(t), \dots, a^{(d)}(t)\right) \in \mathbb{R}^d_+$$

such that

$$\mathbb{P}^n \left(\frac{\mathbf{X} - \mathbf{b}(n)}{\mathbf{a}(n)} \leq \mathbf{x} \right) = F^n(\mathbf{a}(n)\mathbf{x} + \mathbf{b}(n))$$

$$= \left(\mathbb{P}\left(\frac{X^{(i)} - b^{(i)}(n)}{a^{(i)}(n)} \leq x^{(i)}; i = 1, \dots, d \right) \right)^n \rightarrow G(\mathbf{x}), \tag{8.5}$$

where G is a non-degenerate distribution called a max-stable or extreme value distribution. The marginal distributions $G^{(i)}_{\gamma^{(i)}}, i = 1, \dots, d$ of G are one-dimensional extreme value distributions of the type

$$G^{(i)}_{\gamma^{(i)}}\left(x^{(i)}\right) = \exp\left\{ -\left(1 + \gamma^{(i)}x^{(i)}\right)^{-1/\gamma^{(i)}} \right\}, \quad 1 + \gamma^{(i)}x^{(i)} > 0,$$

and $G^{(i)}_{\gamma^{(i)}}(x^{(i)})$ concentrates on $\{u \in \mathbb{R} : 1 + \gamma^{(i)}u > 0\}$ (see de Haan and Ferreira (2006), Embrechts *et al.* (1997), and Resnick (1987)).

Klüppelberg and Resnick (2008) discuss aggregation of risks when (8.5) holds with $\gamma^{(i)} = 0$ for $i = 1, \dots, d$ so that each marginal is in the domain of attraction of the Gumbel distribution. If one chooses a function $e^{(i)}(t)$ as the mean excess function, it could be proved that the following limit holds true:

$$\frac{\overline{F}^{(i)}\left(t + xe^{(i)}(t)\right)}{\overline{F}^{(i)}(t)} \rightarrow e^{-x}, \quad x \in \mathbb{R}, \tag{8.6}$$

as t converges to the right endpoint of $F^{(i)}$. Then we may take:

$$b^{(i)}(t) = \left(\frac{1}{1 - F^{(i)}(.)} \right)^{-1}(t), \quad a^{(i)}(t) = e^{(i)}\left(b^{(i)}(t)\right), \quad i = 1, \dots, d. \tag{8.7}$$

If one also assumes homogeneous risks, marginals of F are the same and we can proceed under the assumption

$$F^{(i)}(.) = F^{(1)}(.), \quad i = 1, \dots, d. \tag{8.8}$$

We now report the main result concerning risk aggregation

Proposition 8.2.1. *Suppose (8.5) holds where all marginals of $F(x)$ are equal and all marginals of $G(x)$ are Gumbel and (8.6) and (8.8) hold. Suppose F does not possess asymptotic independence and define $v(.)$ by*

$$v([-\infty, \mathbf{x}]^c) = -\log(G(\mathbf{x})), \quad \mathbf{x} \neq -\infty.$$

Then

$$N_n := \sum_{j=1}^{n} \epsilon_{(\sum_{i=1}^{d} X^{(i)}_j - db(n))/a(n)} \Rightarrow \sum_k \mathbf{1}_{[j_k > -\infty]} \epsilon_{\sum_{i=1}^{d} j^{(i)}_k},$$

in the space of Radon point measures on $(-\infty, \infty]$, i.e. $M_p(-\infty, \infty]$, where ϵ_x is the probability measure consisting of all mass at x and where the limit N_∞ is a Poisson random measure *with mean measure*

$$\nu\left\{\boldsymbol{x} \in (-\infty, \infty] : \sum_{i=1}^{d} x^{(i)} \in .\right\}$$

and as $n \to \infty$ (Resnick, 2007)

$$n\mathbb{P}\left\{\frac{\sum_{i=1}^{d} X^{(i)} - db(n))}{a(n)} > y\right\} \to \nu\left\{\boldsymbol{x} \in (-\infty, \infty] : \sum_{i=1}^{d} x^{(i)} > y\right\}.$$

To make the argument operational, it is now sufficient to notice the following corollary.

Corollary 8.2.2.   *Under the conditions of Proposition 8.2.1, we have that*

$$\lim_{t \to \infty} \frac{\mathbb{P}\{\sum_{i=1}^{d} X^{(i)} > dt\}}{\mathbb{P}\{X^{(1)} > t\}} = \nu\{\mathbf{x} \in (-\infty, \infty] : \sum_{i=1}^{d} x^{(i)} > 0\}.$$

So, under the homogeneity assumption, evaluating a basket of d risks can be accomplished by simply multiplying the risk of an element of the basket times a random measure.

8.3 SEMI-MARTINGALE COPULAS

Assume a set of stochastic processes $Y^1, \ldots, Y^n$ such that the distribution of each Y^i is given. Consider the problem of constructing a multidimensional random vector $X = (X^1, \ldots, X^n)$ such that the distribution of each X^i is the same as that of Y^i.

Bielecki *et al.* (2008) and Vidozzi A. (2009) address this problem in full generality. They follow the route of working with infinitesimal characteristics when they are defined. The case of Lévy copulas addressed above is actually a special instance of this approach in which the characteristics considered are the Lévy measures.

Formally, the problem addressed is the following: given two real-valued semi-martingales $Y^i, i = 1, 2$ defined on the probability spaces $(\Omega_i, \mathcal{F}^{Y^i}, \mathbb{P}^i)$ and endowed with the canonical filtration $\mathbb{F}^{Y^i} = \{\mathcal{F}_t^{Y^i}\}_{t \geq 0}$, we would like to construct a probability space $(\Omega, \mathcal{F}^X, \mathbb{P})$, with $\Omega = \Omega_1 \times \Omega_2$, such that the finite-dimensional distributions of the components of the canonical process $X = (X^1, X^2)$ on that space are identical to those of Y^1, Y^2. Intuitively, the use of canonical processes is motivated by the need to refer to the most general space embedding all possible processes.

The technique introduced by this branch of literature proposes using the projections of the characteristics of each component with respect to its own filtration. In this way, we actually sterilize the characteristics of the process from the information set which is not specific to the component.

The recipe for constructing a bivariate semi-martingale with given margins works as follows. One constructs $(\Omega, \mathcal{F}^X, \mathbb{F}^X, \mathbb{P})$ in such a way that the $\mathbb{F}^{X^i}$ characteristics of their coordinate processes $X^i, i = 1, 2$, are identical (as functions of trajectories) to the $\mathbb{F}^{Y^i}$ characteristics of $Y^i, i = 1, 2$. This implies that for $i = 1, 2, X^i$ and Y^i are equal in law. Indeed, X^i and Y^i live in the same canonical space Ω_i (this means that the canonical σ-algebras $\mathcal{F}^{Y^i}$ and $\mathcal{F}^{X^i}$ contains the same events), and the $\mathbb{F}^{X^i}$ characteristics of X^i and the $\mathbb{F}^{Y^i}$ characteristics of Y^i coincide (as functions of trajectories). Remember that two processes having the same characteristics share the same finite-dimensional distributions. This fact implies that X^i and Y^i have the same finite dimensional distributions. Here we assume that the finite-dimensional distributions of each Y^i are uniquely determined by its characteristic triple (B^i, C^i, v^i).

As an example of this representation in full generality, we report here the case of a semi-martingale copula.

2-dimensional semi-martingale copula   A triple (B, C, v) defined on the basis $(\Omega, \mathcal{F}^X, \mathbb{F}^X)$ is a semi-martingale copula for $Y^i, i = 1, 2$, if the following conditions hold:

(i) There is a unique probability measure $\mathbb{P}$ on $\mathcal{F}^X$ such that the canonical process on the stochastic basis $(\Omega, \mathcal{F}^X, \mathbb{F}^X)$ is a semi-martingale with characteristic triple (B, C, v).
(ii) Under $\mathbb{P}$, the $\mathbb{F}^{X^i}$ characteristics of X^i, say $(\widetilde{B^i}, \widetilde{C^i}, \widetilde{v^i})$, are equal to the characteristics of Y^i, say $(\widehat{B^i}, \widehat{C^i}, \widehat{v^i})$, for $i = 1, 2$.

It would be easy to check that (see Vidozzi A. (2009) and Vidozzi L. (2009)), given the $\mathbb{F}^X$ characteristic triple of a multivariate process X, the computation of the $\mathbb{F}^{X^i}$ characteristics of the components X^i can be made much simpler if one constructs the multivariate process in such a way that the $\mathbb{F}^X$ characteristic triple of X^i is $\mathbb{F}^{X^i}$-predictable because in this case the computation of projections, which is possible only in very simple and special cases, can be avoided. These observations suggest the following definition.

Consistency with respect to a filtration   We say that a 2-dimensional semi-martingale $X = (X^1, X^2)$ defined on a stochastic basis $(\Omega, \mathcal{F}^X, \mathbb{F}^X, \mathbb{P})$ is consistent with respect to $\mathbb{F}^{X^i}$ if the $\mathbb{F}^X$ characteristic triple (B^i, C^i, v^i) and the $\mathbb{F}^{X^i}$ characteristic triple $(\widetilde{B^i}, \widetilde{C^i}, \widetilde{v^i})$ of the component process X^i coincide (as a function of trajectories).

2-dimensional consistent semi-martingale copula   A triple (B, C, v) defined on the basis $(\Omega, \mathcal{F}^X, \mathbb{F}^X)$ is a consistent semi-martingale copula for $Y^i, i = 1, 2$, if the following conditions hold:

(i) There is a unique probability measure $\mathbb{P}$ on $\mathcal{F}^X$ such that the canonical process on the stochastic basis $(\Omega, \mathcal{F}^X, \mathbb{F}^X)$ is a semi-martingale with characteristic triple (B, C, v).
(ii) Under $\mathbb{P}$, X is consistent with respect to $\mathbb{F}^{X^i}$ and (B^i, C^i, v^i) are equal to $(\widehat{B^i}, \widehat{C^i}, \widehat{v^i})$ for $i = 1, 2$.

The requirement for consistency allows us to avoid the computation of projections of the characteristics on smaller filtrations. We report an example of consistent semi-martingale copulas for an important class of semi-martingales, i.e. pure jump Lévy processes.

Example 8.3.1. *Copulas between pure jump Lévy processes.   We provide an example of a semi-martingale copula that is also a Lévy copula. We know that there is a one-to-one correspondence between a homogeneous Poisson process with values in $\mathbb{R}^2$ and a homogeneous Poisson measure on $E = \{0, 1\}^2 \backslash \{(0, 0)\}$. The $\mathbb{F}$ dual predictable projection of a Poisson measure μ, which we denote v, is uniquely determined by its values on the atoms in E and then a Poisson process X in $\mathbb{R}^2$ is uniquely determined by:*

$$v(dt, \{1, 0\}) = \lambda_{10}dt, \quad v(dt, \{0, 1\}) = \lambda_{01}dt, \quad v(dt, \{1, 1\}) = \lambda_{11}dt,$$

for some positive constants $\lambda_{10}, \lambda_{01},$ and λ_{11}.

Let Y^1 and Y^2 be two Poisson processes in $\mathbb{R}$ with intensities λ_1 and λ_2 respectively. Then the measure v defined before is a semi-martingale copula for X^1 and X^2 if $\lambda_{10}, \lambda_{01},$ and λ_{11}

satisfy the following conditions:

$$\lambda_1 = \lambda_{10} + \lambda_{11},$$
$$\lambda_2 = \lambda_{01} + \lambda_{11},$$
$$\lambda_{11} \in [0, \lambda_1 \wedge \lambda_2].$$

In fact, these conditions imply that v is positive and defines uniquely the probability law of a Poisson random measure on $\{0, 1\}^2$. Finally, since

$$v^1(dt, \{1\}) = v(dt, \{1, 0\}) + v(dt, \{1, 1\})$$
$$= \lambda_{10}dt + \lambda_{11}dt = \lambda_1 dt,$$
$$v^1(dt, \{x\}) = 0, \quad \forall x \neq 1$$

and similarly for $v^2(dt, \{1\})$, it's clear that $v(dt, dx)$ is a semi-martingale copula (also consistent since the characteristics are deterministic).

Assume we want to apply the semi-martingale approach above to the study of multivariate Markov processes. The task is to consider a multivariate Markov process as a collection of Markov processes themselves: this way one would naturally be led to redefine the consistency condition required for semi-martingale copulas in terms of Markov invariance with respect to enlargements of the filtration. To make this approach operational, one is naturally led to work on the relationships between the infinitesimal generators of each component and the overall Markov process. Vidozzi A. (2009) and Bielecki *et al.* (2008) study the consistency conditions among such infinitesimal generators and define a class of Markov copulas accordingly.

Example 8.3.2. *Diffusion processes.* *We assume that the marginal processes Y^i, generated by the infinitesimal generator A^i, are $\mathbb{R}$-valued diffusion processes and we construct a Markov process X with respect to its natural filtration.*

Each Markov process is characterized by an infinitesimal generator

$$A^i f(x_i) = b_i(x_i)\partial_{x_i} f(x_i) + \frac{1}{2}\sigma_i(x_i)^2 \partial_{x_i}^2 f(x_i), \qquad (8.9)$$

where the coefficients $b_i(x_i)$ and $\sigma_i(x_i)$ are suitable smooth functions.

The linear operator A is defined as

$$A f(\mathbf{x}) := \sum_{i=1}^{d} I^{\{i\}^c} \widehat{\otimes} A^i f(\mathbf{x}) + \sum_{i,j=1, i\neq j}^{d} \frac{1}{2} a_{ij}(x_i, x_j)\partial_{x_i}\partial_{x_j} f(\mathbf{x}),$$

where A^i is defined in (8.9), the notation $I^{\{i\}^c} \widehat{\otimes} A^i$ stands for $I^1 \widehat{\otimes} \ldots \widehat{\otimes} A^i \widehat{\otimes} \ldots \widehat{\otimes} I^d$, and I^i denotes the identity operator ($\widehat{\otimes}$ denotes the closure of the tensor product); $a_{ij}(x_i, x_j)$ is a covariance matrix consistent with (8.9). Then the operator A is a Markov copula for $\{(A^i, \mathcal{D}(A^i)) : i = 1, \ldots, d\}$ as defined in Bielecki et al. (2008).

Appendix A

Elements of Probability

A.1 ELEMENTS OF MEASURE THEORY

Definition A.1.1. Given a set Ω, a family $\mathcal{F}$ of subsets of Ω is a σ-algebra if

1. $\emptyset, \Omega \in \mathcal{F}$.
2. If $A \in \mathcal{F}$ then $A^c \in \mathcal{F}$ (where $A^c = \Omega \setminus A$).
3. If $\{A_n\}_{n \geq 1}$ is such that $A_n \in \mathcal{F}$ for all $n \geq 1$, then $\bigcup_{n \geq 1} A_n \in \mathcal{F}$.

The elements of $\mathcal{F}$ are called *measurable sets* and the pair $(\Omega, \mathcal{F})$ a *measurable space*. By the De Morgan formula and 2 and 3:

$$\bigcap_{n \geq 1} A_n = \left(\bigcup_{n \geq 1} A_n^c \right)^c \in \mathcal{F}.$$

Obviously the family $\mathcal{P}(\Omega)$ of all subsets of Ω is always a σ-algebra.

If $\mathcal{A}$ is a family of subsets of Ω, the smallest σ-algebra containing $\mathcal{A}$ is called the σ-*algebra generated by* $\mathcal{A}$. If $\Omega = \mathbb{R}$, the smallest σ-algebra $\mathcal{B}$ containing all subintervals of $\mathbb{R}$ is called the *Borel σ-algebra* and its elements *Borel sets*.

A measure is a function that associates to each measurable set a real number. More precisely

Definition A.1.2. Let $(\Omega, \mathcal{F})$ be a measurable space. A measure is a function $\mu : \mathcal{F} \to [0, +\infty]$ such that:

1. $\mu(\emptyset) = 0$.
2. If $\{A_n\}_{n \geq 1}$ is a disjoint sequence of sets in $\mathcal{F}$ then

$$\mu \left(\bigcup_{n \geq 1} A_n \right) = \sum_{n \geq 1} \mu(A_n).$$

The measure μ is *finite* or *infinite* (say also *integrable* or *not integrable*) as $\mu(\Omega) < +\infty$ or $\mu(\Omega) = +\infty$. Any measurable set $A \in \mathcal{F}$ such that $\mu(A) = 0$ is called *negligible* or the *null* set.

Let $\Omega = \mathbb{R}$. A measure μ on the Borel sets $\mathcal{B}$ is said to be *locally finite* if for every compact $B \in \mathcal{B}$, $\mu(B) < +\infty$. Locally finite measures are also called *Radon measures*. The Lebesgue measure is an example of a Radon measure.

Definition A.1.3. Let $(\Omega, \mathcal{F})$ be a measurable space. A real-valued function $f : \Omega \to \mathbb{R}$ is called measurable if for every Borel set $A \in \mathcal{B}$,

$$f^{-1}(A) = \{\omega \in \Omega : f(\omega) \in A\} \in \mathcal{F}.$$

Let $(\Omega, \mathcal{F})$ be a measurable space provided by the measure μ; two measurable functions f and g are said to be *equal μ-almost everywhere* if and only if $\mu(\{x \in \Omega : f(x) \neq g(x)\}) = 0$ (and we shall write $f = g$ μ-a.e.).

If $\mu(\Omega) = 1$, μ is a *probability* measure and is denoted by $\mathbb{P}$. In particular, if $\mathcal{F}$ is a σ-algebra in Ω and $\mathbb{P}$ a probability on $\mathcal{F}$, the triple $(\Omega, \mathcal{F}, \mathbb{P})$ is called a *probability space*. A *support* of $\mathbb{P}$ is any event $A \in \mathcal{F}$ such that $\mathbb{P}(A) = 1$.

The following are trivial consequences of the above definitions. Let $(\Omega, \mathcal{F}, \mathbb{P})$ be a probability space, then:

1. If $A \in \mathcal{F}$, then $\mathbb{P}(A^c) = 1 - \mathbb{P}(A)$;
2. If $A, B \in \mathcal{F}$, $A \subset B$, then $\mathbb{P}(A) \le \mathbb{P}(B)$;
3. If $A, B \in \mathcal{F}$, then $\mathbb{P}(A \cup B) = \mathbb{P}(A) + \mathbb{P}(B) - \mathbb{P}(A \cap B)$.

Definition A.1.4. Let $(\Omega, \mathcal{F}, \mathbb{P})$ be a probability space. If $A, B \in \mathcal{F}$ with $\mathbb{P}(A) > 0$, the quantity

$$\mathbb{P}(B|A) = \frac{\mathbb{P}(A \cap B)}{\mathbb{P}(A)}$$

is called the "conditional probability of B with respect to A".

Intuitively, the conditional probability $\mathbb{P}(B|A)$ is the probability that B occurs knowing that A has occurred.

Definition A.1.5. Let $(\Omega, \mathcal{F}, \mathbb{P})$ be a probability space. $A, B \in \mathcal{F}$ are "independent" if and only if

$$\mathbb{P}(A \cap B) = \mathbb{P}(A)\mathbb{P}(B).$$

Definition A.1.6. Let $(\Omega, \mathcal{F}, \mathbb{P})$ be a probability space. $A_1, A_2, \ldots \in \mathcal{F}$ are "independent" if and only if for every k and every $i_1, i_2, \ldots, i_k$,

$$\mathbb{P}(A_{i_1} \cap \ldots \cap A_{i_k}) = \mathbb{P}(A_{i_1}) \cdots \mathbb{P}(A_{i_k}).$$

Let $(\Omega, \mathcal{F}, \mathbb{P})$ be a probability space. A real-valued *random variable* is a measurable function $X : \Omega \to \mathbb{R}$. By definition, the function

$$A \to \mathbb{P}(\{\omega \in \Omega : X(\omega) \in A\})$$

is well defined for every real Borel set A. This function is called the *law* or the *distribution* of X. In the sequel we shall use the more concise notation $\mathbb{P}(X \in A)$ to denote $\mathbb{P}(\{\omega \in \Omega : X(\omega) \in A\})$.

The function $F_X : \mathbb{R} \to [0, 1]$ defined as $F_X(t) = \mathbb{P}(X \le t) = \mathbb{P}(X \in (-\infty, t])$ is called the *cumulative distribution function of X*.

Let X and Y be two random variables defined on the same probability space $(\Omega, \mathcal{F}, \mathbb{P})$; we shall say that they are *equal $\mathbb{P}$-almost surely* (and we shall write $X = Y$ a.s.) if and only if $\mathbb{P}(X = Y) = 1$.

The random variables $X_1, \ldots, X_n$ defined on the same probability space $(\Omega, \mathcal{F}, \mathbb{P})$ are *independent* if and only if

$$\mathbb{P}(X_1 \in A_1, \ldots, X_n \in A_n) = \mathbb{P}(X_1 \in A_1) \cdots \mathbb{P}(X_n \in A_n)$$

for all $A_1, \ldots, A_n$ real Borel sets.

A.2 INTEGRATION

Let $(\Omega, \mathcal{F})$ be a measurable space provided by the measure μ. If f is a non-negative simple function, that is $f = \sum_{i=1}^{n} x_i \mathbf{1}_{A_i}$ with $x_i \ge 0$ for all $i = 1, \ldots, n$ and $\{A_i\}$ being a finite decomposition of Ω into elements of $\mathcal{F}$, then the integral is defined by

$$\int_\Omega f(x)\mu(dx) = \sum_{i=1}^n x_i\mu(A_i).$$

Now, if f is a non-negative measurable function, the integral is defined by

$$\int_\Omega f(x)\mu(dx) = \sup\left\{\int_\Omega g(x)\mu(dx) : g \text{ simple function } g \leq f\right\}.$$

For a general measurable function f, consider its positive part

$$f^+(x) = \begin{cases} f(x) & \text{if } f(x) \geq 0 \\ 0 & \text{otherwise} \end{cases}$$

and its negative part

$$f^-(x) = \begin{cases} -f(x) & \text{if } f(x) \leq 0 \\ 0 & \text{otherwise} \end{cases}.$$

These functions are non-negative and measurable and $f = f^+ - f^-$. The general integral is defined by

$$\int_\Omega f(x)\mu(dx) = \int_\Omega f^+(x)\mu(dx) - \int_\Omega f^-(x)\mu(dx) \qquad (A.1)$$

unless $\int_\Omega f^+(x)\mu(dx) = \int_\Omega f^-(x)\mu(dx) = +\infty$, in which case f has no integral.

If $\int_\Omega f^+(x)\mu(dx)$ and $\int_\Omega f^-(x)\mu(dx)$ are both finite (that is $\int_\Omega |f(x)|\mu(dx) < +\infty$), then f is integrable and has (A.1) as its *definite integral*. If $\int_\Omega f^+(x)\mu(dx) = +\infty$ and $\int_\Omega f^-(x)\mu(dx) < +\infty$ (or $\int_\Omega f^-(x)\mu(dx) = +\infty$ and $\int_\Omega f^+(x)\mu(dx) < +\infty$), then f is not integrable but is in accordance with (A.1) assigned $+\infty$ (or $-\infty$), as its definite integral.

Main properties of the integral:

1. If f and g are two integrable functions, then, if $f = g$ μ-a.e., $\int_\Omega f(x)\mu(dx) = \int_\Omega g(x)\mu(dx)$.
2. *Monotonicity.* If f and g are two integrable functions, with $f \geq g$ μ-a.e., then $\int_\Omega f(x)\mu(dx) \geq \int_\Omega g(x)\mu(dx)$.
3. *Linearity.* If f and g are two integrable functions and $a, b \in \mathbb{R}$, then $af + bg$ is integrable and $\int_\Omega(af(x) + bg(x))\mu(dx) = a\int_\Omega f(x)\mu(dx) + b\int_\Omega g(x)\mu(dx)$.
4. $\left|\int_\Omega f(x)\mu(dx) - \int_\Omega g(x)\mu(dx)\right| \leq \int_\Omega |f(x) - g(x)|\mu(dx)$.

A.2.1 Expected Values and Moments

If $(\Omega, \mathcal{F}, \mathbb{P})$ is a probability space and X a random variable, the integral $\int_\Omega X(\omega)\mathbb{P}(d\omega)$, when defined, is called the *expected value of X* and is denoted by $\mathbb{E}^\mathbb{P}[X]$ (when there is no ambiguity the probability can be removed and the expected value indicated by $\mathbb{E}[X]$). Clearly, the expected value only depends on the distribution of a random variable: so random variables with the same distribution share the same expected value.

If, for a positive integer k, $\mathbb{E}[|X|^k] < +\infty$, the expected value $\mathbb{E}[X^k]$ is called the k-moment of X. Since $|x|^j \leq 1 + |x|^k$ for $j \leq k$, if X has a finite k-moment, it has a finite j-moment as well. Clearly, moments, being expectations, are uniquely determined by distributions. The moments of a random variable may or may not exist, depending on how fast the distribution of X decays at infinity.

Definition A.2.1. The set of all random variables for which $\mathbb{E}[|X|^p] < +\infty$ is denoted by L^p and is equipped with the distance $d(X, Y) = (\mathbb{E}[|X - Y|^p])^{1/p}$.

The k-centered moment of a random variable X is defined as the k-moment of $X - \mathbb{E}[X]$, that is

$$\mathbb{E}[(X - \mathbb{E}[X])^k].$$

The second moment of a random variable X is called the variance, $Var(X) = \mathbb{E}[(X - \mathbb{E}[X])^2] = \mathbb{E}[X^2] - \mathbb{E}[X]^2$.

Scale-free versions of centered moments can be obtained by suitably normalizing. For example

$$s(X) = \frac{\mathbb{E}[(X - \mathbb{E}[X])^3]}{Var^{3/2}(X)}$$

is called the *skewness* of X: if $s(X) > 0(< 0)$, X is said to be positively (negatively) skewed.

$$k(X) = \frac{\mathbb{E}[(X - \mathbb{E}[X])^4] - 3Var(X)}{Var^2(X)}$$

is called the excess kurtosis of X. X is leptokurtic (that is fat-tailed) if $k(X) > 0$. By definition, the skewness and the kurtosis are invariant with respect to a change of scale:

$$\forall c > 0, \qquad s(cX) = s(X), \quad k(cX) = k(X).$$

Since for X normally distributed we have $s(X) = k(X) = 0$, these quantities can be seen as measures of deviation from normality.

A.3   THE MOMENT-GENERATING FUNCTION OR LAPLACE TRANSFORM

Definition A.3.1. The moment-generating function of a random variable X is defined as

$$M_X(t) = \mathbb{E}[e^{-tX}]$$

for all t for which this is finite.

The moment-generating function in non-probabilistic contexts is called the Laplace transform.

Notice that $M_X(0) = 1$. If X is non-negative, $M_X(t)$ is well defined for all $t \geq 0$ (if X is non-positive the converse holds). If $M_X(t)$ is defined throughout an interval $(-t_0, t_0)$, with $t_0 > 0$, it can be proved that (Billingsley, 1986) X has finite moments of all orders and

$$M_X(t) = \sum_{k=0}^{+\infty} (-1)^k \frac{t^k}{k!} \mathbb{E}[X^k].$$

As a consequence,

$$\frac{d^k}{dx^k} M_X(0) = (-1)^k \mathbb{E}[X^k].$$

If X and Y are two independent random variables such that M_X and M_Y are both defined on the same t, it can easily be checked that

$$M_{X+Y}(t) = M_X(t)M_Y(t).$$

A.4   THE CHARACTERISTIC FUNCTION

Definition A.4.1. The characteristic function of a random variable X is defined for real t by

$$\phi_X(t) = \mathbb{E}[e^{itX}].$$

The characteristic function in non-probabilistic contexts is called the Fourier transform. The characteristic function has three fundamental properties:

1. If X and Y are independent random variables, $\phi_{X+Y}(t) = \phi_X(t)\phi_Y(t)$.
2. The characteristic function uniquely determines the distribution.
3. From the pointwise convergence of a sequence of characteristic functions, the convergence of the corresponding distributions follows; more precisely, $X_n \xrightarrow{d} X$ if and only if $\phi_{X_n}(t) \to \phi_X(t)$ for all t.

The moments of a random variable are related to the derivatives at 0 of its characteristic function: if $\mathbb{E}[|X|^n] < +\infty$, then ϕ_X has n continuous derivatives at 0 and

$$\mathbb{E}[X^k] = \frac{1}{i^k} \frac{\partial^k \phi_X(0)}{\partial t^k}, \qquad \forall k = 1, \ldots, n. \tag{A.2}$$

On the other hand, if ϕ_X has n continuous derivatives at 0, then $\mathbb{E}[|X|^n] < +\infty$ and (A.2) holds.

Definition A.4.2. The characteristic function of a random vector $X = (X_1, \ldots, X_d)$ is defined for $z \in \mathbb{R}^d$ by

$$\phi_X(t) = \mathbb{E}[e^{iz \cdot X}],$$

where for $y, z \in \mathbb{R}^d$ with $z \cdot y$ we denote the inner product, $z \cdot y = \sum_{i=1}^d z_i y_i$.

As for the case of random variables, the characteristic function identifies the multivariate distribution of the random vector.

A.5   RELEVANT PROBABILITY DISTRIBUTIONS

Binomial Distribution

The binomial distribution with parameters $n \in \mathbb{N} \setminus \{0\}$ and $p \in [0, 1]$ is defined through its density

$$p(x) = \begin{cases} \dbinom{n}{x} p^x (1-p)^{n-x} & x = 0, 1, \ldots, n \\ 0 & \text{otherwise} \end{cases}.$$

It is denoted $B(n,p)$ and is the distribution of a random variable X with values on $\{0, 1, \ldots, n\}$. If $\{X_k\}_{k=1,\ldots,n}$ is a family of independent and identically distributed random variables with $\mathbb{P}(X_k = 1) = p$ and $\mathbb{P}(X_k = 0) = 1 - p$, then $X = \sum_{k=1}^n X_k$ is binomially distributed. Moreover, $\mathbb{E}[X] = np$ and $Var(X) = np(1-p)$.

Poisson Distribution

The Poisson distribution with parameter $\lambda > 0$ is defined through its density

$$p(x) = \begin{cases} e^{-\lambda}\frac{\lambda^x}{x!} & x = 0, 1, \ldots \\ 0 & \text{otherwise} \end{cases}.$$

It is denoted $Poi(\lambda)$ and is the distribution of a random variable with values on $\mathbb{N}$. It is obtained as the limit of binomial distributions $B(n, \frac{\lambda}{n})$ as $n \to +\infty$. For $\lambda > 0$, $\mathbb{E}[X] = \lambda$ and $Var(X) = \lambda$.

The characteristic function of the Poisson distribution with parameter λ is

$$\phi(t) = \exp(\lambda(e^{it} - 1)).$$

If $X_1 \overset{d}{=} Poi(\lambda_1)$, $X_2 \overset{d}{=} Poi(\lambda_2)$ are independent then

$$\phi_{X_1+X_2}(t) = \phi_{X_1}(t)\phi_{X_2}(t) = \exp(\lambda_1(e^{it} - 1))\exp(\lambda_2(e^{it} - 1))$$
$$= \exp((\lambda_1 + \lambda_2)(e^{it} - 1))$$

and $X_1 + X_2 \overset{d}{=} Poi(\lambda_1 + \lambda_2)$. As a consequence, if $X \overset{d}{=} Poi(\lambda)$, for every integer $n \geq 1$, $X = X_1 + X_2 + \cdots + X_n$ where $X_1, \ldots, X_n$ are i.i.d. random variables with law $Poi(\frac{\lambda}{n})$.

The moment-generating function is

$$M(t) = \exp(\lambda(e^{-t} - 1)).$$

Normal Distribution

The normal distribution with parameters $\mu \in \mathbb{R}$ and $\sigma^2 > 0$ is defined through its density

$$f(x) = \frac{1}{\sqrt{2\pi}\sigma}e^{-\frac{(x-\mu)^2}{2\sigma^2}}, \qquad x \in \mathbb{R}.$$

It is denoted $N(\mu, \sigma^2)$ and is the distribution of a random variable with values on $\mathbb{R}$. If $X \overset{d}{=} N(0, 1)$, then $\sigma X + \mu \overset{d}{=} N(\mu, \sigma^2)$.

Since $e^{-\frac{x^2}{2}}$ is a symmetric function around $x = 0$, if $X \overset{d}{=} N(0, 1)$ then also $-X \overset{d}{=} N(0, 1)$ and this can be expressed saying that the distribution $N(0, 1)$ is symmetric. As a consequence of $xe^{-\frac{x^2}{2}}$ being an odd function, if $X \overset{d}{=} N(0, 1)$

$$\mathbb{E}[X] = \int_{-\infty}^{+\infty} \frac{1}{\sqrt{2\pi}}xe^{-\frac{x^2}{2}}dx = 0.$$

On the other hand, integrating by parts,

$$\mathbb{E}[X^2] = \int_{-\infty}^{+\infty} \frac{1}{\sqrt{2\pi}}x^2 e^{-\frac{x^2}{2}}dx = 1,$$

and so $Var(X) = 1$.

By the linearity of expectations, if $X \overset{d}{=} N(\mu, \sigma^2)$ then

$$\mathbb{E}[X] = \mu, \qquad Var(X) = \sigma^2.$$

More generally if $X \overset{d}{=} N(\mu, \sigma^2)$ all odd-order moments are zero, while the even ones are

$$\mathbb{E}[(X - \mu)^{2k}] = \sigma^k \frac{(2k)!}{2^k k!}.$$

The characteristic function of the normal distribution with parameters μ and σ^2 is

$$\phi(t) = \exp\left(-\frac{\sigma^2 t^2}{2} + i\mu t\right).$$

If $X_1 \overset{d}{=} N(\mu_1, \sigma_1^2)$, $X_2 \overset{d}{=} N(\mu_2, \sigma_2^2)$ are independent then

$$\phi_{X_1+X_2}(t) = \phi_{X_1}(t)\phi_{X_2}(t) = \exp\left(-\frac{\sigma_1^2 t^2}{2} + i\mu_1 t\right)\exp\left(-\frac{\sigma_2^2 t^2}{2} + i\mu_2 t\right)$$

$$= \exp\left(-\frac{(\sigma_1^2 + \sigma_2^2)t^2}{2} + i(\mu_1 + \mu_2)t\right)$$

and $X_1 + X_2 \overset{d}{=} N(\mu_1 + \mu_2, \sigma_1^2 + \sigma_2^2)$. As a consequence, if $X \overset{d}{=} N(\mu, \sigma^2)$, for every integer $n \geq 1$, $X = X_1 + X_2 + \cdots + X_n$ where $X_1, \ldots, X_n$ are i.i.d. random variables with law $N(\frac{\mu}{n}, (\frac{\sigma}{\sqrt{n}})^2)$.

The moment-generating function of a random variable normally distributed with mean μ and variance σ^2 is

$$M(t) = \exp\left(\frac{\sigma^2 t^2}{2} - \mu t\right).$$

Exponential Distribution

The exponential distribution with parameter $\lambda > 0$ is defined through its density

$$f(x) = \begin{cases} \lambda e^{-\lambda x} & x > 0 \\ 0 & \text{otherwise} \end{cases}.$$

It is denoted $\mathcal{E}(\lambda)$ and is the distribution of a random variable with values on $[0, +\infty)$.

This distribution is characterized by the *lack of memory property*, meaning that if $X \overset{d}{=} \mathcal{E}(\lambda)$,

$$\mathbb{P}(X > t + s | X > t) = \frac{\mathbb{P}(X > t + s, X > t)}{\mathbb{P}(X > t)} = \frac{\mathbb{P}(X > t + s)}{\mathbb{P}(X > t)}$$

$$= \frac{\int_{t+s}^{+\infty} \lambda e^{-\lambda x} dx}{\int_{t}^{+\infty} \lambda e^{-\lambda x} dx} = \frac{e^{-\lambda(t+s)}}{e^{-\lambda t}}$$

$$= e^{-\lambda s} = \mathbb{P}(X > s).$$

Moreover, if $X \overset{d}{=} \mathcal{E}(\lambda)$,

$$\mathbb{E}[X] = \int_0^{+\infty} x \lambda e^{-\lambda x} dx = \frac{1}{\lambda},$$

$$Var(X) = \mathbb{E}[(X - \frac{1}{\lambda})^2] = \int_0^{+\infty} (x - \frac{1}{\lambda}) \lambda e^{-\lambda x} dx = \frac{1}{\lambda^2}.$$

Gamma Distribution

The *gamma function* is a function $\Gamma : \mathbb{R}^+ \to \mathbb{R}^+$ defined as

$$\Gamma(\alpha) = \int_0^{+\infty} x^{\alpha-1} e^{-x} dx.$$

Except for some special cases, the above integral is not explicitly computable. Nevertheless, integrating by parts:

$$\Gamma(\alpha + 1) = \int_0^{+\infty} x^{\alpha} e^{-x} dx = -x^{\alpha} e^{-x} \big|_0^{+\infty} + \alpha \int_0^{+\infty} x^{\alpha-1} e^{-x} dx = \alpha\Gamma(\alpha),$$

and for integers α, inductively we get

$$\Gamma(n) = (n-1)!.$$

The gamma distribution with parameters $\alpha,\ \lambda > 0$ is defined through its density

$$f(x) = \begin{cases} \frac{\lambda^{\alpha}}{\Gamma(\alpha)} x^{\alpha-1} e^{-\lambda x} & x > 0 \\ 0 & \text{otherwise} \end{cases}.$$

It is denoted $\Gamma(\alpha, \lambda)$ and is the distribution of a random variable with values on $[0, +\infty)$. If $\alpha = 1$ we recover the exponential distribution with parameter λ. If $X \stackrel{d}{=} N(0, \sigma^2)$, then $X^2 \stackrel{d}{=} \Gamma(\frac{1}{2}, \frac{1}{2\sigma^2})$. Gamma distributions with integer α are also called *Erlang laws*. On the other hand, distributions of type $\Gamma(\frac{n}{2}, \frac{1}{2})$ are called *chi-square laws with n degrees of freedom* and denoted $\chi^2(n)$.

If $\beta > 0$ and $X \stackrel{d}{=} \Gamma(\alpha, \lambda)$,

$$\mathbb{E}\left[X^{\beta}\right] = \frac{\lambda^{\alpha}}{\Gamma(\alpha)} \int_0^{+\infty} x^{\beta+\alpha-1} e^{-\lambda x} dx$$

$$= \frac{\lambda^{\alpha}}{\Gamma(\alpha)} \frac{\Gamma(\alpha+\beta)}{\lambda^{\alpha+\beta}} \left\{ \frac{\lambda^{\alpha+\beta}}{\Gamma(\alpha+\beta)} \int_0^{+\infty} x^{\beta+\alpha-1} e^{-\lambda x} dx \right\} = \frac{\Gamma(\alpha+\beta)}{\lambda^{\beta}\Gamma(\alpha)},$$

since the quantity inside $\{\}$ is 1, being the integral of the density of the distribution $\Gamma(\alpha + \beta, \lambda)$. Assuming $\beta = 1$ and $\beta = 2$, we get

$$\mathbb{E}[X] = \frac{\Gamma(\alpha + 1)}{\lambda\Gamma(\alpha)} = \frac{\alpha}{\lambda},$$

$$\mathbb{E}[X^2] = \frac{\Gamma(\alpha + 2)}{\lambda^2\Gamma(\alpha)} = \frac{\alpha(\alpha + 1)}{\lambda^2},$$

and

$$Var(X) = \mathbb{E}[X^2] - (\mathbb{E}[X])^2 = \frac{\alpha}{\lambda^2}.$$

In particular, if $X \stackrel{d}{=} \chi^2(n)$,

$$\mathbb{E}[X] = n \text{ and } Var(X) = 2n.$$

The characteristic function of a random variable $X \stackrel{d}{=} \Gamma(\alpha, \lambda)$ is

$$\phi(t) = \frac{\lambda^{\alpha}}{\Gamma(\alpha)} \int_0^{+\infty} v^{\alpha-1} e^{-(-it+\lambda)v} dv.$$

The last complex-valued integral has to be computed applying contour complex integration techniques, through which we get

$$\phi(t) = \frac{1}{(1 - i\lambda^{-1}t)^\alpha}.$$

If $X_1 \overset{d}{=} \Gamma(\alpha_1, \lambda)$, $X_2 \overset{d}{=} \Gamma(\alpha_1, \lambda)$ are independent, then

$$\phi_{X_1+X_2}(t) = \phi_{X_1}(t)\phi_{X_2}(t) = \frac{1}{(1 - i\lambda^{-1}t)^{\alpha_1}} \frac{1}{(1 - i\lambda^{-1}t)^{\alpha_2}} = \frac{1}{(1 - i\lambda^{-1}t)^{\alpha_1+\alpha_2}}$$

and $X_1 + X_2 \overset{d}{=} \Gamma(\alpha_1 + \alpha_2, \lambda)$. As a consequence, if $X \overset{d}{=} \Gamma(\alpha, \lambda)$, for every integer $n \geq 1$, $X = X_1 + X_2 + \cdots + X_n$ where $X_1, \ldots, X_n$ are i.i.d. random variables with law $\Gamma(\frac{\alpha}{n}, \lambda)$.

Moreover, by the above results, the distribution $\Gamma(\frac{n}{2}, \frac{1}{2})$ is the law of a random variable Y of type $Y = X_1^2 + \cdots + X_n^2$, where $X_1, \ldots, X_n$ are i.i.d. $N(0, 1)$ random variables.

The moment-generating function of the Gamma distribution is

$$M(t) = \frac{1}{(1 + \lambda^{-1}t)^\alpha}.$$

Stable Distributions

Definition A.5.1. A random variable Y is said to have a stable distribution if for all $n \geq 1$ it satisfies

$$Y_1 + \ldots + Y_n \overset{d}{=} a_n Y + b_n, \tag{A.3}$$

where $Y_1, \ldots, Y_n$ are i.i.d. copies of Y, $a_n > 0$ and $b_n \in \mathbb{R}$.

Getting to the characteristic function, Y has a stable distribution if and only if

$$\phi_Y(u) = e^{iu\eta - c|u|^\alpha(1 - i\beta sgn(u)g(u))}, \tag{A.4}$$

with

$$g(u) = \begin{cases} tan\frac{\pi\alpha}{2} & \text{for } \alpha \in (0, 1) \cup (1, 2] \\ \frac{2}{\pi}log|u| & \text{for } \alpha = 1 \end{cases}, \tag{A.5}$$

where $\alpha \in (0, 2]$, $\beta \in [-1, 1]$, $c > 0$, and $\eta \in \mathbb{R}$. We shall indicate $Y \overset{d}{=} Stable_\alpha(c, \beta, \eta)$. In this representation c is the scale parameter (note that it has nothing to do with the Gaussian component if $\alpha < 2$), η is the location parameter, α determines the shape of the distribution and is called the *index* of the stable distribution, and β is the skewness parameter.

When $\beta = 0$ and $\eta = 0$, Y is said to have a symmetric stable distribution and the characteristic function is given by

$$\phi_Y(u) = e^{-c^\alpha|u|^\alpha}.$$

Furthermore, when $\alpha = 2$, $Y \overset{d}{=} N(\eta, c^2)$. It can be proved that these distributions are heavy-tailed:

- $\mathbb{E}[|Y|^p] < +\infty$, if $p \in (0, \alpha)$,
- $\mathbb{E}[|Y|^p] = +\infty$, if $p \in [\alpha, 2]$.

The probability density of an α-stable distribution law is not known in closed form except in the following cases (plus the degenerate case of a constant random variable):

1. $\alpha = 2$ and $\eta = 0$ corresponds to the normal distribution.
2. $\alpha = 1$, $\beta = 0$, and $\eta = 0$ corresponds to the Cauchy distribution whose density is given by $\frac{c}{\pi}\frac{1}{(x^2+c^2)}$, for $x \in \mathbb{R}$.
3. $\alpha = \frac{1}{2}$, $\beta = 1$, and $\eta = 0$ corresponds to the inverse Gaussian distribution whose density is given by $\frac{c}{\sqrt{2\pi x^3}}e^{-\frac{c^2}{2x}}$, for $x > 0$. It is the distribution of the r.v.

$$\tau_c = \inf\{t > 0 : B_t > c\},$$

where B is a Brownian motion.

If we consider a set of independent random variables $Y_i \overset{d}{=} Stable_\alpha(c_i, \beta, \eta_i)$, $i = 1, \ldots, N$, it is trivial to show through the characteristic function that $Y_1 + \cdots + Y_N \overset{d}{=} Stable_\alpha(c_1 + \cdots + c_N, \beta, \eta_1 + \cdots + \eta_N)$. Moreover, by subtracting b_n/n from each of the terms on the left-hand side of (A.3) and dividing by a_n, we can see that any stable random variable can be written as the sum of suitable stable random variables with the same skewness parameter.

Pareto Distribution

The Pareto distribution is characterized by its cumulative distribution function

$$F(x) = 1 - \left(\frac{c}{x}\right)^\alpha, \qquad x > c,$$

with $\alpha > 0$ and $c > 0$. The density function can be written as

$$f(x) = \alpha c^\alpha \frac{1}{x^{\alpha+1}}.$$

The expectation exists if $\alpha > 1$

$$\mathbb{E}[X] = \frac{\alpha}{\alpha - 1}c,$$

while the variance exists if $\alpha > 2$

$$Var(X) = \frac{\alpha}{(\alpha - 1)^2(\alpha - 2)}c^2.$$

As for the characteristic function, it has a very complicated expression that we present just for the sake of completeness:

$$\phi(t) = \alpha(-ict)^\alpha \int_{-ict}^{+\infty} z^{-\alpha-1}e^{-z}dz.$$

A.6 RANDOM VECTORS AND MULTIVARIATE DISTRIBUTIONS

Let $(\Omega, \mathcal{F}, \mathbb{P})$ be a probability space. A function $X = (X^1, \ldots, X^m) : \Omega \to \mathbb{R}^m$, such that each X^i is a random variable, is called a *random vector*.

The multivariate cumulative distribution function of the random vector X is the function $F_{X^1,\ldots,X^m} : \mathbb{R}^m \to [0, 1]$ defined as

$$F_{X^1,\ldots,X^m}(x_1, \ldots, x_m) = \mathbb{P}(X^1 \le x_1, \ldots, X^m \le x_m).$$

The cumulative distribution function of each X^i is called the *marginal* cumulative distribution and is given by

$$F_{X^i}(x_i) = \lim_{x_j \to +\infty, j \neq i} F_{X^1, \dots, X^m}(x_1, \dots, x_m).$$

If the cumulative distribution $F_{X^1, \dots, X^m}(x_1, \dots, x_m)$ admits a density, that is there exists $f : \mathbb{R}^m \to [0, +\infty)$ such that

$$F_{X^1, \dots, X^m}(x_1, \dots, x_m) = \int_{-\infty}^{x_1} \cdots \int_{-\infty}^{x_m} f(z_1, \dots, z_m) dz_1 \dots dz_m,$$

the density of each random variable X^i can be recovered through

$$f_{X^i}(z_i) = \int_{-\infty}^{+\infty} \cdots \int_{-\infty}^{+\infty} f(z_1, \dots, z_{i-1}, z_i, z_{i+1}, \dots, z_m) dz_1 \cdots dz_{i-1} dz_{i+1} \cdots dz_m.$$

If the random variables $X^1, \dots, X^m$ are independent, we have

$$F_{X^1, \dots, X^m}(x_1, \dots, x_m) = \prod_{i=1}^{m} F_{X^i}(x_i)$$

and

$$f_{X^1, \dots, X^m}(z_1, \dots, z_m) = \prod_{i=1}^{m} f_{X^i}(z_i).$$

Let $g : \mathbb{R}^m \to \mathbb{R}^m$ be a one-to-one mapping and suppose that its inverse function is continuously differentiable. If $J(z_1, \dots, z_m)$ is the Jacobian of g^{-1}, we have that the density of the random variable $Y = g(X)$ is

$$f(g^{-1}(z_1, \dots, z_m)) \cdot |J(z_1, \dots, z_m)|.$$

Definition A.6.1. The characteristic function of a random variable $X = (X^1, \dots, X^m)$ is defined for real $v \in \mathbb{R}^m$ by

$$\phi_X(v) = \mathbb{E}[e^{it \cdot X}],$$

where $v \cdot X = \sum_{u=1}^{m} t_u X^u$ denotes the inner product.

The characteristic function of random vectors satisfies the same properties as random variables, and uniquely determines the distribution of X.

A.6.1   The Multivariate Normal Distribution

Let us consider the random vector $X = (X^1, \dots, X^m)$ with independent components each having the standard normal distribution. Since each X^i has density $\frac{1}{\sqrt{2\pi}} e^{-\frac{x_i^2}{2}}$, X has density

$$f_{X^1, \dots, X^m}(x_1, \dots, x_m) = \prod_{i=1}^{m} \frac{1}{\sqrt{2\pi}} e^{-\frac{x_i^2}{2}} = \frac{1}{(2\pi)^{m/2}} e^{-\frac{|x|^2}{2}},$$

where $|x| = \sqrt{x_1^2 + \cdots + x_m^2}$. Its characteristic function is

$$\mathbb{E}[e^{iv \cdot X}] = \mathbb{E}[\prod_{u=1}^{m} e^{iv_u X^u}] = \prod_{u=1}^{m} e^{-iv_u^2/2} = e^{-\frac{|v|^2}{2}}.$$

Let A be an $m \times m$ matrix and put $Y = AX$, where X is viewed as a column vector. Since $\mathbb{E}[X^i X^j] = \mathbf{1}_{\{i=j\}}$, the matrix Σ of the covariances of Y has entries $\sigma_{ij} = \mathbb{E}[Y^i Y^j] = \sum_{k=1}^{m} a_{ik} a_{jk}$. Thus $\Sigma = AA'$, where the prime denotes transpose. The matrix Σ is symmetric and non-negative definite.

The characteristic function of $Y = AX$ is

$$\mathbb{E}[e^{iv'(AX)}] = \mathbb{E}[e^{i(A'v)'X}] = e^{-\frac{|A'v|^2}{2}} = e^{-\frac{v'\Sigma v}{2}}.$$

The distribution of Y is the *Centered normal distribution with covariance* Σ.

If Σ is symmetric and non-negative definite, then for an appropriate orthogonal matrix U, $U'\Sigma U = D$ is a diagonal matrix whose diagonal elements are the eigenvalues of Σ and hence are non-negative. If D_0 is the diagonal matrix whose elements are the square roots of those of D and if $A = UD_0$, then $\Sigma = AA'$. Thus, for every non-negative definite matrix Σ, there exists a unique centered normal distribution (namely the distribution of AX) with covariance matrix Σ.

If Σ is non-singular, so is A. Since X has density, by the Jacobian transformation formula, $Y = AX$ has density

$$f(A^{-1}x)|\det A^{-1}|.$$

Since $\Sigma = AA'$, $|\det A^{-1}| = (\det \Sigma)^{-1/2}$. Moreover, $\Sigma^{-1} = (A')^{-1} A^{-1}$, so that $|A^{-1}x|^2 = x'\Sigma^{-1}x$. Thus, the centered normal distribution has density

$$\frac{1}{2\pi^{m/2}} \frac{1}{(\det \Sigma)^{1/2}} e^{-\frac{1}{2} x'\Sigma x}.$$

If Σ is singular, the distribution can have no density.

If C is a $k \times m$ matrix and Y is an $\mathbb{R}^m$-valued random vector with centered normal distribution with covariance matrix Σ, CY is an $\mathbb{R}^k$-valued random vector with characteristic function

$$\exp\left(-\frac{1}{2}(C'v)'\Sigma(C'v)\right) = \exp\left(-\frac{1}{2}v'(C\Sigma C')v\right),$$

with $v \in \mathbb{R}^k$. Hence, CY has the centered normal distribution in $\mathbb{R}^k$ with covariance matrix $C\Sigma C'$.

The general normal distribution is a translation of one of these centered distributions and has characteristic function

$$e^{i\mu \cdot v - \frac{v'\Sigma v}{2}}.$$

A.7 INFINITE DIVISIBILITY

In this section we provide basic notions of infinite divisibility. We refer the interested reader to Sato (1999).

Definition A.7.1. A random variable Y is said to have an *infinitely divisible distribution* if for every $m \geq 1$,

$$Y \overset{d}{=} Y_1^{(m)} + \ldots + Y_m^{(m)}$$

for some i.i.d. random variables $Y_1^{(m)}, \ldots, Y_m^{(m)}$.

Notice that the distribution of $Y_j^{(m)}$ may vary as m varies, but not as j varies. It is an immediate consequence of the definition that for every $m \geq 1$,

$$\phi_Y(t) = \prod_{u=1}^{m} \phi_{Y_u^{(m)}}(t).$$

Many known distributions are infinitely divisible, some are not. The normal, Poisson, Gamma and stable distributions are infinitely divisible. This follows from the fact, observed above, that sums of independent random variables distributed as each of the above distributions are again of the same type with adequate parameters. The Pareto distribution is infinitely divisible as well, but this is very difficult to prove since, in this case, the random variables $Y_1^{(m)}, \ldots, Y_m^{(m)}$ of the definition are not of Pareto type (see Thorin (1977)).

Infinite divisible distributions are characterized by the following key result.

Theorem A.7.1. *Lévy–Khintchine theorem. A real-valued random variable X has an infinitely divisible distribution if there are parameters $a \in \mathbb{R}$, $\sigma^2 \geq 0$ and a Radon measure v on $\mathbb{R} \setminus \{0\}$ with $\int_{-\infty}^{+\infty}(1 \wedge x^2)v(dx) < +\infty$ such that*

$$\phi_X(\lambda) = \mathbb{E}\left[e^{i\lambda X}\right] = e^{-\psi(\lambda)},$$

where

$$\psi(\lambda) = -ia\lambda + \frac{1}{2}\sigma^2\lambda^2 - \int_{-\infty}^{+\infty}\left(e^{i\lambda x} - 1 - i\lambda x \boldsymbol{I}_{\{|x|\leq 1\}}\right)v(dx), \qquad \lambda \in \mathbb{R}.$$

Infinite divisible distributions are parameterized by their Lévy–Khintchine characteristics (a, σ^2, v). $\psi(\lambda)$ is called the *characteristic exponent*, and v the *Lévy measure*.

Example A.7.1. *1. For the normal distribution, $v = 0$ and $a = 0$.*

2. If $\sigma = 0$ and with the Lévy measure

$$v(dx) = \begin{cases} c_1 x^{-1-\alpha}dx, & \text{for } x > 0 \\ c_2|x|^{-1-\alpha}dx, & \text{for } x < 0 \end{cases} \tag{A.6}$$

with $c_1, c_2 > 0$ and $0 < \alpha < 2$, we get the α-stable distribution.

3. For the Poisson distribution with parameter $\lambda > 0$, $a = 0$, $\sigma = 0$ and $v(dx) = \lambda\delta_1(dx)$.

4. If $a = \sigma = 0$ and $v(dx) = \alpha x^{-1}e^{-\lambda x}$, with $x > 0$ we recover the gamma distribution with parameters α and λ.

The definition of infinite divisibility trivially extends to the multivariate case.

Definition A.7.2. A random vector $X = (X_1, \ldots, X_d)$ is said to have an *infinitely divisible distribution* if for every $m \geq 1$,

$$X \overset{d}{=} X_1^{(m)} + \ldots + X_m^{(m)}$$

for some i.i.d. random variables $X_1^{(m)}, \ldots, X_m^{(m)}$.

The following is the multivariate version of the Lévy–Khintchine theorem.

Theorem A.7.2. *A random vector X with values in $\mathbb{R}^d$ has an infinitely divisible distribution if there are a positive definite $d \times d$-matrix Σ, a vector $a \in \mathbb{R}^d$, and a measure v on $\mathbb{R}^d \setminus \{0\}$ with $\int_{-\infty}^{+\infty}(1 \wedge x^2)v(dx) < +\infty$ such that*

$$\phi_X(\lambda) = \mathbb{E}[e^{i\lambda X}] = e^{-\psi(\lambda)},$$

where

$$\psi(\lambda) = -ia \cdot \lambda + \frac{1}{2}\lambda'\Sigma\lambda - \int_{-\infty}^{+\infty}\left(e^{i\lambda \cdot x} - 1 - i\lambda \cdot x\boldsymbol{I}_{\{|x|\leq 1\}}\right)v(dx), \qquad \lambda \in \mathbb{R}^d.$$

The multivariate normal distribution introduced above is infinite divisible.

A.8   CONVERGENCE OF SEQUENCES OF RANDOM VARIABLES

Let $\{X_n\}_n$ be a sequence of random variables defined on the same probability space as the random variable X. X_n is said to *converge almost surely* to X ($X_n \to X$ a.s.) if

$$\mathbb{P}\left(\lim_{n\to+\infty} X_n = X\right) = 1.$$

X_n is said to *converge in L^p* to X ($X_n \xrightarrow{L^p} X$) if

$$\lim_{n\to+\infty} \mathbb{E}\left[|X_n - X|^p\right) = 0.$$

The above type of convergence is the convergence in the metric space L^p. This space is *complete* (more precisely it is a Banach space), meaning that every Cauchy sequence of random variables X_n (that is such that for all $\epsilon > 0$ there exists an $\bar{n}$ such that for all $n, m > \bar{n}$, $d(X^n, X^m) < \epsilon$) converges in L^p to a random variable $X \in L^p$.

X_n is said to *converge in probability* to X ($X_n \xrightarrow{\mathbb{P}} X$) if for each $\epsilon > 0$

$$\lim_{n\to+\infty} \mathbb{P}(|X_n - X| > \epsilon) = 0.$$

X_n is said to *converge in distribution or in law* to X ($X_n \xrightarrow{d} X$) if

$$\lim_{n\to+\infty} F_{X_n}(t) = F_X(t)$$

for every t in which F_X is continuous. Unlike the other notion of convergence, it does not require the random variables to be defined on a common probability space.

This definition induces the concept of weak convergence of probabilities on $\mathcal{B}(\mathbb{R})$. In fact, if $\mathbb{P}_n$ and $\mathbb{P}$ are probabilities on $\mathcal{B}(\mathbb{R})$, we say that $\mathbb{P}_n$ *weakly converges to* $\mathbb{P}$ if

$$\lim_{n\to+\infty} \mathbb{P}_n((-\infty, x]) = \mathbb{P}((-\infty, x])$$

for all x such that $\mathbb{P}(\{x\}) = 0$.

The last definition can be extended to finite measures on $\mathcal{B}(\mathbb{R})$. We say that the sequence of measures on $\mathcal{B}(\mathbb{R})$, μ_n, *vaguely converges* to the measure μ on $\mathcal{B}(\mathbb{R})$ if

$$\lim_{n \to +\infty} \mu_n((-\infty, x]) = \mu((-\infty, x])$$

for all x such that $\mu(\{x\}) = 0$.

A.8.1   The Strong Law of Large Numbers

Let $X_1, X_2, \ldots, X_n, \ldots$ be a sequence of independent random variables defined on the same probability space $(\Omega, \mathcal{F}, \mathbb{P})$. We assume that they are identically distributed with finite mean μ. If $S_n = \sum_{i=1}^{n} X_i$, then

$$\mathbb{P}\left(\lim_{n \to +\infty} \frac{S_n}{n} = \mu\right) = 1.$$

A.9   THE RADON–NIKODYM DERIVATIVE

Let $(\Omega, \mathcal{F})$ be a measurable space with two measures μ and ν. If, for every $A \in \mathcal{F}$, $\mu(A) = 0 \Rightarrow \nu(A) = 0$, then ν is said to be *absolutely continuous with respect to μ*. This means that all negligible sets for μ are negligible sets for ν as well. The following result, known as the *Radon–Nikodym theorem*, characterizes absolute continuity.

Theorem A.9.1. *If ν is absolutely continuous with respect to μ, there exists a measurable function $Z : \Omega \to [0, +\infty)$ such that for any $A \in \mathcal{F}$*

$$\nu(A) = \int_A Z(\omega)\mu(d\omega).$$

Z is called the density or Radon–Nikodym derivative of ν with respect to μ and is usually denoted $\frac{d\nu}{d\mu}$. For every function f integrable with respect to the measure ν,

$$\int_\Omega f(\omega)\nu(d\omega) = \int_\Omega f(\omega)Z(\omega)\mu(d\omega) = \int_\Omega f(\omega)\frac{d\nu}{d\mu}(\omega)\mu(d\omega).$$

If μ is also absolutely continuous with respect to ν, then μ and ν are said to be *equivalent*, meaning that they share the same negligible sets. This is obviously equivalent to $\frac{d\nu}{d\mu} > 0$.

A.10   CONDITIONAL EXPECTATION

Definition A.10.1. Let $(\Omega, \mathcal{F}, \mathbb{P})$ be a probability space and $\mathcal{A} \subset \mathcal{F}$ a σ-algebra. There exists a random variable denoted $\mathbb{E}[X|\mathcal{A}]$ called the "conditional expected value of X given $\mathcal{A}$", which has these two properties:

1. $\mathbb{E}[X|\mathcal{A}]$ is measurable $\mathcal{A}$ and integrable.
2. $\mathbb{E}[X|\mathcal{A}]$ satisfies the functional equation

$$\int_A \mathbb{E}[X|\mathcal{A}]d\mathbb{P} = \int_A Xd\mathbb{P}, \qquad A \in \mathcal{A}.$$

To prove the existence of such a random variable, consider first the case of non-negative X. Define a measure ν on $\mathcal{A}$ by $\nu(A) = \int_A X\mathbb{P}(d\omega)$. This measure is finite because X is integrable and it is absolutely continuous with respect to $\mathbb{P}$. By the Radon–Nikodym theorem there exists a function Y, measurable $\mathcal{A}$, such that $v(A) = \int_A Y(\omega)\mathbb{P}(d\omega)$. This Y has properties $1.$ and $2.$ If X is not necessarily non-negative, $\mathbb{E}[X^+|\mathcal{A}] - \mathbb{E}[X^-|\mathcal{A}]$ clearly has the required properties.

In general, there will be many such random variables $\mathbb{E}[X|\mathcal{A}]$; any one of them is called a *version* of the conditional expected value. Any two versions are equal with probability 1. Obviously, $\mathbb{E}[X|\{\emptyset, \Omega\}] = \mathbb{E}[X]$ and $\mathbb{E}[X|\mathcal{F}] = X$ with probability 1. As $\mathcal{A}$ increases, condition $1.$ becomes weaker and condition $2.$ becomes stronger.

The value $\mathbb{E}[X|\mathcal{A}](\omega)$ is to be interpreted as the expected value of X for someone who knows for each $A \in \mathcal{A}$ whether or not it contains the point ω, which in general remains unknown itself. Condition $1.$ ensures that $\mathbb{E}[X|\mathcal{A}]$ can in principle be calculated from this partial information alone. Condition $2.$ can be restated as $\int_A (\mathbb{E}[X|\mathcal{A}] - X)\,d\mathbb{P} = 0$; if the observer, in possession of the partial information contained in $\mathcal{A}$, is offered the opportunity to bet, paying an entry fee of $\mathcal{E}[X|\mathcal{A}]$ and being returned the amount X, and if he adopts the strategy of betting if A occurs, this equation says that the game is fair.

Properties of the Conditional Expectation

Suppose that X, Y, and X_n are integrable.

1. If $X = a$ with probability 1, then $\mathbb{E}[X|\mathcal{A}] = a$.
2. For constant a and b, $\mathbb{E}[aX + bY|\mathcal{A}] = a\mathbb{E}[X|\mathcal{A}] + b\mathbb{E}[Y|\mathcal{A}]$.
3. If $X \leq Y$ with probability 1, then $\mathbb{E}[X|\mathcal{A}] \leq \mathbb{E}[Y|\mathcal{A}]$.
4. $|\mathbb{E}[X|\mathcal{A}]| \leq \mathbb{E}[|X||\mathcal{A}]$.
5. $\lim_{n\to+\infty} \mathbb{E}[X_n|\mathcal{A}] = \mathbb{E}[X|\mathcal{A}]$ with probability 1.
6. If X is measurable $\mathcal{A}$ and if XY is integrable, then $\mathbb{E}[XY|\mathcal{A}] = X\mathbb{E}[Y|\mathcal{A}]$ with probability 1.
7. If $\mathcal{A}_1 \subset \mathcal{A}_2$ are σ-algebras, then $\mathbb{E}[\mathbb{E}[X|\mathcal{A}_2]|\mathcal{A}_1] = \mathbb{E}[X|\mathcal{A}_1]$ with probability 1.
8. If X is independent on the partial information provided by $\mathcal{A}$, then $\mathbb{E}[X|\mathcal{A}] = \mathbb{E}[X]$.
9. *Jensen's inequality*: if ϕ is a convex function on the real line and $\phi(X)$ is integrable, then $\phi(\mathbb{E}[X|\mathcal{A}]) \leq \mathbb{E}[\phi(X)|\mathcal{A}]$.

Appendix B
Elements of Stochastic Processes Theory

B.1 STOCHASTIC PROCESSES

Let $(\Omega, \mathcal{F}, \mathbb{P})$ be a probability space. A stochastic process is a collection of random variables $(X_t)_{t \in \mathcal{T}}$ with values in a common state space, which we will choose specifically as $\mathbb{R}$ (or $\mathbb{R}^d$). In this book we interpret the index t as time. If $\mathcal{T} = [0, +\infty)$ (or $\mathcal{T} = [0, T]$), we are dealing with a *continuous time* stochastic process, while if $\mathcal{T} = \mathbb{N}$ (or $\mathcal{T} = \{0, 1, \ldots, N\}$) we are dealing with a *discrete time* stochastic process.

The functions of time $t \to X_t(\omega)$ are the paths or trajectories of the process: thus a stochastic process can be viewed as a random function that is a random variable taking values in a function space.

Remark B.1.1. In the case of a continuous time stochastic process, the trajectories can be continuous or with jumps for some $t \geq 0$:

$$\Delta X_t = X_{t^+} - X_{t^-} = \lim_{h \downarrow 0} X_{t+h} - \lim_{h \downarrow 0} X_{t-h}. \tag{B.1}$$

If all trajectories are continuous functions of time apart from a negligible set, the process is said to be continuous and its state space is $C(\mathcal{T})$, the space of all real-valued continuous functions. In the case of discontinuous paths, we anyway suppose that limits in (B.1) always exist for all $t \geq 0$ and that $X_{t^+} = X_t$, i.e. that the paths are almost surely right-continuous with left limits. These paths are called *cadlag*, which is a French acronym for *continu á droite, limite á gauche* which means "right-continuous with left limit". The jump at t is denoted by $\Delta X_t = X_t - X_{t^-}$. However, cadlag trajectories cannot jump too wildly. In fact, as a consequence of the existence of limits, in any interval $[0, T]$, for every $b > 0$ the number of jumps greater than b must be finite and the number of jumps is at most countable. This way, in $[0, T]$ every cadlag trajectory has a finite number of large jumps and a possibly infinite but countable set of small jumps. The space of all real-valued cadlag functions is denoted by $D(\mathcal{T})$.

The choice, among all others, of assuming cadlag trajectories for financial modeling is justified by the following arguments. If a cadlag trajectory has a jump at time t, then the value of $X_t(\omega)$ is unknown before t following the trajectory up to time t: the discontinuity is a sudden event at time t. By contrast, if the left limit coincides with $X_t(\omega)$ then, an observer following the trajectory up to time t will approach the value of $X_t(\omega)$. It is natural, in a concrete financial context, to assume jumps to be sudden and unforeseeable events.

B.1.1 Filtrations

While time t is elapsing, the observer increases his/her endowment of information. In fact, some events that are random at time 0 may not be random anymore at a certain time $t > 0$. In fact, the set of information available at time t can be sufficient to reveal if the event has occurred or not. In order to model the flow of information, we introduce the notion of filtration.

Definition B.1.1. Given a probability space $(\Omega, \mathcal{F}, \mathbb{P})$, a filtration is an increasing family of σ-algebras $(\mathcal{F}_t)_{t \in \mathcal{T}}$, such that for all $t \geq s$, $t, s \in \mathcal{T}$, $\mathcal{F}_s \subseteq \mathcal{F}_t \subseteq \mathcal{F}$.

A probability space equipped with a filtration is called a *filtered probability space*.

$\mathcal{F}_t$ can be interpreted as the set of all events occurring within time t and so represents the information known at time t. By definition, an $\mathcal{F}_t$-measurable random variable is a random variable whose value is known at time t.

Given a stochastic process $(X_t)_{t \in \mathcal{T}}$, if X_t is $\mathcal{F}_t$-measurable for every $t \in \mathcal{T}$ we say that the stochastic process is $(\mathcal{F}_t)_{t \in \mathcal{T}}$-adapted. This means that the values of the process at time t are revealed by the known information $\mathcal{F}_t$.

Clearly, the values of the process at time t, X_t, are revealed by the σ-algebra generated by X_t. We shall call the *natural filtration* generated by the process X_t the filtration $\left(\mathcal{F}_t^X\right)_{t \in \mathcal{T}}$ such that $\mathcal{F}_t^X$ is the smallest σ-algebra with respect to which X_t is adapted completely by the null sets.

The assumption that all negligible sets are contained in each $\mathcal{F}_t$ implies, in particular, that all null sets are in $\mathcal{F}_0$, meaning that the fact that a certain evolution for the process is impossible is already known at time 0.

If $\mathcal{T} = \mathbb{N}$ (or $\mathcal{T} = \{0, 1, \ldots, N\}$), a stochastic process is said to be *predictable* if X_t is $\mathcal{F}_{t-1}$-measurable for every $t \in \mathcal{T}$.

If $\mathcal{T} = [0, +\infty)$ (or $\mathcal{T} = [0, T]$), the notion of predictability is more complicated. In this framework, the *predictable filtration* is the smallest filtration with respect to which all *cag* (left-continuous) processes are adapted and a general stochastic process is called predictable if it is adapted to this filtration.

B.1.2 Stopping Times

In a stochastic setting it is natural to deal with events happening at random times. For example, given a stochastic process $(X_t)_{t \in \mathcal{T}}$, we may be interested in the first time at which the value of the process exceeds a given bound b; more precisely, if

$$\tau_b = \inf\{t \in \mathcal{T} : X_t > b\} \tag{B.2}$$

and if $X_0 < b$, τ_b is a random variable.

A *random time* τ is a random variable with values in the set of times $\mathcal{T}$. It represents the time at which some event is going to occur. Given a filtration $(\mathcal{F}_t)_{t \in \mathcal{T}}$ one can ask if the information $\mathcal{F}_t$ available at time t is sufficient to state if the event has already happened ($\tau \leq t$) or not ($\tau > t$).

Definition B.1.2. Given a filtered probability space, a random variable τ with values in $\mathcal{T}$ is a "stopping time" if for all $t \in \mathcal{T}$, $\{\tau \leq t\} \in \mathcal{F}_t$.

The name "stopping time" is due to the notion of a *stopped process*: given an adapted stochastic process $(X_t)_{t \in \mathcal{T}}$ and a stopping time τ, the process stopped at τ is defined by

$$X_{t \wedge \tau} = \begin{cases} X_t & \text{if } t < \tau \\ X_\tau & \text{if } t \geq \tau \end{cases}.$$

The random time τ_b defined in (B.2) is indeed a stopping time.

Given a filtration $(\mathcal{F}_t)_{t\in T}$ and a stopping time τ, the known information at time τ is the σ-algebra generated by all adapted processes observed up to time τ. More precisely,

$$\mathcal{F}_\tau = \{A \in \mathcal{F} : \forall t \in T, A \cap \{\tau \le t\} \in \mathcal{F}_t\}.$$

B.2 MARTINGALES

Let $(\Omega, \mathcal{F}, \mathbb{P})$ be equipped with a filtration $(\mathcal{F}_t)_{t\in T}$.

Definition B.2.1. A process $(X_t)_{t\in T}$ is a *martingale* if it is $(\mathcal{F}_t)_{t\in T}$-adapted, $\mathbb{E}[|X_t|]$ is finite for any $t \in T$, and

$$\forall s < t, \ s, t \in T \qquad \mathbb{E}[X_t | \mathcal{F}_s] = X_s. \tag{B.3}$$

In other words, the best prediction of a martingale future value is its current value.

An obvious consequence of (B.3) is that a martingale has constant expectation: $\forall t \in T$, $\mathbb{E}[X_t] = \mathbb{E}[X_0]$.

The stochastic process $(X_t)_{t\in T}$ is defined to be a martingale if it is a martingale relative to some filtration $(\mathcal{F}_t)_{t\in T}$. If $(\mathcal{G}_t)_{t\in T}$ is another filtration with respect to which $(X_t)_{t\in T}$ is again adapted and such that $\mathcal{G}_t \subset \mathcal{F}_t$ for all $t \in T$, then

$$\mathbb{E}[X_t | \mathcal{G}_s] = \mathbb{E}[\mathbb{E}[X_t | \mathcal{F}_s] | \mathcal{G}_s] = \mathbb{E}[X_s | \mathcal{G}_s] = X_s$$

and $(X_t)_{t\in T}$ is a martingale relative to $(\mathcal{G}_t)_{t\in T}$ as well. This is *a fortiori* true for the natural filtration $\left(\mathcal{F}_t^X\right)_{t\in T}$. In this special case (B.3) reduces to

$$\forall i_1, \ldots, i_n, t \in T \text{ with } i_1 < \cdots < i_n < t \qquad \mathbb{E}[X_t | X_{i_n}, \ldots, X_{i_1}] = X_{i_n}.$$

In the case of a discrete time martingale, that is $T = \mathbb{N}$ or $T = \{0, 1, \ldots, N\}$, the martingale property (B.3) is equivalent to

$$\mathbb{E}[X_{n+1} | \mathcal{F}_n] = X_n \qquad \forall n \in T. \tag{B.4}$$

In this case the defining conditions for a martingale can also be given in terms of the differences

$$\Delta_n = X_n - X_{n-1}$$

$(\Delta_1 = X_1)$. (B.4) is equivalent to

$$\mathbb{E}[\Delta_{n+1} | \mathcal{F}_n] = 0.$$

Note that, since $X_n = \Delta_1 + \cdots + \Delta_n$, the stochastic processes $(X_n)_{n\in T}$ and $(\Delta_n)_{n\in T}$ generate the same filtration, that is $\mathcal{F}_n^X = \mathcal{F}_n^\Delta$ for all n.

Example B.2.1. *Suppose that Z is an integrable random variable on $(\Omega, \mathcal{F}, \mathbb{P})$ and $(\mathcal{F}_t)_{t\in T}$ a filtration. If*

$$X_t = \mathbb{E}[Z | \mathcal{F}_t]$$

then it is easy to check that $(X_t)_{t\in T}$ is a martingale relative to $(\mathcal{F}_t)_{t\in T}$.

Definition B.2.2. The stochastic process $(X_t)_{t\in T}$ is a *submartingale* if it is $(\mathcal{F}_t)_{t\in T}$-adapted, $\mathbb{E}[|X_t|]$ is finite for any $t \in T$, and

$$\forall s < t, \ s, t \in T \qquad \mathbb{E}[X_t | \mathcal{F}_s] \ge X_s. \tag{B.5}$$

The stochastic process is called a *supermartingale* if (B.5) is replaced by

$$\forall s < t, \ s, t \in T \qquad \mathbb{E}\left[X_t | \mathcal{F}_s \right] \leq X_s.$$

Convex functions of martingales are submartingales.

Theorem B.2.1. *If $(X_t)_{t \in T}$ is a martingale with respect to the filtration $(\mathcal{F}_t)_{t \in T}$, if ψ is convex and if $Y_t = \psi(X_t)$ are integrable, then $(Y_t)_{t \in T}$ is a submartingale relative to $(\mathcal{F}_t)_{t \in T}$.*

The following closure property of a martingale is due to Doob.

Theorem B.2.2. *Closure. Let $(M_t)_{t \geq 0}$ be a martingale such that $\sup_{t \geq 0} \mathbb{E}[M_t] < +\infty$, then there exists a random variable H with $\mathbb{E}[|Y|] < +\infty$ such that $M_t \underset{t \to +\infty}{\to} H$ almost surely.*

A generalization of the concept of martingale is that of local martingale.

Definition B.2.3. A stochastic process $(Y_t)_{t \geq 0}$ is a *local martingale* with respect to a given filtration $(\mathcal{F}_t)_{t \geq 0}$ if there exists a sequence of stopping times $\tau_1 \leq \tau_2 \leq \cdots$ with $\tau_n \to +\infty$ a.s. such that, for all n, the stopped process $Y_{t \wedge \tau_n}$ is an $(\mathcal{F}_t)_{t \geq 0}$-martingale.

Of course a martingale is a local martingale.

B.3 MARKOV PROCESSES

This section is devoted to the basics on Markov processes. We refer the interested reader to Ethier and Kurtz (1985).

Let E be a metric space, $\mathcal{B}(E)$ be the collection of all Borel sets of E, and $B(E)$ the set of all measurable bounded functions with the $||f||_\infty = \sup_{x \in E} |f|$ norm. Even if all the definitions and results of this section work for a general metric space E, we shall think of E as $\mathbb{R}^m$ since this is the case in financial appblications.

Definition B.3.1. Let $(X_t)_{t \in T}$ be a stochastic process defined on a probability space $(\Omega, \mathcal{F}, \mathbb{P})$ with values in E and let $\left(\mathcal{F}_t^X \right)_{t \in T}$ be its natural filtration. Then $(X_t)_{t \in T}$ is a Markov process if

$$\mathbb{P}\left(X_t \in C | \mathcal{F}_s^X \right) = \mathbb{P}\left(X_t \in C | X_s \right), \qquad (\text{B.6})$$

for all $s, t \in T$, $s < t$ and $C \in E$.

If $(\mathcal{G}_t)_{t \in T}$ is a filtration with $F_t^X \subset \mathcal{G}_t$ (that is such that the stochastic process is $\mathcal{G}_t$-adapted), then $(X_t)_{t \in T}$ is a *Markov process with respect to* $(\mathcal{G}_t)_{t \in T}$ if

$$\mathbb{P}\left(X_t \in C | \mathcal{G}_s \right) = \mathbb{P}\left(X_t \in C | X_s \right). \qquad (\text{B.7})$$

Obviously, if $(X_t)_{t \in T}$ is a Markov process with respect to $(\mathcal{G}_t)_{t \in T}$ it is a Markov process as well.

In what follows we will consider a generic filtration $(\mathcal{G}_t)_{t \in T}$ with respect to which $(X_t)_{t \in T}$ is adapted. If we assume $E = \mathbb{R}^m$, (B.7) is equivalent to

$$\mathbb{P}\left(X_t^1 \leq x_1, \ldots, X_t^m \leq x_m \big| \mathcal{G}_s \right) = \mathbb{P}\left(X_t \leq x_1, \ldots, X_t^m \leq x_m \big| X_s \right),$$

for all $(x_1, \ldots, x_m) \in \mathbb{R}^m$.

Moreover, (B.7) implies

$$\mathbb{E}\left[f(X_t) | \mathcal{G}_s \right] = \mathbb{E}\left[f(X_t) | X_s \right]$$

for all $s, t \in T$, $s < t$ and for all $f \in B(E)$.

Definition B.3.2. A function $P(t, x, C)$ defined on $T \times E \times \mathcal{B}(E)$ is a time homogeneous transition function if

$$P(t, x, \cdot) \text{ is a probability on } \mathcal{B}(E) \text{ for all } (t, x) \in T \times E$$

$$P(0, x, \cdot) = \mathbf{1}_x \text{ unit mass at } x \text{ for all } x \in E$$

$$P(\cdot, \cdot, C) \in B(T \times E), \qquad C \in \mathcal{B}(E)$$

and

$$P(t, x, C) = \int P(t - s, y, C)P(s, x, dy), \quad s, t \in T, s < t, x \in E, C \in \mathcal{B}(E). \tag{B.8}$$

(B.8) is called the *Chapman–Kolmogorov equation*.

Definition B.3.3. A transition function $P(t, x, C)$ is a transition function for a time homogeneous Markov process $(X_t)_{t \in T}$ with respect to the filtration $(\mathcal{G}_t)_{t \in T}$ if

$$\mathbb{P}(X_t \in C | \mathcal{G}_s) = P(t - s, X_s, C) \tag{B.9}$$

for all $s, t \in T, s < t$ and $C \in \mathcal{B}(E)$, or equivalently if

$$\mathbb{E}[f(X_t) | \mathcal{G}_s] = \int f(y)P(t - s, X_s, dy)$$

for all $s, t \in T, s < t$ and $f \in B(E)$.

To see that the Chapmann–Kolmogorov equation (B.8) is satisfied given (B.9), notice that

$$\begin{aligned}
P(t, X_u, C) &= \mathbb{P}(X_{t+u} \in C | \mathcal{G}_u) \\
&= \mathbb{E}[\mathbb{P}(X_{t+u} \in C | \mathcal{G}_{u+s}) | \mathcal{G}_u] \\
&= \mathbb{E}[P(t - s, X_{u+s}, C) | \mathcal{G}_u] \\
&= \int P(t - s, y, C)P(s, X_u, dy)
\end{aligned}$$

for all $s, t, u \in T, s < t, x \in E, C \in \mathcal{B}(E)$. The distribution of X_0 is called the *initial distribution of* $(X_t)_{t \in T}$.

A transition function for $(X_t)_{t \in T}$ and a given initial distribution ν on $\mathcal{B}(E)$ determine the finite-dimensional distributions of $(X_t)_{t \in T}$ by

$$\mathbb{P}(X_0 \in C_0, X_{t_1} \in C_1, \ldots, X_{t_n} \in C_n) =$$

$$\int_{C_0} \cdots \int_{C_1} P(t_n - t_{n-1}, y_{n-1}, C_n)P(t_{n-1} - t_{n-2}, y_{n-2}, dy_{n-1}) \cdots P(t_1, y_0, dy_1)\nu(dy_0), \tag{B.10}$$

for all $t_1, t_2, \ldots, t_n \in T, 0 < t_1 < \cdots < t_n$. More precisely, if E is complete and separable (and this is the case if $E = \mathbb{R}^m$), the following result holds.

Theorem B.3.1. *Let $P(t, x, C)$ be a homogeneous transition function and let ν be a probability measure on $\mathcal{B}(E)$. Then there exists a Markov process with respect to its natural filtration $(X_t)_{t \in T}$, whose finite-dimensional distributions are uniquely determined by (B.10).*

In the case $T = \mathbb{N}$ (or $T = \{0, 1, \ldots, N\}$), it is sufficient to know the transition probabilities for the unitary time interval. More precisely, in this case, a transition function is a function

$P(x,C)$ defined on $E \times \mathcal{B}(E)$ such that

$$P(x, \cdot) \text{ is a probability on } \mathcal{B}(E) \text{ for all } x \in E$$

and

$$P(\cdot, C) \in B(E) \text{ for all } C \in \mathcal{B}(E).$$

In the considered discrete times setting, $P(x; C)$ is a transition function for a time homogeneous Markov process with respect to the filtration $(\mathcal{G}_n)_{n \in \mathcal{T}}$, if

$$\mathbb{P}(X_{n+1} \in C | \mathcal{G}_n) = P(X_n, C)$$

with $C \in \mathcal{B}(E)$. Notice that

$$\mathbb{P}(X_{n+m} \in C | \mathcal{G}_n) \int \cdots \int P(y_{m-1}, C) P(y_{m-2}, dy_{m-1}) \cdots P(y_n, dy_1).$$

Definition B.3.4. Let $(X_t)_{t \in \mathcal{T}}$ be a Markov process with respect to the filtration $(\mathcal{G}_t)_{t \in \mathcal{T}}$ with transition function $P(t, x, C)$ and τ a stopping time with respect to the filtration $(\mathcal{G}_t)_{t \in \mathcal{T}}$, with $\tau < +\infty$ a.s. Then $(X_t)_{t \in \mathcal{T}}$ is *strong Markov at τ* if

$$\mathbb{P}(X_{\tau+t} \in C | \mathcal{G}_\tau) = \mathbb{P}(X_{\tau+t} \in C | X_\tau),$$

that is

$$\mathbb{P}(X_{\tau+t} \in C | \mathcal{G}_\tau) = P(t, X_\tau, C).$$

$(X_t)_{t \in \mathcal{T}}$ is a *strong Markov process* if it is a strong Markov process at all $(\mathcal{G}_t)_{t \in \mathcal{T}}$ stopping times τ with $\tau < +\infty$.

Let us now concentrate on the case $\mathcal{T} = [0, +\infty)$. Unfortunately, formulas for transition functions cannot be obtained apart from the one-dimensional Brownian motion. Consequently, directly defining a transition function is usually not a useful method of specifying a Markov process. One alternative way to do this is through semi-groups.

Definition B.3.5. A one-parameter family $\{T(t) : t \geq 0\}$ of bounded linear operators on a Banach space L is called a *semi-group* of operators if

1. $T(0) = Identity$,
2. $T(s + t) = T(s)T(t)$ for all $s, t \geq 0$.

Definition B.3.6. The infinitesimal generator of a semi-group $\{T(t) : t \geq 0\}$ on L is the linear operator A defined by

$$Af = \lim_{t \to 0} \frac{T(t)f - f}{t},$$

$$\mathcal{D}(A) = \{f \in L \text{ for which the above integral exists}\}.$$

Let now $T(t)$ be a semi-group on a closed subspace $L(E) \subset B(E)$. We say that an E-valued Markov process X corresponds to $T(t)$ if

$$E[f(X_{t+s})|\mathcal{F}_t^X] = T(s)f(X_s),$$

for all $s, t \geq 0$ and $f \in L(E)$.

It can be proved that under suitable (but satisfied in most common examples) assumptions, the semi-group $T(t)$ and an initial distribution ν determine the finite-dimensional distributions of X. The relation between the semi-group $T(t)$ and the transition function is given by

$$T(t)f(x) = \int f(y)P(t, x, dy),$$

which defines a semi-group thanks to the Chapman–Kolmogorov property.

B.4 LÉVY PROCESSES

For details on Lévy processes we refer the interested reader to, among others, Cherubini *et al.* (2011), Sato (1999), or Cont and Tankov (2004). In this section we just sketch the main definitions and properties.

Definition B.4.1. An $\mathbb{R}^d$-valued stochastic process $X = (X_t)_{t \geq 0}$ on $(\Omega, \mathcal{F}, \mathbb{P})$ is called a *Lévy process* if it has

1. **Independent increments.** The random variables $X_{t_0}, X_{t_1} - X_{t_0}, \ldots, X_{t_n} - X_{t_{n-1}}$ are independent for all $n \geq 1$ and $0 \leq t_0 < t_1 < \ldots < t_n$.
2. **Stationary increments.** $X_{t+h} - X_t$ has the same distribution as X_h for all $h, t \geq 0$.
3. It is stochastically continuous: for every $t \geq 0$ and $\epsilon > 0$

$$\lim_{s \to t} \mathbb{P}[|X_s - X_t| > \epsilon] = 0.$$

4. The paths $t \to X_t$ are right-continuous with left limits with probability 1.

It is implicit in 2. that $\mathbb{P}(X_0 = 0) = 1$. Moreover, it is an immediate consequence of 1. that a Lévy process is a Markov process.

The above definition implies that increments of Lévy processes are infinitely divisible. In particular, if $(X_t)_{t \geq 0}$ is a Lévy process, the distribution of X_t necessarily has to be of the infinitely divisible type.

Theorem B.4.1. *If $(X_t)_{t \geq 0}$ is a Lévy process, then, for any $t \geq 0$, the distribution of X_t is infinitely divisible and, if $\phi(u) = \phi_{X_1}(u)$ for all $u \in \mathbb{R}^d$, we have $\phi_{X_t}(u) = \phi_{X_1}^t(u)$. Conversely, if $\phi(u)$ is the characteristic function of an infinitely divisible distribution on $\mathbb{R}^d$, then there is a Lévy process $(X_t)_{t \geq 0}$ such that $\phi_{X_1}(u) = \phi(u)$. Moreover, if $(X_t)_{t \geq 0}$ and $(X_t')_{t \geq 0}$ are Lévy processes such that $\phi_{X_1}(u) = \phi_{X_1'}(u)$, then they are identical in law.*

By the above theorem and the Lévy–Khintchine theorem for infinitely divisible distributions, we can state the following fundamental result representing the characteristic function of a Lévy process in terms of its characteristic triplet (a, Σ, ν).

Theorem B.4.2. *Lévy–Khintchine representation. Let X_t be a Lévy process, then there exist a vector $a \in \mathbb{R}^d$, a positive definite $d \times d$-matrix Σ, and a Radon measure ν on $\mathbb{R}^d \setminus \{0\}$ with $\int_{-\infty}^{+\infty}(1 \wedge x^2)\nu(dx) < +\infty$ such that*

$$\phi_{X_t}(\lambda) = \mathbb{E}\left[e^{i\lambda \cdot X_t}\right] = e^{-t\psi(\lambda)},$$

where

$$\psi(\lambda) = -ia \cdot \lambda + \frac{1}{2}\lambda'\Sigma\lambda - \int_{-\infty}^{+\infty} \left(e^{i\lambda \cdot x} - 1 - i\lambda \cdot x\mathbf{1}_{\{|x|\leq 1\}}\right)v(dx), \qquad \lambda \in \mathbb{R}^d.$$

Lévy processes are characterized by their Lévy–Khintchine characteristics (a, Σ, v), where we call a the *drift coefficient*, Σ the *Brownian covariance matrix*, and v the *Lévy measure* or *jump measure*. $\psi(\lambda)$ is called the *characteristic exponent* of the Lévy process. For real-valued Lévy processes the above formulas take the form

$$\phi_{X_t}(\lambda) = \mathbb{E}\left[e^{i\lambda \cdot X_t}\right] = e^{-t\psi(\lambda)} \qquad \lambda \in \mathbb{R}$$

and

$$\psi(\lambda) = -ia\lambda + \frac{1}{2}\sigma^2\lambda^2 - \int_{-\infty}^{+\infty} \left(e^{i\lambda x} - 1 - i\lambda x\mathbf{1}_{\{|x|\leq 1\}}\right)v(dx) \qquad \lambda \in \mathbb{R}.$$

The Lévy measure v is called the "jump measure" since it is defined as

$$v(A) = \mathbb{E}[\#\{t \in [0, 1] : \Delta X_t \neq 0, \Delta X_t \in A\}], \qquad A \in \mathcal{B}(\mathbb{R}^d).$$

That is, $v(A)$ is the expected number, per unit time, of jumps whose size belongs to A.

It can be proved that any Lévy process is made up of a drift, a diffusion, a set of finite jumps of large size, and a set of infinite jumps of infinitesimal size. This is the content of the following decomposition theorem.

Theorem B.4.3. *Lévy–Ito decomposition theorem.* Let $(Z_t)_{t\geq 0}$ *be a Lévy process and* v *its Lévy measure. Then there exist a triplet* (a, Σ, v) *and a d-dimensional Brownian motion* $(B_t)_{t\geq 0}$ *such that*

$$Z_t = at + \sigma B_t + M_t + C_t,$$

where

$$C_t = \sum_{s\leq t} \Delta X_s \mathbf{1}_{\{|\Delta X_s|>1\}}$$

(big jumps) and

$$M_t = \lim_{\epsilon \downarrow 0} \left(\sum_{s\leq t} \Delta X_s \mathbb{I}_{\{\epsilon < |\Delta X_s|\leq 1\}} - t\int_{\{x\in\mathbb{R}:\epsilon<|x|\leq 1\}} xv(dx)\right)$$

is a martingale (of small jumps compensated by a linear drift).

The d-dimensional Brownian motion with drift is the only Lévy process with continuous trajectories and it represents the continuous part of the general Lévy process. As for the purely jumps component, if $v(\mathbb{R} \setminus \{0\}) < +\infty$, the characteristic triplet $(0, 0, v)$ defines a piecewise constant process and the number of jumps in each bounded time interval is finite.

If $v(\mathbb{R} \setminus \{0\}) = +\infty$ the set of small jumps of every trajectory of the Lévy process associated with the characteristic triplet $(0, 0, v)$ is countably infinite.

The following is a list of univariate Lévy processes and their characteristics.

1. **Brownian motion.** The Brownian motion is a Lévy process with continuous paths such that

$$X_t \overset{d}{=} N(0, t)$$

for all $t > 0$. It is parameterized by the characteristics

$$a = 0, \quad \sigma^2 > 0, \quad v = 0.$$

Brownian motion has the *scaling property* $\left(\sqrt{c} X_{\frac{t}{c}} \right)_{t \geq 0} \overset{d}{=} X$.

2. **Poisson process.** The Poisson process with rate $\lambda \in (0, +\infty)$ is a Lévy process with

$$\mathbb{P}(X_t = k) = \frac{(\lambda t)^k}{k!} e^{-\lambda t},$$

$k \geq 0, t \geq 0$ (Poisson distribution). It is parameterized by the characteristics

$$a = 0, \quad \sigma = 0, \quad v(dx) = \lambda \delta_1(x).$$

3. **Compound Poisson process.** The compound Poisson process is defined as

$$C_t = \sum_{k=1}^{N_t} Z_k, \quad t \geq 0$$

for a Poisson process $(N_t)_{t \geq 0}$ with rate λ and independent (also of N_t) identically distributed jumps $Z_k, k \geq 1$. It is parameterized by the characteristics

$$a = \sigma^2 = 0, \quad v(dx) = \lambda \mu(dx)$$

where μ is the law of the i.i.d. random variables Z_k and λ is the intensity of the Poisson process counting the jumps.

4. **Stable process.** The α-stable process for $0 < \alpha < 2$ is parameterized by the characteristics

$$\sigma = 0, \text{ and } v(dx) = \begin{cases} c_1 x^{-1-\alpha} dx, & \text{for } x > 0 \\ c_2 |x|^{-1-\alpha} dx, & \text{for } x < 0 \end{cases}$$

with $c_1, c_2 > 0$. Since it can be proved that in this case the characteristic exponent is

$$\psi(u) = iua - \gamma |u|^\alpha (1 - i\beta sgn(u)g(u))$$

with

$$g(u) = \begin{cases} tan\frac{\pi\alpha}{2} \text{ for } \alpha \in (0, 1) \cup (1, 2] \\ \frac{2}{\pi} log|u| \text{ for } \alpha = 1 \end{cases}$$

it is easy to verify that the *scaling property* stated for the Brownian motion extends to all α-stable processes with $a = 0$, that is

$$(X_{ct})_{t \geq 0} \overset{d}{=} \left(c^{\frac{1}{\alpha}} X_t \right)_{t \geq 0}.$$

5. **Gamma process.** The Gamma process, where $X_t \overset{d}{=} Gamma(\alpha t, \beta)$, is characterized by the characteristics

$$a = \sigma^2 = 0, v(dx) = \alpha x^{-1} e^{-\beta x} dx, x > 0.$$

6. **Variance Gamma process.** The variance Gamma process is defined as the difference $X = G - H$ of two independent Gamma processes G and H. Let $G \overset{d}{=} Gamma(\alpha_+, \beta_+)$ and $H \overset{d}{=} Gamma(\alpha_-, \beta_-)$. The Lévy–Khintchine characteristics of the variance Gamma process are

$$a = \sigma = 0, \text{ and } v(dx) = \begin{cases} \alpha_+ |x|^{-1} e^{-\beta_+|x|} dx, & x > 0 \\ \alpha_- |x|^{-1} e^{-\beta_-|x|} dx, & x < 0 \end{cases}.$$

7. **CGMY process.** As a natural generalization of the variance Gamma process, Carr, Geman, Madan, and Yor (CGMY) suggested the following Lévy measure for financial price processes:

$$v(dx) = \begin{cases} C_+ e^{-G|x|} |x|^{-Y-1} dx, & x > 0 \\ C_- e^{-M|x|} |x|^{-Y-1} dx, & x < 0 \end{cases}$$

for parameters $C > 0, G \geq 0, M \geq 0, Y < 2$.

B.4.1 Subordinators

Definition B.4.2. An increasing $\mathbb{R}$-valued Lévy process (that is, for $t \geq s$ we have almost surely that $X_t \geq X_s$) is called a *subordinator*.

Subordinators are used extensively as stochastic clocks to apply the time-change technique. Clearly a non-negative Lévy process, i.e. one where $X_t \geq 0$ a.s., for all $t \geq 0$, is automatically increasing, since every increment $X_{s+t} - X_s$ has the same distribution as X_t and therefore it is also non-negative.

Main examples of subordinators are:

1. **The gamma process.** The gamma process is an increasing Lévy process by definition since the support of the marginal distributions is the positive real line.
2. **The Poisson process.** The Poisson process is an increasing Lévy process by definition since the support of the marginal distributions is the set of natural numbers.
3. **The increasing compound process.** The compound Poisson process $C_t = \sum_{k=1}^{N_t}$ for a Poisson process N_t and identically distributed non-negative $Z_1, Z_2, \ldots$ with probability density function concentrated in $(0, +\infty)$. We can add a drift and consider $\tilde{C}_t = at + C_t$ for some $a > 0$ to get *a compound Poisson process with drift*.
4. **The stable subordinator.** The stable subordinator is best defined in terms of its Lévy–Khintchine characteristics $a = 0$ and $v(dx) = x^{-\alpha-1} dx$, for $x > 0$. This gives

$$\mathbb{E}(exp(i\gamma X_t)) = exp\left\{ t \int_0^{+\infty} (e^{i\gamma x} - 1) x^{-\alpha-1} dx \right\}.$$

More generally we can also consider tempered stable processes with $v(dx) = x^{-\alpha-1} e^{-\rho x}$, $\rho > 0, x > 0$.

B.5   SEMI-MARTINGALES

Here, we briefly introduce (for the sake of simplicity, in some cases we will skip the formal hypotheses) the main definitions and properties of semi-martingales related to the results presented in Chapter 8 of this book. For details and a complete analysis of this topic we refer to Jacod and Shiryaev (2003).

Given $(\Omega, \mathcal{F}, \mathbb{P})$ and a filtration $(\mathcal{F}_t)_{t \geq 0}$, a semi-martingale is a process with values in $\mathbb{R}^d$ of the form

$$X = X_0 + M + A,$$

where X_0 is $\mathcal{F}_0$-measurable, M is a local martingale, and A is a cadlag process of finite variation (that is, its paths are a.s. of finite variation on compact sets of $[0, +\infty)$). The decomposition of a semi-martingale is not unique, however; there is at most one such decomposition where the process A can be chosen predictably.

Definition B.5.1. A special semi-martingale is a semi-martingale X which admits a unique decomposition

$$X = X_0 + M + A$$

where the process A is predictable.

Such a decomposition is called canonical decomposition. The following result can be found in Jacod and Shiryaev (2003), Lemma I.4.24.

Lemma B.5.1. *If a semi-martingale X satisfies $|\Delta X_t| \leq a$ for some $a > 0$ for all t, then it is special and its canonical decomposition satisfies $|\Delta A_t| \leq a$ and $|M_t| \leq 2a$.*

Definition B.5.2. Let X be a real-valued semi-martingale. The *quadratic variation process* of X, denoted $[X, X]$, is defined by

$$[X, X] = X^2 - 2 \int X_- dX.$$

For a real-valued semi-martingale X, the process $[X, X]^c$, denotes the *path-by-path* continuous part of $[X, X]$ that satisfies

$$[X, X]_t = [X, X]_t^c + X_0^2 + \sum_{0 < s \leq t} (\Delta X_s)^2.$$

The behavior of a semi-martingale is determined by its characteristic triplet. Before introducing these concepts, we need the following notions.

In this section, given two σ-algebras $\mathcal{F}$ and $\mathcal{G}$ with $\mathcal{F} \otimes \mathcal{F}$, we denote the smallest σ-algebra containing $\mathcal{F} \times \mathcal{G}$.

Definition B.5.3. A real-valued function ν defined on $\Omega \times \mathcal{B}([0, +\infty)) \otimes \mathcal{B}(\mathbb{R}^d)$ is called a random measure if, for all $\omega \in \Omega$, $\nu(\omega, \cdot, \cdot)$ is a measure on $\mathcal{B}([0, +\infty)) \otimes \mathcal{B}(\mathbb{R}^d)$.

A random measure is said to be predictable if for any predictable random function W such that the integral

$$W * \nu_t(\omega) = \int_{[0,t] \times \mathbb{R}^d} W_s(x) \nu(\omega, ds, dx)$$

exists, $W * \nu_t$ is a predictable process.

Let X be an $\mathbb{R}^d$-valued semi-martingale. By Proposition II.1.16 in Jacod and Shiryaev (2003),

$$\mu(\omega, dt, dx) = \sum_{s>0} \mathbf{1}_{\{\Delta X_s \neq 0\}} \delta_{(s, \Delta X_s)}(dt, dx)$$

defines an integer-valued random measure on $\mathcal{B}([0, +\infty)) \bigotimes \mathcal{B}(\mathbb{R}^d)$, where $\delta_{(s, \Delta X_s)}$ denotes the Dirac delta function centered at $(s, \Delta X_s)$ and the summation is taken over a countable index set since a semi-martingale has almost surely right-continuous paths. The measure ν counts the jumps of the semi-martingale X. The next theorem is the counterpart of the Doob–Meyer decomposition for random measures.

Theorem B.5.2. *Let μ be a σ-finite random measure. There exists a random measure, called the compensator or dual predictable projection, $(\mu)^p$, which is characterized as being a predictable random measure, satisfying one of the following equivalent conditions:*

1. *$\mathbb{E}[W * \mu_\infty] = [W * \mu_\infty^p]$ for every measurable function Q such that the process $|W| * \mu$ is integrable.*
2. *For every measurable function W such that the process $|W| * \mu$ is locally integrable, then $|W| * \mu^p$ is locally integrable, $W * \mu^p$ is the compensator of $W * \mu$, i.e. the process $W * \mu - W * \mu^p$ is a local martingale.*

We are now ready to introduce the notion of semi-martingale characteristics. This notion extends the three terms, namely the drift, the variance of the Brownian part, and the Lévy measure, that characterizes Lévy processes.

We first need the following concept:

Definition B.5.4. We call a function $h : \mathbb{R}^d \to \mathbb{R}^d$ a truncation function, if it is bounded with compact support and $h(x) = x$ in a neighborhood of 0.

If h is a truncation function, $\Delta X_s - h(\Delta X_s) \neq 0$ only if $\Delta X_s| > b$ for some $b > 0$. If we let

$$X_t(h) = X_t - \sum_{s \leq t} (\Delta X_s - h(\Delta X_s)),$$

then $X(h)$ is a semi-martingale with bounded jumps and in view of Lemma B.5.1 it is special, therefore it admits a canonical decomposition

$$X(h) = X_0 + M(h) + B(h),$$

where $M(h)$ is an $\mathbb{R}^d$-valued local martingale and $B(h)$ is a predictable process of finite variation.

Definition B.5.5. Let h be fixed. We call the characteristics of X the triplet (b, C, ν) consisting of:

1. b, an $\mathbb{R}^d$-valued predictable of finite variation process of the canonical decomposition of $X(h)$, i.e. $b = B(h)$.
2. C, an $\mathbb{R}^d \times \mathbb{R}^d$-valued continuous process of finite variation (which is the continuous part of the quadratic variation of the local martingale M, i.e. $C_{i,j} = [M^i, M^j]^c$).
3. ν, a predictable random measure on $\mathcal{B}([0, +\infty)) \bigotimes \mathcal{B}(\mathbb{R}^d)$. In particular, ν is the predictable projection of the random measure μ counting the jumps of X, i.e. $\nu = \mu^p$.

Because the compensator and the canonical decomposition of a special semi-martingale are unique a.s., the characteristic triplets also uniquely defined a.s. Moreover, notice that the characteristics depend on the filtration one is working with. In the case of a Lévy process, we have.

Theorem B.5.3. *A d-dimensional process is a Lévy process if and only if it is a semi-martingale admitting as characteristics*

$$b_t = at, \ C_t = \Sigma t, \ \nu(\omega, dt, dx) = dt K(dx),$$

where (a, Σ, K) is the characteristic triplet of the Lévy process given by the Lévy–Khintchine theorem.

References

Aas, K., Czado, C., Frigessi, A., and Bakken, H. (2009) Pair copula constructions of multiple dependence. *Insurance Mathematics and Economics*, **44**(2), 182–198.

Alexander C. (2001) Orthogonal GARCH. In *Mastering Risk* Vol 2, Financial Times. Prentice-Hall, London, **2**, 21–38.

Ang, A. and Chen, J. (2002) Asymmetric correlations of equity portfolios. *Journal of Financial Economics*, **63**(3), 443–494.

Asharaf, S., Narasimha Murty, M., and Shevade, S.K. (2006) Rough set based incremental clustering of interval data. *Pattern Recognition Letters*, **27**, 515–519.

Azzalini, A. and Capitanio, A. (1999) Statistical applications of the multivariate skew normal distribution. *Journal of the Royal Statistical Society: Series B (Statistical Methodology)*, **61**(3), 579–602.

Baba, Y., Engle, R.F., Kraft, D., and Kroner, K. (1990) Multivariate Simultaneous Generalized ARCH, UCSD, Department of Economics, unpublished manuscript.

Baglioni A. and Cherubini, U. (2010) Marking-to-market government guarantees to financial systems: an empirical analysis of Europe. Working paper.

Barbe, P., Genest, C., Ghoudi, K., and Rémillard, B. (1996) On Kendall's process. *Journal of Multivariate Analysis*, **58**, 197–229.

Beare, B. (2009) Copulas and temporal dependence. UCD, Department of, Economics.

Bedford, T. and Cooke, R.M. (2002) Vines: a new graphical model for dependent random variables. *Annals of Statistics*, **30**(4), 1031–1068.

Bielecki, T., Jakubowski, J., Vidozzi, A., and Vidozzi, L. (2008) Study of dependence for some stochastic processes. *Stochastic Analysis and Applications*, **26**, 903–924.

Bielecki, T., Vidozzi, A., and Vidozzi, L. (2008) Multivariate Markov processes with given Markovian margins and Markov copulae. Working paper, Illinois Institute of Technology.

Billingsley, P. (1986) *Probability and Measure*, 2nd edn. John Wiley and, Sons, New York.

Black, F. and Cox, J.C. (1976) Valuing corporate securities: some effects of bond indenture provisions. *Journal of Finance*, **31**, 351–367.

Boero, G., Silvapulle, P., and Tursunalieva, A. (2010) Modelling the bivariate dependence structure of exchange rates before and after the introduction of the euro: a semi-parametric approach. *International Journal of Finance and Economics*, in press (available online).

Bouyè, E. and Salmon, M. (2009) Dynamic copula quantile regressions and tail area dynamic dependence in forex markets. *The European Journal of Finance*, **15**(78), 721–750.

Bradley, R.C. (2007) *Introduction to Strong Mixing Conditions*. Kendrick, Press, Heber City.

Breeden, D.T. and Litzenberger, R.H. (1978) Prices of state-contingent claims implicit in option prices. *Journal of Business*, **51**, 621–651.

Brèmaud, P. and Yor, M. (1978) Changes of filtrations and of probability measures. *Z. Wahrscheinl*, **45**, 269–295.

Burtschell, X., Gregory, J., and Laurent, J.P. (2007) Beyond the Gaussian copula: stochastic and local correlation. *Journal of Credit Risk*, **3**(1), 31–62.

Calinski, T. and Harabasz, J. (1974) A dendrite method for cluster analysis. *Communications in Statistics*, **3**, 1–27.

Carr, P. and Wu, L. (2004) Time changed Lèvy processes and option pricing. *Journal of Financial Economics*, **17**, 113–141.

Chan, K. and Tong, H. (2001) *Chaos: A Statistical Perspective*. Springer-Verlag, New York.

Chen, X. and Fan, Y. (2006) Estimation of copula-based semiparametric time series models. *Journal of Econometrics*, **130**, 307–335.

Chen, X., Koenker, R., and Xiao, Z. (2009a) Copula-based nonlinear quantile autoregression. *Econometrics Journal*, **12**, S50–S67.

Chen, X., Wu, W.B., and Yi, Y. (2009b) Efficient estimation of copula-based semiparametric Markov models. *Annals of Statistics*, **37**(6B), 4214–4253.

Cherubini, U. and Luciano, E. (2002) Bivariate option pricing with copulas. *Applied Mathematical Finance*, **8**, 69–85.

Cherubini, U. and Romagnol, S. (2009) Computing copula volume in *n*-dimensions. *Applied Mathematical Finance*, **16**(4), 307–314.

Cherubini, U., Gobbi, F., and Mulinacci, S. (2010a) Semi parametric estimation and simulation of actively managed portfolios. Working paper.

Cherubini, U., Gobbi, F., Mulinacci, S., and Romagnoli, S. (2010b) On the term structure of multivariate equity derivatives. Working paper.

Cherubini, U., Gobbi, F., Mulinacci, S., and Romagnoli, S. (2010c) A copula-based model for spatial and temporal dependence of equity markets. In F. Durante, W. Haerdle, P. Jaworski, and T. Rychlik (eds). *Workshop on Copula Theory and its Applications*. Lecture Notes in Statistics-Proceedings. Springer, Berlin.

Cherubini, U., Luciano, E., and Vecchiato, W. (2004) *Copula Methods in Finance*. John Wiley & Sons, London.

Cherubini, U., Mulinacci, S., and Romagnoli, S. (2011) A copula-based model of speculative price dynamics in discrete time. *Journal of Multivariate Analysis*, **102**, 1047–1063.

Cont, R. and Tankov, P. (2004) *Financial Modelling with Jump Processes*. Chapman & Hall, New York.

CreditMetrics Manual. RiskMetrics Group (available at www.riskmetrics.com).

Dambis, K.E. (1965) On the decomposition of continuous submartingales. *Theory of Probability and its Applications*, **10**, 401–410.

Darsow, W.F., Nguyen, B., and Olsen, E.T. (1992) Copulas and Markov processes. *Illinois Journal of Mathematics*, **36**, 600–642.

Davydov, Y. (1973) Mixing conditions for Markov chains. *Theory of Probability and its Applications*, **18**, 312–328.

de Haan, L. and Ferreira, A. (2006) *Extreme Value Theory: An Introduction*, Springer-Verlag, New York.

Deheuvels, P. (1979) La function de dépendance empirique et ses propriétés. Un test non paramétriquen d'indépendace. *Académie Royale de Belgique Bulletin Cl Sci.*, **65**(5), 274–292.

Deheuvels, P. (1981) A non-parametric test for independence. Université de Paris. Institut de Statistique.

Dubins, L.E. and Schwarz, G. (1965) On continuous martingales. *Proceedings of the National Academy of Sciences USA*, **53**, 913–916.

Duffie, D. and Singleton, K. (1999) Simulating correlated defaults. Stanford University, working paper.

Durante, F., Hofert, M., and Scherer, M. (2009) Multivariate hierarchical copulas with shocks. *Methods in Computing and Applied Probability*, **12**(4), 681–694.

Embrechts, P., Kluppenberg, P., and Mikosch, T. (1997) *Modeling Extremal Event for Insurance and Finance*. Springer, Berlin.

Embrechts, P., McNeil, A., and Strauman., D. (2002) Correlation and dependence in risk management: properties and pitfalls. In M.A.H. Dempster (ed.), *Risk Management: Value at Risk and Beyond*. Cambridge University Press, Cambridge, pp. 176–223.

Emmer, S. and Tasche, D. (2003) Calculating credit risk capital charges with the one-factor model. Working paper.

Engle, R.F. (2002) Dynamic conditional correlation. *Journal of Business & Economic Statistics*, **20**(3), 339–350.

Engle, R.F., Ng, V., and Rothschild, M. (1990) Asset pricing with a Factor ARCH covariance structure: empirical estimates for Treasury Bills. *Journal of Econometrics*, **45**, 213–237.

Erb, C.B., Harvey, C.R., and Viskanta, T.E. (1994) Forecasting international equity correlations. *Financial Analysts Journal*, **50**(6), 32–45.

Ethier, N.E. and Kurtz, T.G. (1985) *Markov Processes: Characterization and Convergence*. John Wiley & Sons, New York.

Fang, K.T., Kots, S., and Ng, W. (1990) *Symmetric Multivariate and Related Distributions*. Chapman & Hall, London.

Fengler, M.R., Herwartz, H., and Werner, C. (2010) Option pricing under time varying correlation with conditional dependence: A copula based approach to recover the index skew from the constituent dynamics, working paper.

Fermanian, J.D. and Wegkamp, M. (2004) Time dependent copulas. Preprint INSEE. Paris, France.

Florens, J.P. and Fougere, D. (1991) Non-causality in continuous time: applications to counting processes. Cahier du GREMAQ 91.b, Universit des Sciences Sociales de Toulouse.

Francois, D., Wertz, V., and Verleysen, M. (2005) Non-Euclidean metrics for similarity search in noisy datasets. *Proceedings of the European Symposium on Artificial Neural Networks*, ESANN'2005.

Gagliardini, P. and Gourièroux, C. (2007) An efficient nonparametric estimator for models with nonlinear dependence. *Journal of Econometrics*, **137**, 189–229.

Gordy, M. (2003) A risk-factor model foundation for ratings-based bank capital rules. *Journal of Financial Intermediation*, **12**(3), 199–232.

Gordy, M. (2004) Granularity adjustment in portfolio credit risk measurement. In G. Szego (ed.), *Risk Measures for the 21st Century*. John Wiley & Sons, New York.

Gourièroux, C., Laurent J.P., and Scaillet, O. (2000) Sensitivity analysis of values at risk. *Journal of Empirical Finance*, **7**, 225–245.

Gourièroux, C. and Robert, C.Y. (2006) Stochastic unit root models. *Econometric Theory*, **22**, 1052–1090.

Granger, C.J.W. (2003) Time series concepts for conditional distributions. *Oxford Bulletin for Economics and Statistics*, **65**, 689–701.

Granger, C.J.W., Maasoumi, E., and Racine, J. (2004) A dependence metric for possibly nonlinear processes. *Journal of Time Series Analysis*, **25**, 649–669.

Hartigan, P.M. (1985) Computation of the dip statistic to test for unimodality. *Applied Statistics (JRSS C)*, **34**, 320–325.

Henriksson, R.D. and Merton, R.C. (1981) On market timing and investment performance. II. Statistical procedures for evaluating forecasting skills. *Journal of Business*, **54**(4), 513–533.

Hiemstra, C. and Jones, J.D. (1994) Testing for linear and non-linear Grange causality in the stock price–volume relation. *Journal of Finance*, **49**, 1639–1664.

Hofert, M. (2008) Sampling Archimedean copulas. *Computational Statistics and Data Analysis*, **52**(12), 5163–5174.

Ibragimov, R. (2005) Copula-based dependence characterization and modeling for time series. Harvard Institute of Economic Research, discussion paper no. 2094.

Ibragimov, R. (2009) Copula based characterizations for higher-order Markov processes. *Econometric Theory*, **25**(3), 819–846.

Ibragimov, R. and Lentzas, G. (2009) Copulas and long memory. Harvard Institute of Economic Research, discussion paper no. 2160 (Revised November 2009).

Jacod, J. and Shiryaev, A.N. (2003) *Limit Theorems For Stochastic Processes*, 2nd edn. Springer-Verlag, Berlin.

Joe, H. (1989) Relative entropy measures of multivariate dependence. *Journal of the American Statistical Association*, **84**(405), 157–164.

Joe, H. (1990) Multivariate concordance. *Journal of Multivariate Analysis*, **35**, 12–30.

Joe, H. (1997) *Multivariate Models and Dependence Concepts*, Chapman &, Hall, London.

Kallsen, J. and Tankov, P. (2006) Characterization of dependence of multidimensional Lévy processes using Lévy copulas. *Journal of Multivariate Analysis*, **97**, 1551–1572.

Kendall, M.G. (1970) *Rank Correlation Methods*. Griffin, London.

Khoudraji, A. (1995) Contribution letude des copules et la modèlisation de valeurs extrêmes multivarièes. PhD Thesis, Universit Lavae, Quèbec.

Klement, E.P., Mesiar, R., and Pap, E. (2002) Invariant copulas. Kybernetica (Prague), 275–285.

Klüppelberg, C. and Resnick, S. (2008) The Pareto copula aggregation of risks and Emperor's socks. *Journal of Applied Probabilities*, **45**(1), 67–84.

Koenker, R. and Basset Jr., G. (1978) Regression quantiles. *Econometrica*, **46**(1), 33–50.

Koenker, R. and Xiao, Z. (2006) Quantile autoregression. *Journal of the American Statistical Association*, **101**, 980–990.

Kohonen, T. (1982) Clustering taxonomy, and topological maps of patterns. In *Proceedings 6ICPR, International Conference on Pattern Recognition*. IEEE Computer Society Press, Washington, DC, pp. 114–128.

Kolmogorov, A.N. and Rozanov, Y.A. (1960) On the strong mixing conditions for stationary Gaussian sequences. *Theory of Probability and its Applications*, **5**, 204–207.

Krzanowski, W.J. and Lai, Y.T. (1985) A criterion for determining the number of groups in a data set using sum of squared clustering. *Biometrics*, **44**, 23–34.

Li, D.X. (2000) On default correlation: a copula function approach. *Journal of Fixed Income*, **9**(4), 43–54.

Longin, F. and Solnik, B. (2001) Extreme correlation of international equity markets. *The Journal of Finance*, **56**(2), 649–676.

Luciano, E. and Schoutens, W. (2006) A multivariate jump-driven financial asset model. *Quantitative Finance*, **6**(5), 385–402.

Mai, J.F. and Scherer, M. (2009a) A tractable multivariate default model based on a stochastic time-change. *International Journal of Theoretical and Applied Finance*, **12**(2), 227–249.

Mai, J.F. and Scherer, M. (2009b) Pricing k-th to default swaps in a Lèvy-time framework. *Journal of Credit Risk*, **5**(3), 55–70.

Mai, J.F. and Scherer, M. (2011) Reparameterizing Marshall–Olkin copulas with applications to sampling. *Journal of Statistical Computation & Simulation*, **81**(1), 59–78.

Marshall, A.W. and Olkin, I. (1967a) A generalized bivariate exponential distribution. *Journal of Applied Probability*, **4**, 291–302.

Marshall, A.W. and Olkin, I. (1967b) A multivariate exponential distribution. *Journal of American Statistics Association*, **62**, 30–44.

Martin, R. and Wilde, T. (2003) Unsystematic credit risk. *Risk*, **15**(3), 26–32.

McNeil, A.J., Frey, R., and Embrechts, P. (2005) *Quantitative Risk Management. Concepts, Techniques and Tools*. Princeton Series in Finance. Princeton University Press, Princeton, NJ.

McNeil, A.J. (2008) Sampling nested Archimedean copulas. *Journal of Statistical Computational Simulation*, **78**(6), 567–581.

McNeil, A.J. and Neslehova, J. (2009) Multivariate Archimedean, copulas, d-monotone functions and l_1-norm symmetric distributions. *Annals of Statistics*, **37**(5B), 3059–3097.

Merton, R.C. (1974) On the pricing of corporate debt: the risk structure of interest rates. *Journal of Finance*, **29**, 449–470.

Merton, R.C. (1981) On market timing and investment performance. II. An equilibrium theory of value for market forecasts. *Journal of Business*, **54**(3), 363–406.

Mohebi, E. and Sap, M.N.M. (2009) Hybrid Kohonen self organizing map for the uncertainty involved in overlapping clusters using simulated annealing. In *Proceedings of the UKSim 2009: 11th International Conference on Computer Modelling and Simulation*, pp. 53–58.

Monroe, I. (1978) Processes that can be embedded in Brownian motion. *Annals of Applied Probability*, **6**(1), 42–56.

Muhr, M. and Granitzer, M. (2009) Automatic cluster number selection using a split and merge K-means approach, DEXA Workshops 2009, pp. 363–367.

Munakata, T. (2008) *Fundamentals of the New Artificial Intelligence*. Springer-Verlag, New York.

Nelsen, R.N. (2006) *Introduction to Copulas*, 2nd edn. Springer-Verlag, Heidelberg.

Patton, A.J. (2001) Modelling time-varying exchange rate dependence using the conditional copula. UCSD Department of Economics, working paper.

Patton, A.J. (2006a) Modelling asymmetric exchange rate dependence. *International Economic Review*, **47**(2), 527–556.

Patton, A.J. (2006b) Estimation of multivariate models for time series of possibly different lengths. *Journal of Applied Econometrics*, **21**, 147–173.

Pawlak, Z. (1982) Rough sets. *International Journal of Computer Information Science*, **11**, 341–356.

Pelleg, D. and Moore, A. (2000) X-means: extending K-means with efficient estimation of the number of clusters. In ICML-2000.

Rau-Breedow, H. (2002) Credit portfolio modelling, marginal risk contributions and granularity adjustment. Working paper.

Resnick, S.I. (1987) *Extreme Values, Regular Variation and Point Processes*. Springer-Verlag, New York.

Resnick, S.I. (2007) *Heavy Tail Phenomena: Probabilistic and Statistical Modeling*. Springer Series in Operations Research and Financial Engineering. Springer-Verlag, New York.

Rockinger, M. and Jondeau, E. (2001) Conditional dependency of financial series: an application of copulas. Les Cahiers de Recherche 723, HEC Paris.

Rosenblatt, M. (1956) Remarks on some nonparametric estimates of a density function. *Annals of Mathematical Statistics*, **27**, 832–837.

Ruymgaart, F.H. and van Zuijlen, M.C.A. (1978) Asymptotic normality of multivariate linear rank statistics in the non-i.i.d. case. *Annals of Statistics*, **6**(3), 588–602.

Sato, K. (1999) *Lévy Processes and Infinitely Divisible Distributions*. Cambridge University Press, Cambridge.

Savu, C. and Trede, M. (2008) Goodness-of-fit tests for parametric families of Archimedean copulas. *Quantitative Finance*, **8**, 109–116.

Schönbucher, P. (2003) *Credit Derivatives Pricing Models: Model, Pricing and Implementation*. John Wiley & Sons, New York.

Schmitz, V. (2003) Copulas and stochastic processes. Aachen University, PhD dissertation.

Schweizer, B. and Wolff, E. (1976) Sur une mesure de dépendance pour les variables aléatoires. *Comptes Rendus de l'Académie des Sciences de Paris*, **283**, 659–661.

Schweizer, B. and Wolff, E. (1981) On non-parametric measures of dependence for random variables. *Annals of Statistics*, **9**, 879–885.

Silvapulle, P. and Moosa, I.A. (1999) The relationship between spot and futures prices: evidence for the crude oil market. *Journal of Futures Markets*, **19**(2), 175–193.

Sklar, A. (1959) Fonctions de repartition à *n* dimensions et leurs marges. *Publication Inst. Statist. Univ. Paris*, **8**, 229–231.

Su, M.C. and Chou, C.H. (2001) A modified version of the K-means algorithm with a distance based on cluster symmetry. *IEEE Tran. Pattern Anal. Mach. Intell.* **23**(6), 674–680.

Tankov, P. (2003) Dependence structure of spectrally positive multidimensional Lévy processes. Working paper.

Tasche, D. (1999) Risk contributions and performance measurement. Working paper.

Thorin, O. (1977) On the infinite divisibility of the Pareto distribution. *Scandinavian Actuarial Journal*, 31–40.

van der Goorberg, R., Genest, C., and Werker, B. (2005) Bivariate option pricing using dynamic copula models. *Insurance, Mathematics and Economics*, **37**, 101–114.

van der Vaart, A. and Wellner, J. (1996) Weak convergence and empirical processes: with applications to statistics (Springer Series in Statistics).

Vasicek, O. (1991) Limiting loan loss distributions. KMV working paper.

Vidozzi, A. (2009a) Two essays in mathematical finance. Doctoral dissertation Illinois Institute of Technology.

Vidozzi, L. (2009b) Two essays on multivariate stochastic processes and applications to credit risk modeling. Doctoral dissertation, Illinois Institute of Technology.

Wang, L., Bo, L., and Jiao, L. (2006) A modified K-means clustering with a density-sensitive distance metric. *Rough Sets and Knowledge Technology, Lecture notes in Computer Science*, **4062**, 544–551.

Wolff, E.F. (1981) *N*-dimensional measures of dependence. *Stochastica*, **4**, 175–188.

Xing, E.P., Ng, A.Y., Jordan, M.I., and Russell, S. (2001) Distance metric learning, with application to clustering with sideinformation, Advances in Neural Information Processing System 2003, Whistler, British Columbia, Canada, pp. 505–512.

Yu, J. and Phillips, P.C.B. (2001) Gaussian approach for continuous time models of the short term interest rate. *Econometrics Journal*, **4**, 1–17.

Extra Reading

Amsler, C., Prokhorov, A., and Schmidt, P. (2010) Using copulas to model time dependence in stochastic frontier models. Working paper.

Bernardi, E. and Romagnoli, S. (2011) Computing the volume of a high-dimensional semi-unsupervised hierarchical copula. *International Journal of Computer Mathematics*, **88**(12), 2591–2607.

Berrada, T., Dupuis, D.J., Jacquier, E., Papageorgiou, N., and Rémillard, B. (2006) Credit migration and derivatives pricing using copulas. *Journal of Computing in Finance*, **10**, 43–68.

Bielecki, T., Jakubowski, J., and Nieweglowski, M. (2010) Dynamic modeling of dependence in finance via copulae between stochastic processes, In P. Jaworski, F. Durante, W. Haerdle, and T. Rychlik (eds). *Copula Theory and its Applications*. Lecture Notes in Statistics-Proceedings. Springer, Berlin, pp. 33–76.

Bingham, N.H., Goldie C.M., and Teugels, J.L. (1987) *Regular Variation*. Cambridge University Press, Cambridge.

Black, F. and Scholes, M. (1973) The pricing of options and corporate liabilities. *Journal of Political Economy*, **81**, 637–654.

Block, H.W. and Ting, M.L. (1981) Some concepts of multivariate dependence. *Communications in Statistics A – Theory Methods*, **10**, 749–762.

Block, H., Savits, T., and Shaked, M. (1982) Some concepts of negative dependence. *Annals of Probability*, **10**, 765–772.

Blomqvist, N. (1950) On a measure of dependence between two random variables. *Annals of Mathematical Statistics*, **21**, 593–600.

Bouyè, E., Durrleman, V., Nikeghbali, A., Riboulet, G., and Roncalli, T. (2000) Copulas for finance – a reading guide and some applications. Groupe de Recherche Opèrationelle, Crédit Lyonnais, working paper.

Brindley, E.C. and Thompson Jr., W.A. (1972) Dependence and aging aspects of multivariate survival. *Journal of the American Statistical Association*, **67**, 822–830.

Buchinsky, M. (1998) Recent advances in quantile regression models: a practical guideline for empirical research. *Journal of Human Resources*, **33**(1), 88–126.

Cambanis, S., Huang, S., and Simons, G. (1981) On the theory of elliptically contoured distributions. *Journal of Multivariate Analysis*, **11**, 368–385.

Cameron, A.C., Li, T., Trivedi, P.K., and Zimmer, D.M. (2004) Modelling the differences in counted outcomes using bivariate copula models with application to mismeasured counts. *Econometrics Journal*, **7**, 566–584.

Cazzulani, L., Meneguzzo, D., and Vecchiato, W. (2001) Copulas: a statistical perspective with empirical applications. IntesaBCI Risk Management, Department, working paper.

Cherubini, U. and Luciano, E. (2001) Value at risk trade-off and capital allocation with copulas. *Economic Notes*, **30**(2), 235–256.

Cherubini, U., Mulinacci, S., and Romagnoli, S. (2008) A copula-based model of the term structure of CDO tranches. In W.K. Härdle, N. Hautsch, and L. Overbeck (eds), *Applied Quantitative Finance*, 2nd, edn. Springer, Berlin.

Cherubini, U. and Romagnoli, S. (2010) The dependence structure of running maxima and minima: results and option pricing applications. *Mathematical Finance*, **20**(1), 35–58.

Cherubini, U., Mulinacci, S., and Romagnoli, S. (2011) On the distribution of the (un)bounded sum of random variables. *Insurance Mathematics and Economics*, **48**(1), 56–63.

Clark, P. (1973) A subordinated stochastic process with finite variance for speculative prices. *Econometrica*, **41**, 135–155.

Clayton, D.G. (1978) A model for association in bivariate life tables and its application in epidemiological studies of familial tendency in chronic disease incidence. *Biometrika*, **65**, 141–151.

Coles, S., Currie, J., and Tawn, J. (1999) *Dependence measures for extreme value analysis*. Department of Mathematics and Statistics, Lancaster University, working paper.

Cont, R. (2001) Empirical properties of asset returns: stylized facts and statistical issue. *Quantitative Finance*, **1**, 223–236.

Cook, R.D. and Johnson, M.E. (1981) A family of distributions for modeling non-elliptically symmetric multivariate data. *Journal of the Royal Statistical Society, Series B*, **43**, 210–218.

Cook, R.D. and Johnson, M.E. (1986) Generalized Burr–Pareto–logistic distributions with applications to a uranium exploration data set. *Technometrics*, **28**, 123–131.

Cox, J.C., Ingersoll, J., and Ross, S. (1985) A theory of the term structure of interest rates. *Econometrica*, **53**, 385–407.

Cox, D.R. and Oakes, D. (1984) *Analysis of Survival Data*. Chapman & Hall. London.

Davidson, R. and MacKinnon, J. (1993) *Estimation and Inference in Econometrics*. Oxford University Press, Oxford.

de Haan, L. (1970) On regular variation and its application to the weak convergence of sample extremes. Mathematics Centrum, Amsterdam.

de Haan, L. and Resnick, S.I. (1977) Limit theory for multivariate sample extremes. *Z. Wahrscheinlichkeitstheorie und Verw. Gebiete*, **40**(4), 317–337.

Deheuvels, P. (1978) Caractérisation complète des Lois Extrèmes Multivariées et de la Convergence des Types Extrèmes. *Publications de l'Institut de Statistique de l'Université de Paris*, **23**, 1–36.

Devroye, L. (1986) *Non-Uniform Random Variate Generation*. Springer-Verlag, New York.

Dhaene, J., Denuit, M., Goovaerts, M.J., Kaas, R., and Vyncke, D. (2002) The concept of comonoticity in actuarial science and finance: theory. *Insurance: Mathematics and Economics*, **31**, 3–33.

Diebold, F.X., Gunther, T., and Tay, A.S. (1998) Evaluating density forecasts with applications to financial risk management. *International Economic Review*, **39**, 863–883.

Diebold, F.X., Hahn, J., and Tay, A.S. (1999) Multivariate density forecast evaluation and calibration in financial risk management. *Review of Economics and Statistics*, **81**(4), 661–673.

Dobric, J. and Schmidt, F. (2002) Nonparametric estimation of the lower tail dependence λ_L in bivariate copula. *Journal of Applied Statistics*, **32**(4), 387–407.

Drouet-Mari, D. and Kotz, S. (2001) *Correlation and Dependence*. Imperial College Press, London.

Durante, F. (2009) Construction of non-exchangeable bivariate distribution functions. *Statistical Papers*. **50**(2), 383–391.

Durbin, J. and Stuart, A. (1951) Inversions and rank correlations. *Journal of the Royal Statistical Society, Series B*, **2**, 303–309.

Durrleman, V., Nikeghbali, A., and Roncalli, T. (2000a) Copulas approximation and new families. Groupe de Recherche Opèrationelle, Crédit Lyonnais, working paper.

Durrleman, V., Nikeghbali, A., and Roncalli, T. (2000b) Which copula is the right one? Groupe de Recherche Opèrationelle, Crédit Lyonnais, working paper.

Efron, B. and Tibshirani, R.J. (1993) *An Introduction to Bootstrap*. Chapman & Hall. London.

El Karoui, N., Jeanblanc, M., and Jiao, Y. (2009) Successive default times. Working paper.

Embrechts, P. and Hofert, M. (2011) A note on generalized inverses. Available at http://www.math.ethz.ch/baltes/ftp/inversesPEMH.pdf.

Embrechts, P., Hoing, A., and Juri, A. (2003) Using copulae to bound the value-at-risk for functions of dependent risks. *Finance and Stochastics*, **7**, 145–167.

Embrechts, P., Lindskog, F., and McNeil, A. (2003) Modelling dependence with copulas and applications to risk management. In S. Rachev (ed.) *Handbook of Heavy Tailed Distributions in Finance*. Elsevier, Amsterdam, pp. 329–384.

Embrechts, P. and Puccetti, G. (2006a) Bounds for functions of dependent risks. *Finance and Stochastics*, **10**, 341–352.

Embrechts, P. and Puccetti, G. (2006b) Bounds for functions of multivariate risks. *Journal of Multivariate Analysis*, **27**, 526–547.

Embrechts, P. and Puccetti, G. (2006c) Aggregating risk capital, with an application to operational risk. *Geneva Risk and Insurance Review*, **31**, 71–90.

Engle, R.F. (ed.) (1996) *ARCH Selected Readings*. Oxford University Press, Oxford.

Engle, R.F. and Manganelli, S. (1999) CAViaR: Conditional Autoregressive Value at Risk by regression quantiles. UCSD Department of Economics, working paper.

Engle, R.F. and Russell, J. (1998) The autoregressive conditional duration model. *Econometrica*, **66**, 1127–1162.

Engle, R.F. (2008) *Anticipating Correlations: A New Paradigm for Risk Management*. Princeton University Press, Princeton, NJ.

Evans, M., Hastings, N., and Peacock, B. (1993) *Statistical Distributions*. John Wiley & Sons, New York.

Feller, W. (1968) *An Introduction to Probability Theory and Its Applications*, Vol. I. John Wiley & Sons, New York.

Feller, W. (1971) *An Introduction to Probability Theory and Its Applications*, Vol. II. John Wiley & Sons, New York.

Ferguson, T.S. (1995) A class of symmetric bivariate uniform distributions. *Statistics Papers*, **36**, 31–40.

Féron, R. (1956) Sur les tableaux de corrélation dont les marges sont données, cas de l'espace à trois dimensions. *Publ. Inst. Statist. Univ. Paris*, **5**, 3–12.

Finger, C. (ed.) (2003) CreditGrades Technical Document.

Fisher, M., Kock, C., Schlouter, S., and Weigert, F. (2009) An empirical analysis of multivariate copula models. *Quantitative Finance*, **9**(7), 839–854.

Frank, M.J. (1979) On the simultaneous associativity of $F(x,y)$ and $x + y - F(x, y)$. *Aequationes Mathematicae*, **19**, 194–226.

Frank, M.J. and Schweizer, B. (1979) On the duality of general infimal and supremal convolutions. *Rend. Math.*, **12**, 1–23.

Frank, M.J., Nelsen, R., and Schweizer, B. (1987) Best-poss bounds for the distribution of a sum – a problem of Kolmogorov. *Probab. Th. Rel. Fields*, **74**, 199–211.

Fréchet, M. (1935) Généralisations du théorème des probabilités totales. *Fund. Math.*, **25**, 379–387.

Fréchet, M. (1951) Sur le tableaux de corrélation dont les marges sont données. *Ann. Univ. Lyon*, **9**(A), 53–77.

Fréchet, M. (1958) Remarques au sujet de la note précédente. *C. R. Acad. Sci. Paris*, **246**, 2719–2720.

Frees, E.W. and Valdez, E. (1998) Understanding relationship using copulas. *North American Actuarial Journal*, **2**, 1–25.

Frey, R. and McNeil, A.J. (2001) Modeling dependent defaults. Deptartment of Mathematics ETH Zurich, working paper.

Geluk, J.L. and de Haan, L. (1987) Regular variation, extensions and Tauberian theorems. *CWI Tract. Stichting Mathematisch Centrum*, Vol. 40. Centrum voor Wiskunde en Informatica, Amsterdam.

Genest, C. (1987) Frank's family of bivariate distributions. *Biometrika*, **74**, 549–555.

Genest, C., Ghoudi, K., and Rivest, L. (1995) A semiparametric estimation procedure of dependence parameters in multivariate families of distributions. *Biometrika*, **82**(3), 543–552.

Genest, C. and MacKay, J. (1986) The joy of copulas: bivariate distributions with uniform marginals. *American Statistician*, **40**, 280–283.

Genest, C. and Rivest, L. (1993) Statistical inference procedures for bivariate Archimedean copulas. *Journal of the American Statistical Association*, **88**, 1034–1043.

Genest, C., Ghoudi, K., and Rivest, L.P. (1995) A semiparametric estimation procedure of dependence parameters in multivariate families of distributions. *Biometrika*, **82**(3), 543–552.

Genest, C. and Neslehova, J. (2007) A primer on copulas for count data. *ASTIN Bulletin*, **37**, 475–515.

Georges, P., Lamy, A., Nicolas, E., Quibel, G., and Roncalli, T. (2001) Multivariate survival modelling: a unified approach with copulas. Groupe de Recherche Opèrationelle, Crédit Lyonnais, working paper.

Ghysel, E., Gourièroux, C., and Jasiak, J. (2004) Stochastic volatility duration models. *Journal of Econometrics*, **119**, 413–433.

Gibbons, J.D. (1992) *Nonparametric Statistical Inference*. Marcel Dekker, New York.

Gourièroux, C. and Monfort, A. (2002) Equidependence in qualitative and duration models with application to credit risk. CREST, working paper.

Greco, S., Matarazzo, B., and Slowinski, R. (2001) Rought sets theory for multicriteria decision analysis. *European Journal of Operational Researches*, **129**(1), 1–47.

Gumbel, E.J. (1960) Bivariate exponential distributions. *Journal of the American Statistical Association*, **55**, 698–707.

Hardle, W. (1990) *Applied Nonparametric Regression*. ESM Cambridge University Press, Cambridge.

Hering, C., Hofert, M., Mai, J.F., and Scherer, M. (2009) Constructing hierarchical Archimedean copulas with Lèvy subordinators. *Journal of Multivariate Analysis*, **101**(6), 1428–1433.

Heston, S.L. (1993) A closed form solution for options with stochastic volatility with applications to bond and currency options. *Review of Financial Studies*, **6**, 327–343.

Hoeffding, W. (1940) Masstabinvariante Korrelationstheorie. *Schriften des Mathematischen Instituts und des Instituts fur Angewandte Mathematik der Universitat Berlin*, **5**, 179–233.

Hofert, M. (2010) Sampling nested Archimedean copulas with applications to CDO pricing. Südwestdeutscher Verlag für Hochschulschriften AG & Co. KG, PhD thesis.

Hofert, M. (2011a) Efficiently sampling nested Archimedean copulas. *Computational Statistics and Data Analysis*, **55**, 57–70.

Hofert, M. (2011b) A stochastic representation and sampling algorithm for nested Archimedean copulas. *Journal of Statistical Computation and Simulation*, **55**, 57–70.

Hofert, M. and Mächler, M. (2011) Nested Archimedean copulas meet R: the nacopula package. *Journal of Statistical Software*, in press.

Hofert, M. and Scherer, M. (2010) CDO pricing with nested Archimedean copulas. *Quantitative Finance*, **10**.

Hougaard, P. (1986) A class of multivariate failure time distributions. *Biometrika*, **73**, 671–678.

Huang, W. and Prokhorov, A. (2008) Goodness-of-fit test for copulas. *Econometric Reviews*.

Hull, J. and White, A. (1998) Value at Risk when daily changes in market variables are not normally distributed. *Journal of Derivatives*, **5**(3), 9–19.

Hutchinson, T.P. and Lai, C.D. (1990) *Continuous Bivariate Distribution, Emphasising Applications*. Rumsby Scientific Publishing, Adelaide.

Jaworski, P., Durante, F., Haerdle, W., and Rychlik T. (eds). (2010) *Copula Methods and its Applications*, Lecture Notes in Statistics-Proceedings, Vol. 198. Springer, Berlin.

Joag-Dev, K. (1984) Measures of dependence, In P.R. Krishnaiak (ed.), *Handbook of Statistics*. North-Holland, New York, pp. 79–88.

Joe, H. and Hu, T. (1996) Multivariate distributions from mixtures of max-infinitely divisible distributions. *Journal of Multivariate Analysis*, **57**, 240–265.

Joe, H. and Xu, J.J. (1996) The estimation method of inference functions for margins for multivariate models. Department of Statistics, University of British Columbia, Technical report 166.

Jouanin, J.F., Rapuch, G., Riboulet, G., and Roncalli, T. (2001) Modelling dependence for credit derivatives with copulas. Groupe de Recherche Opèrationelle, Crédit Lyonnais, working paper.

J.P. Morgan (2000) Guide to Credit Derivatives. *Risk*.

Kendall, M.G. (1938) A new measure of rank correlation. *Biometrika*, **30**, 81–93.

Kim, G., Silvapulle, M., and Silvapulle, P. (2007) Comparison of semiparametric and parametric methods for estimating copulas. *Computational Statistics and Data Analysis*, **51**, 2836–2850.

Kimberling, C.H. (1974) A probabilistic interpretation of complete monotonicity. *Aequationes Mathemati*, **10**, 152–164.

Kimeldorf, G. and Sampson, A.R. (1975) Uniform representation of bivariate distributions. *Communications in Statistics*, **4**, 617–627.

Kim, T.H. and White, H. (2002) Estimation inference and testing for possibly misspecified quantile regression. In T.B. Fomby and R.C. Hill (eds), *Maximum Likelihood Estimation of Misspecified Models: Twenty Years Later*. Advances in Econometrics, Vol. 17. Emerald Group, pp. 107–132.

Koenker, R. and Park, J.B. (1996) An interior point algorithm for nonlinear quantile regression. *Journal of Econometrics*, **71**(1), 265–283.

Konijn, H.S. (1959) Positive and negative dependence of two random variables. *Sankhyà*, **21**, 269–280.

Kurowicka, D. and Joe, H. (2010) *Dependence Modeling. Vine Copulae Handbook*. World Scientific, Singapore.

Laurent, J.P. and Gregory, J. (2002) Basket default swaps. CDOs and factor copulas. BNP Paribas and University of Lyon, working paper

Lee, A.J. (1993) Generating random binary deviates having fixed marginal distributions and specified degrees of association. *American Statistician*, **47**, 209–215.

Lehmann, E. (1966) Some concepts of dependence, *Annals of Mathematical Statistics*, **37**, 1137–1153.

Lehmann, E. and Casella, G. (1998) *Theory of Point Estimation*. Springer-Verlag, New York.

Li, X., Mikusinski, P., Sherwood, H., and Taylor, M.D. (1997) On approximation of copulas. In V. Benes, and J. Stepan (eds) *Distributions with Given Marginals and Moment Problems*. Kluwer, Dordrecht.

Liebscher, E. (2008) Construction of asymmetric multivariate copulas. *Journal of Multivariate Analysis*, **99**(10), 2234–2250.

Lindskog, F. (2000) Linear correlation estimation. RiskLab ETH, working paper.

Lindskog, F., McNeil, A., and Schmock, U. (2001) A note on Kendall's tau for elliptical distribution. ETH Zurich, working paper.

Ling, C.H. (1965) Representation of associative functions. *Publ. Math. Debrecen*, **12**, 189–212.

Luciano, E. and Marena, M. (2002) Portfolio value at risk bounds. *International Transactions in Operational Research*, **9**(5), 629–641.

Luciano, E. and Marena, M. (2003a) Value at risk bounds for portfolios of non-normal returns. In C. Zopoudinis (ed.), *New Trends in Banking Management*. Physica-Verlag, Berlin, pp. 207–222.

Luciano, E. and Marena, M. (2003b) Copulae as a new tool in financial modelling. *Operational Research: An International Journal*, **2**(2), 139–155.

Madan, D. and Seneta, E. (1998) The variance gamma. (VG) model for share market returns. *Journal of Business*, **63**(4), 511–524.

Mai, J.F. (2010) Extendibility of Marshall–Olkin distributions via Lèvy subordinators and an application to portfolio credit risk. Technische Universitat Munchen Zentrum Mathematik HVB-Stiftungsinstitut fur Finanzmathematik, PhD thesis.

Mai, J.F., Scherer, M., and Zagst, R. (2011) CIID default models and implied copulas. Working paper.

Makarov, G.D. (1981) Estimates for the distribution function of a sum of two random variables when the marginal distributions are fixed. *Theory of Probability and its Applications*, **26**, 803–806.

Marshall, A.W. and Olkin, I. (1988) Families of multivariate distributions. *Journal of the American Statistical Association*, **83**, 834–841.

Mashal, R. and Zeevi, A. (2002) Beyond correlation: extreme co-movements between financial assets. Columbia Business School. Working paper.

Meneguzzo, D. and Vecchiato, W. (2000) Improvement on Value at Risk measures by combining conditional autoregressive and extreme value approaches. *Review in International Business and Finance*, **16**(12), 275–324.

Meneguzzo, D. and Vecchiato, W. (2003) Copula sensitivity in collateralized debt obligations and basket default swaps. *The Journal of Future Markets*, **24**(1), 37–70.

Mikusinski, P., Sherwood, H., and Taylor, M.D. (1992) Shuffles of min. *Stochastica*, **13**, 61–74.

Moore, D.S. and Spruill, M.C. (1975) Unified large-sample theory of general chi-squared statistics for tests of fit. *Annals Statistics*, **3**, 599–616.

Moynihan, R., Schweizer, B., and Sklar, A. (1978) Inequalities: among binary operations on probability distribution functions. *General Inequalities*. **1**, 133–149.

Muller, A. and Scarsini, M. (2000) Some remarks on the supermodular order. *Journal of Multivariate Analysis*, **73**(1), 107–119.

Nagpal, K. and Bahar, R. (2001) Measuring default correlation. *Risk*, March, 129–132.

Nelsen, R.B. (1991) Copulas and association. In G. Dall'Aglio, S. Kotz, and G. Salinetti (eds), *Advances in Probability Distributions with Given Marginals*. Kluwer Academic, Dordrecht, pp. 51–74.

Nelsen, R.B. (1996) Nonparametric measures of multivariate association. In *Distibution with Fixed Marginals and Related Topics*. IMS Lecture Notes-Monograph Series, Vol. 28, pp. 223–232.

Oakes, D. (1982) A model for association in bivariate survival data. *Journal of the Royal Statistical Society, Series B*, **44**, 414–422.

Oakes, D. (1986) Semiparametric inference in a model for association in bivariate survival data. *Biometrika*, **73**, 353–361.

O'Kane, D. and Livesey, M. (2004) Base correlation explained. *Quantitative Credit Research*, Lehman Brothers, Q3(4).

Pearson, N.D. and Sun, T.S. (1994) Exploiting the conditional density in estimating the term structure: an application to the Cox–Ingersoll–Ross model. *Journal of Finance*, **49**, 1279–1304.

Plackett, R.L. (1965) A class of bivariate distributions. *Journal of the American Statistical Association*, **60**, 516–522.

Prokhorov, A. and Schmidt, P. (2009) Likelihood-based estimation in a panel setting: robustness redundancy and validity of copulas. *Journal of Econometrics*, **153**, 93–104.

Rémillard, B. (2010) Goodness-of-fit tests for copulas of multivariate time series. SSRN Working Paper Series No. 1729982.

Rémillard, B., Papageorgiou, N., and Soustra, F. (2010) Dynamic copulas. Technical Report G-2010-18, Gerad. 1.

Rényi, A. (1959) On measures of dependence. *Acta. Math. Acad. Sci. Hungar.*, **10**, 441–451.

Robinson, P. (1983) Nonparametric estimators for time series. *Journal of Time Series Analysis*, **4**, 185–207.

Roncalli, T. (2001) Copulas: a tool for dependence in finance. Groupe de Recherche Opèrationelle, Crédit Lyonnais, working paper.

Roncalli, T. (2002) Gestiondes Risques Multiples. Cours ENSAI de 3' année, Groupe de Recherche Opèrationelle, Crédit Lyonnais, working paper.

Roncalli, T. (2003) Gestion des Risques Multiples: Le Risque de Credit 2ème partie, Groupe de Recherche Opèrationelle, Crèdit Lyonnais, working paper.

Rosemberg, J.V. (2000) Nonparametric pricing of multivariate contingent claims. Stern School of Business, working paper.

Rousseuw, P. and Molenberghs, G. (1993) Transformations of non-positive semidefinite correlation matrices. *Communications in Statisitcs – Theory and Methods*, **22**(4), 965–984.

Ruschendorf, L., Schweizer, B., and Taylor, M.D. (1996) *Distributions with Fixed Marginals and Related Topics*. Institute of Mathematical Statistics, Hayward.

Samorodnitsky, G. and Taqqu, M.S. (1995) *Stable Non-Gaussian Random Processes. Stochastic Models with Infinite Variance*. Chapman & Hall, New York.

Savu, C. and Trede, M. (2010) Hierarchical Archimedean copulas. *Quantitative Finance*, **10**, 295–304.

Scaillet, O. (2000) Nonparametric estimation of copulas for time series. IRES working paper.

Scarsini, M. (1984) On measures of concordance. *Stochastica*, **8**, 201–218.

Schönbucher, P. and Schubert, D. (2001) Copula-dependent default risk in intensity models. University of Bonn, Germany, working paper.

Schmid, F. and Schmidt, R. (2007) Multivariate extensions of Spearman's rho and related statistics. *Statistics and Probability Letters*, **77**, 407–416.

Schmid, F. and Schmidt, R. (2006a) Multivariate conditional version of Spearman's rho and related measures of tail dependence. *Journal of Multivariate Analysis*, **98**, 1123–1140.

Schmid, F. and Schmidt, R. (2006b) Bootstrapping Spearman's multivariate rho. In A. Rizzi and M. Vichi (eds), *Proceedings of COMPSTAT 2006*, pp. 759–766.

Schweizer, B. (1991) Thirty years of copulas. In G. Dall'Aglio, S. Kotz, and G. Salinetti (eds), *Advances in Probability Distributions with Given Marginals*. Kluwer Academic, Dordrecht, pp. 13–50.

Schweizer, B. and Sklar, A. (1961) Associative functions and statistical triangle inequalities. *Publ. Math. Debrecen*, **8**, 169–186.

Schweizer, B. and Sklar, A. (1974) Operations on distribution functions not derivable from operations on random variables. *Studia Mathematica*, **52**, 43–52.

Schweizer, B. and Sklar, A. (1983) *Probabilistic Metric Spaces*. Elsevier Science, New York.

Scott, D. (1992) *Multivariate Density Estimation: Theory and Practice and Visualization*. John Wiley & Sons, New York.

Serfling, R.J. (1980) *Approximation Theorems of Mathematical Statistics*. John Wiley & Sons, New York.

Shao, J. (1999) *Mathematical Statistics*. Springer-Verlag, New York.

Shao, J. and Tu, D. (1995) *The Jacknife and Bootstrap*. Springer-Verlag, New York.

Silvapulle, P. and Moosa, I. (1999) The relationship between spot and futures prices: evidence from the crude oil market. *Journal of Futures Markets*, **19**, 147–167.

Silverman, B. (1986) *Density Estimation for Statistics and Data Analysis*. Chapman & Hall, London.

Sklar, A. (1973) Random variables, distribution functions, and copulas. *Kybernetica*, **9**, 449–460.

Sklar, A. (1996) Random variables, distribution functions, and copulas – a personal look backward and forward. In L. Ruschendorf, B. Schweizer, and M.D. Taylor (eds), *Distributions with Fixed Marginals and Related Topics*. Institute of Mathematical Statistics, Hayward, pp. 1–14.

Sklar, A. (1998) personal communication.

Song, P. (2000) Multivariate dispersion models generated from Gaussian copula. *Scandinavian Journal of Statistics*, **27**(2), 305–320.

Stulz, R.M. (1982) Options on the minimum and maximum of two risky assets. *Journal of Finance*, **10**, 161–185.

Tjostheim, D. (1996) Measures of dependence and tests of independence. *Statistics*, **28**, 249–284.

Vasicek, O. (2002) Loan portfolio value. *Risk*, **15**(2), 160–162.

Wang, S.S. (1998) Aggregation of correlated risk portfolios: models & algorithms. CAS Committee on Theory of Risk, working paper.

White, H. (1984) *Asymptotic Theory for Econometricians*. Academic Press, New York.

Wilde, T. (2002) Probing granularity. *Risk*, **14**(8), 103–106.

Williamson, R.C. (1989) Probabilistic arithmetic. University of Queensland, PhD thesis.

Williamson, R.C. and Downs, T. (1990) Probabilistic arithmetic I: Numerical methods for calculating convolutions and dependency bounds. *International Journal of Approximate Reasoning*, **4**, 89–158.

Index

Index compiled by Terry Halliday

Printed and bound by CPI Group (UK) Ltd, Croydon, CR0 4YY

27/10/2024

14580375-0004